Irrigation and Water Management

Irrigation and Water Management

Edited by **Davis Twomey**

SYRAWOOD
PUBLISHING HOUSE

New York

Published by Syrawood Publishing House,
750 Third Avenue, 9th Floor,
New York, NY 10017, USA
www.syrawoodpublishinghouse.com

Irrigation and Water Management
Edited by Davis Twomey

© 2016 Syrawood Publishing House

International Standard Book Number: 978-1-68286-125-7 (Hardback)

Contents

Preface

Irrigation plays a crucial role in influencing the quality and quantity of crop yield. This book fulfills the need for a comprehensive text on irrigation and water management. The text discusses innovative techniques for irrigation, water sources, groundwater management, technological advances and challenges of this field. The scope of the book makes it a useful reference for agricultural engineers, students and professionals.

All of the data presented henceforth, was collaborated in the wake of recent advancements in the field. The aim of this book is to present the diversified developments from across the globe in a comprehensible manner. The opinions expressed in each chapter belong solely to the contributing authors. Their interpretations of the topics are the integral part of this book, which I have carefully compiled for a better understanding of the readers.

At the end, I would like to thank all those who dedicated their time and efforts for the successful completion of this book. I also wish to convey my gratitude towards my friends and family who supported me at every step.

Editor

Effect of dune sand mulch on water recharging into root zone using drip lines

Khumbulani DHAVU[1]* and Hiroshi YASUDA[2]

[1]Bioresources Engineering, School of Engineering, University of KwaZulu-Natal, Private Bag X01, Scottsville, South Africa.
[2]Arid Land Research Center, Tottori University, 1390 Hamasaka, Tottori, 680-0001, Japan.

Surface drip irrigation has been used for agricultural production for more than three decades. Surface drip irrigation (SDI) can maintain sufficient water content in soil for seed germination and emergence, seedling development and plant growth. Insufficient water content in the root zone may cause crop failure. In this study, the effects of Tottori dune sand mulch on drip lines on water recharging into root zone was examined under drip irrigation. Two irrigation levels of 60 and 100% evapotranspiration, and three dune sand mulches of 0, 2 and 5 cm thickness on the drip lines were used in this experiment. The results indicated that water recharging into the root zone under 5 cm dune sand was highest among the three dune sand mulch treatments. The drip lines under the 2 cm dune sand mulch expanded and contracted resulting in protrusion from the sand mulch. An irrigation level of 100% evapotranspiration under the 5 cm dune sand mulch resulted in increase of dry matter yield of sorghum. From this study, it is concluded that to increase water recharging into the root zone (to the depth of 25 cm), the minimum thickness of dune sand mulch on the drip lines was 5 cm.

Key words: irrigation level, root zone, dune sand, surface drip irrigation, water recharging.

INTRODUCTION

Water is one of the most important resources for the growth and development of human life. Water demand for agriculture to meet the increasing demand for food is increasing worldwide. The population of the world benefits from the efficient use of water resources. In order to reduce the severity of water scarcity, water management should be improved. Agriculture has the greatest potential for solving the problem of global water scarcity (Longo and Spears, 2003).

SDI has been used for agricultural production for more than three decades. SDI has higher water use efficiency than other irrigation methods. Surface drip irrigation can keep sufficient water content in the soil and high soil temperature for seed germination and seedling development (Wang et al., 2000).

When the drip lines are covered with sandy soil, the sandy soil surface tends to dry, which would help to reduce soil evaporation. The top 20 cm of sandy soil had lower water content than the deeper soil layers resulting in reduced soil evaporation when drip lines were buried at the depth of 45 cm (Phene et al., 1983; Solomon, 1993).

Lamm and Trooien (2005), and Neelam and Rajput (2007) proposed that placement of drip lines might be practical at depths shallower than 10 cm under sandy soil. Neelam and Rajput (2007) evaluated the effect of placing drip lines at the depths of 0, 5, 10, 15 and 20 cm under sandy loam soil on yield of potato. When drip lines were buried at the depth of 5 cm, movement of water to the soil surface was observed. Their findings showed that sandy loam of 5 cm thickness was not sufficient to effectively restrict soil evaporation.

The effectiveness of vegetative mulches may be limited because their high porosity permits rapid diffusion (Hillel, 1998). In a banana plantation in central Uganda, McIntyre et al. (2000) found that mulching decreased surface soil bulk density and thus led to more rapid recharge of soil

*Corresponding author. E-mail: khumbudhavu@yahoo.co.uk.

water. Water recharging is important in supplying adequate water to the root zone, and mulch materials have been used to reduce soil evaporation and to increase infiltration. Ssali et al. (2003) assessed the effects of maize stovermulch on recharging of water into the soil. They found that there was higher water recharging under mulched surfaces than under unmulched surfaces. They attributed the increase in water recharging to the increased infiltration rather than the increased porosity.

Our study hypothesized that Tottori dune sand could be used as a mulch to increase amount of water recharged into the root zone. The objective of this study was to evaluate the effects of dune sand mulch on drip lines on water recharging into the root zone, and to determine the minimum thickness of dune sand mulch.

MATERIALS AND METHODS

Plots and sub-plots

The experiment was carried out in a 154 m^2 glasshouse at the Arid Land Research Center, Tottori University, Japan (35°32'N; 134°13'E; 23 m above sea level). The length, width and maximum height of the glasshouse were 22, 7 and 4.5 m, respectively. It was unheated and naturally ventilated with a single continuous roof vent. Lateral windows were kept open during daytime. The soil type of the Tottori dune sand was Arenosol (silicious sand, typicUdipsamment) with 96% sand (Qiu et al., 1999).

Soil water pressures (kPa) and the respective soil water contents (m^3/m^3) were measured using a pressure plate. Undisturbed soil samples were collected from the upper soil layer using 100 cm^3 stainless steel cores. The soil water retention curve was determined using the pressure plate laboratory method (Richards, 1948; Klute, 1986). Richrads (1948) and Klute (1986) found that the pressure plate method can reliably measure soil water characteristics when undisturbed soil samples are used. The relationship between soil water pressure and volumetric water content of the Tottori dune sand is shown in Figure 1. The field capacity and permanent wilting point of the dune sand were 0.074 and 0.022 m^3/m^3, respectively, and the corresponding matric potential is -0.006 and -1.5 MPa, respectively (Qiu et al., 1999). Porosity and saturated soil hydraulic conductivity of the dune sand were 0.4 m^3/m^3 and 2.7×10.4 m/s, respectively (Qiu et al., 1999). Some of the physical properties of the dune sand are summarized in Table 1.

Two plots (Plot A and Plot B), each of 4.80 m long and 1.0 m wide, were used in this experiment as shown in Figure 2. Sorghum (Sorghum bicolor) was planted on Plot A and B on June 16, 2008 at a 50 cm row and 30 cm in-row spacing. After sowing, the drip lines were mulched with dune sand. The sorghum was harvested on October 24, 2008. Fertilizer was applied at the rate of 180 kg/ha N, 45 kg/ha P and 80 kg/ha K just before sowing. Top dressing was done at the rate of 100 kg/ha N at the middle growth stage.

Thickness of mulch

Each of the two plots was further divided into three sub-plots. Three drip lines were arranged on each sub-plot. The drip lines were spaced at 50 cm. The spacing within the drip line was 30 cm. This irrigation system was operated at a pressure head of 14 m.

On the first sub-plot of each plot, three drip lines were not mulched, and are referred to as T0. On the second sub-plot of each plot, the three drip lines were mulched with 2 cm of the dune sand, and are referred to as T2. On the third sub-plot of each plot, the

three drip lines were mulched with 5 cm of the dune sand, and are referred to as T5. The section view of the position of the drip lines, sensors and plants under T0, T2 and T5 is shown in Figure 3.

Irrigation level

Two small evaporation pans were placed at random in each sub-plot, and were weighed twice daily, at 08:30 and at 20:30. Irrigation water was applied every other day based on small pan evaporation. Different irrigation levels were applied to Plot A and B. For Plot A, 60% of the estimated evapotranspiration of sorghum was applied and this is referred to as 0.6Ep. For Plot B, 100% of the estimated evapotranspiration of sorghum was applied and this is referred to as 1.0Ep.

Evapotranspiration and amount of irrigation water

Evaporation of the small evaporation pan was converted to the class A pan evaporation by the Agodzo et al. (1997) equation:

$$E_A = a \times E_S^b \tag{1}$$

Where, E_A is the class A pan evaporation (mm), a and b are fitting parameters, and a = 0.17, b = 1.92 and E_S is the evaporation from the small evaporation pan (mm). Measured E_S and E_A (E_S and E_A were measured daily at 08:30) were correlated for 10 days at the Arid Land Research Center, Tottori University, Japan. E_A was measured using a meter rule daily at 08:30. The meter rule measured the depth of water lost in the last 24 h.

The class A pan evaporation was then converted to the potential evapotranspiration by the Doorenbos and Pruit (1977) equation:

$$ETo = Kp \times E_A \tag{2}$$

Where, ETo is the potential evapotranspiration (mm) and Kp is the pan coefficient (dimensionless). Kp = 0.80, and was obtained from Kp = 0.75 as given by Doorenbos and Pruit (1977) based on location, and adjusted by 7.5% to Kp = 0.80 for sorghum (Kp = 1.075 × 0.75 = 0.80). The Kp values relate to evaporation pans located in an open field with no crops taller than one meter (1 m), and depend on general wind and humidity conditions of an area. The Kp values were, therefore, adjusted because the small evaporation pans were placed in a glasshouse (a small enclosure) and surrounded by sorghum. The ETo from Equation (2) was converted to evapotranspiration of sorghum (ETc) (mm) by the Doorenbos and Pruit (1977) equation:

$$ETc = Kc \times ETo \tag{3}$$

Where, Kc is the crop coefficient (dimensionless). The Allen et al. (1998) Kc values of 0.70, 1.10 and 0.55 for early growth stage, middle growth stage and late growth stage, respectively, were used in this experiment. Irrigation time was calculated by the following equations:

$$T = \frac{A \times 0.6 \times ETc}{Q} \tag{4}$$

$$T = \frac{A \times 1.0 \times ETc}{Q} \tag{5}$$

Where, T is the irrigation time (hours), ETc is the evapotranspiration of sorghum (mm), A is the area of the wetted horizontal region (assumed circular and of radius of 15 cm) (m^2) below the emitters and Q is the emitter discharge (l/h).

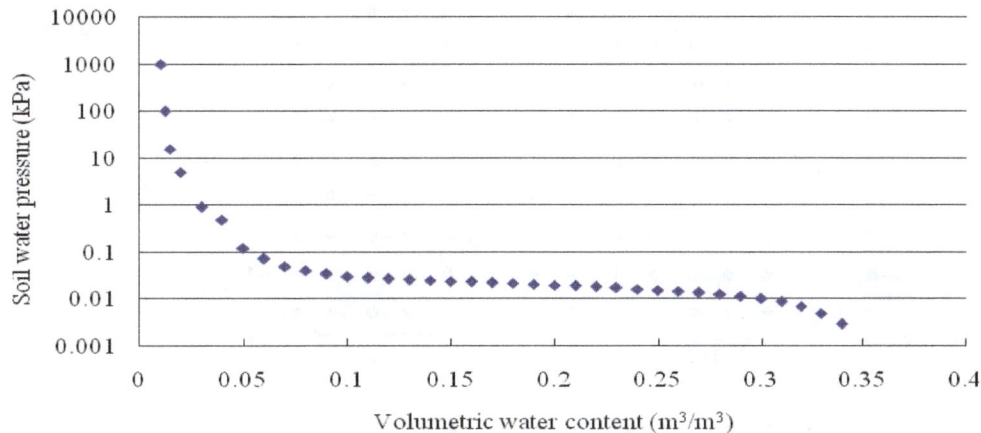

Figure 1. Relation between soil water pressure and volumetric water content of the Tottori sand dune soil.

Table 1. Physical properties of Tottori sand dune soil.

Depth (cm)	Particle size composition (%)			Bulk density (Mg/m^3)
	Sand	Silt	Clay	
0 - 5	96.2	2.0	1.8	1.50
5 -10	95.7	3.0	1.3	1.49
10 - 15	95.2	3.0	1.8	1.49
25 - 30	94.7	3.5	1.8	1.50

Source: Yamanaka and Yonetani (1999).

The total amount of irrigation water during the cropping period was 344 mm for 0.6Ep to each mulch treatment. The total amount of irrigation water during the cropping period was 521 mm for 1.0Ep to each mulch treatment.

Soil water and water recharged into the root zone

Soil water content was measured by ECH$_2$O capacitance probes (Decagon Devices Inc., Pullman, Washington, USA) installed at depths of 5 and 25 cm. The capacitance probes were calibrated using the method described by Cobos and Chambers (2010). The sensor installation depth of 25 cm was based on Yamamoto and Cho (1978) who reported that the most effective root-water uptake zone in the dune sand under surface drip irrigation was the top 25 cm. Soil water content was not measured at the surface because the dune sand dries quickly. Soil water content was recorded every hour by Em50 ECH$_2$O data loggers (Decagon Devices Inc., Pullman, Washington, USA).

The data of soil water content at an hour before irrigation and at 12 h after irrigation was selected from the hourly recorded data. The average soil water content was determined from the soil water content at the depths of 5 and 25 cm at one hour before irrigation and at 12 h after irrigation. Soil water content at one hour before irrigation was selected because that represented the amount of water retained in the soil before irrigation. Soil water content at 12 h after irrigation was selected because gravitational water was expected to have drained. In course textured (sandy) soils, the gravitational water drainage is completed within a period of a few hours while in fine textured (clayey) soil; the drainage may take

some (2 to 3) days (Brouwer et al., 1985). Water recharged (m^3/m^3) into the root zone was the difference between the average soil water content at one hour before irrigation and the average soil water content at 12 h after irrigation. Water recharged in mm into the root zone was the product of water recharged in m^3/m^3 into the root zone and the root zone depth of 20 cm (between the depth of 5 and 25 cm).

$$Ri = (\theta ai - \theta bi) \times D \times 10 \tag{6}$$

Where, Ri is the water recharged into the root zone at ith irrigation (mm), θai is the average soil water content (m^3/m^3) at 12 h after irrigation, θbi is the average soil water content (m^3/m^3) at one hour before irrigation and D is the depth from 5 to 25 cm (cm).

Climatic conditions and plant growth

Air temperature and humidity in the glasshouse were measured at the height of 2 m by ESPEC temperature and humidity sensors (ESPEC MIC Corp., Aichi, Japan). These were installed at the centre of the glasshouse. Air temperature and humidity were recorded every hour by ESPEC data loggers (ESPEC MIC Corp., Aichi, Japan). The climatic conditions in the glasshouse are summarized in Table 2. The maximum air temperature is the highest daily air temperature in each month. The minimum air temperature is the lowest daily air temperature in each month. The average humidity is the average of maximum and minimum daily relative humidity in each month.

Three plants were randomly selected from each sub-plot of Plots

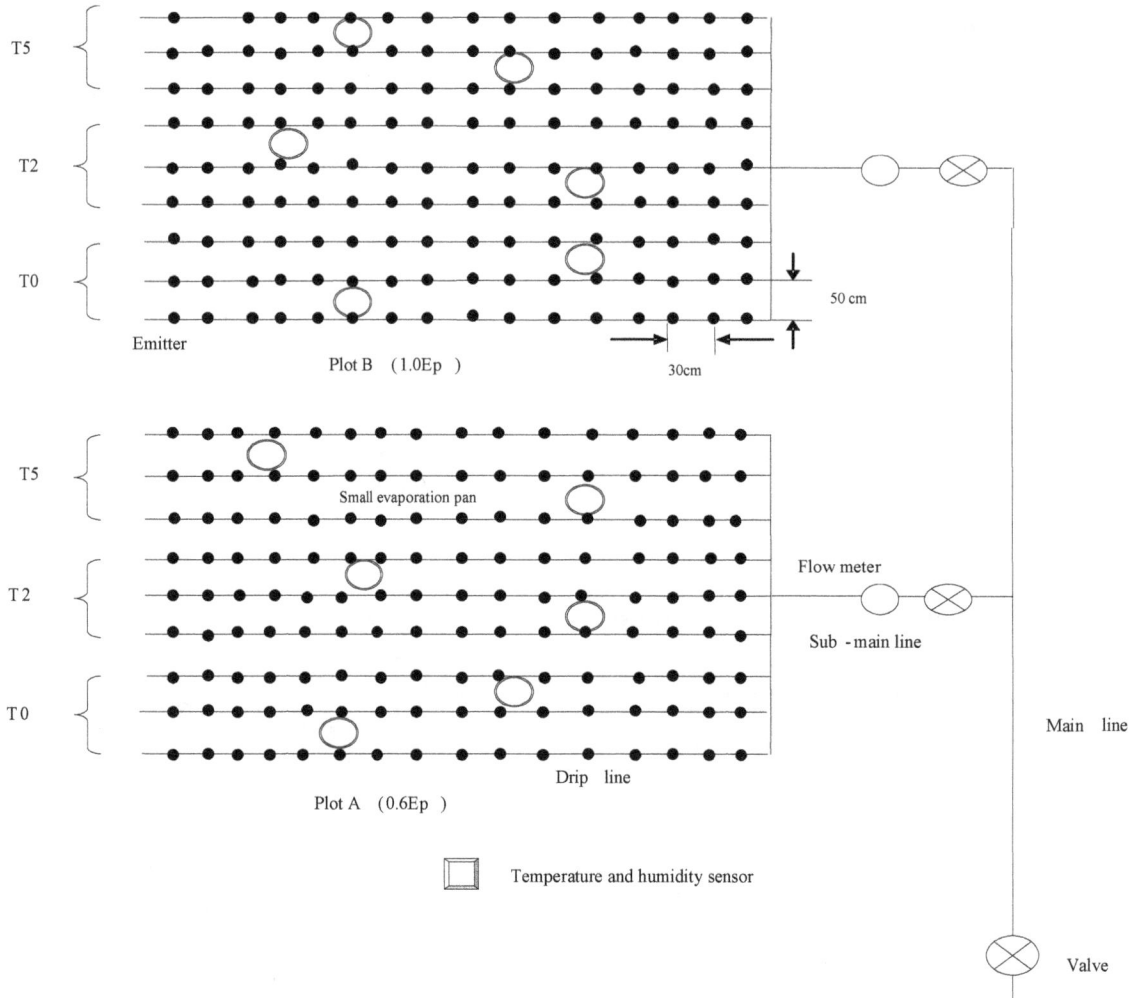

Figure 2. Experiment plots.

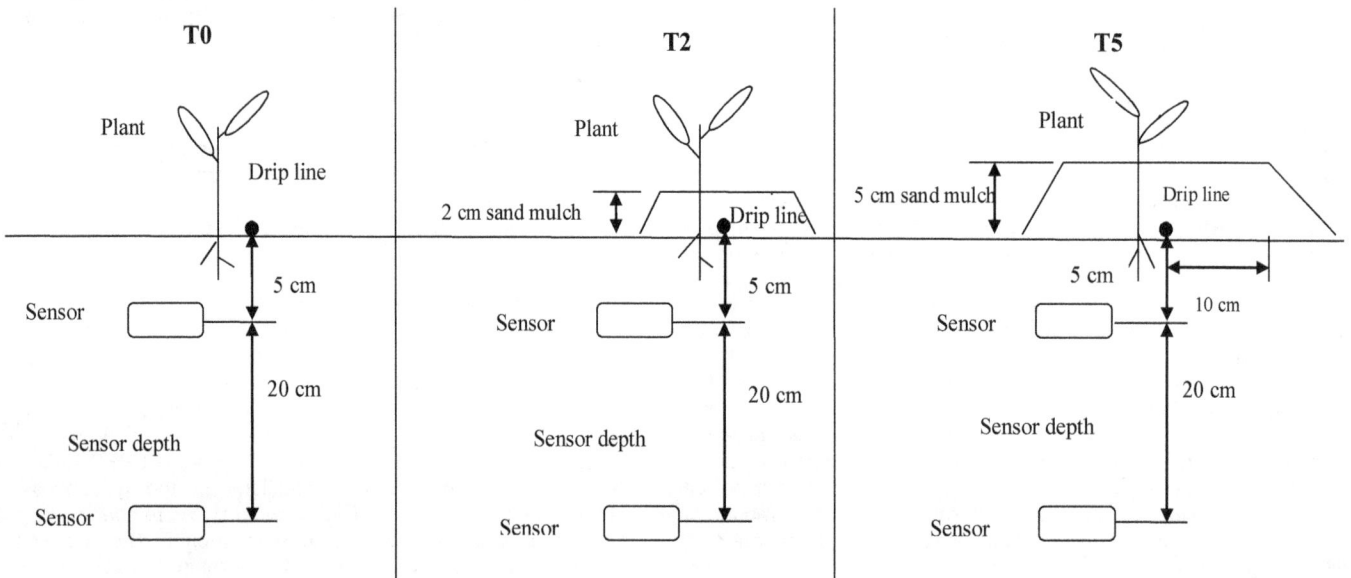

Figure 3. Section view of the position of drip lines and soil moisture sensors under T0, T2 and T5.

Table 2. Climatic conditions in glasshouse.

Month	Air temperature (°C)		Average humidity (%)
	Maximum	Minimum	
June	31.1	12.4	83
July	35.9	16.7	80
August	39.5	18.1	81
September	33.8	13.7	80
October	28.5	9.4	76

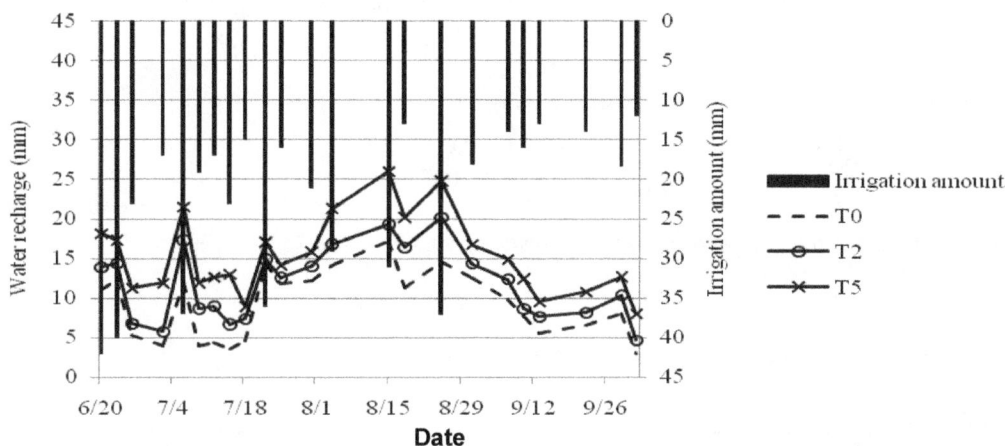

Figure 4. Amount of irrigation water and water recharged under T0, T2 and T5 for 0.6Ep.

A and B for measurement of plant height and dry matter weight. Plant height was measured at the early, middle and late growth stage by a meter rule. Plant height was measured from dune sand surface to the latest leaf. The dry matter weight was measured after drying the samples in an oven at 70°C for 48 h.

RESULTS AND DISCUSSION

Water recharged

The amount of irrigation water and water recharged into the root zone under T0, T2 and T5 for 0.6Ep are shown in Figure 4. The orders of water recharged into the root zone from high to low were T5, T2 and T0. The 5 cm dune sand mulch on drip lines improved the amount of water recharged through the dune sand. A possible explanation for the improved water recharged under 5 cm dune sand mulch could be reduced upward movement of water to the surface. The reduced upward movement of water to the soil surface was due to the dune sand dry layer of 5 cm thickness above the wet layer.

The amount of irrigation water and water recharged into the root zone under T0, T2 and T5 for 1.0Ep are shown in Figure 5. The orders of water recharged into the root zone from high to low were T5, T2 and T0. The 5 cm dune sand mulch on drip lines improved the amount of water recharged through the dune sand to the root zone.

Improved amount of water recharged for 1.0Ep could be for the same reason as for 0.6Ep. Total water recharged into the root zone was the sum of the water recharged calculated after every irrigation event. Mean water recharged into the root zone was the average of the water recharged calculated after every irrigation event.

The total and mean water recharged into the root zone under T0, T2 and T5 for 0.6Ep and 1.0Ep are shown in Table 3. The total amount of water recharged was the highest under T5 for 0.6Ep and 1.0Ep. The 5 cm dune sand mulch on drip lines improved the amount water recharged through dune sand.

The total amount water recharged in the root zone under T0, T2 and T5 for 1.0Ep was higher than those of 0.6Ep. A greater amount of irrigation water was applied at 1.0Ep than 0.6Ep. Applying sufficient amount of irrigation water resulted in higher amount of water recharged into root zone as compared to insufficient amount of irrigation. Under 0.6Ep we practised deficit irrigation which meant insufficient amount of irrigation was replenished. Deficit irrigation reduces the amount of water content in the soil compared to full irrigation.

Plant growth

Average plant heights at the early growth stage, middle

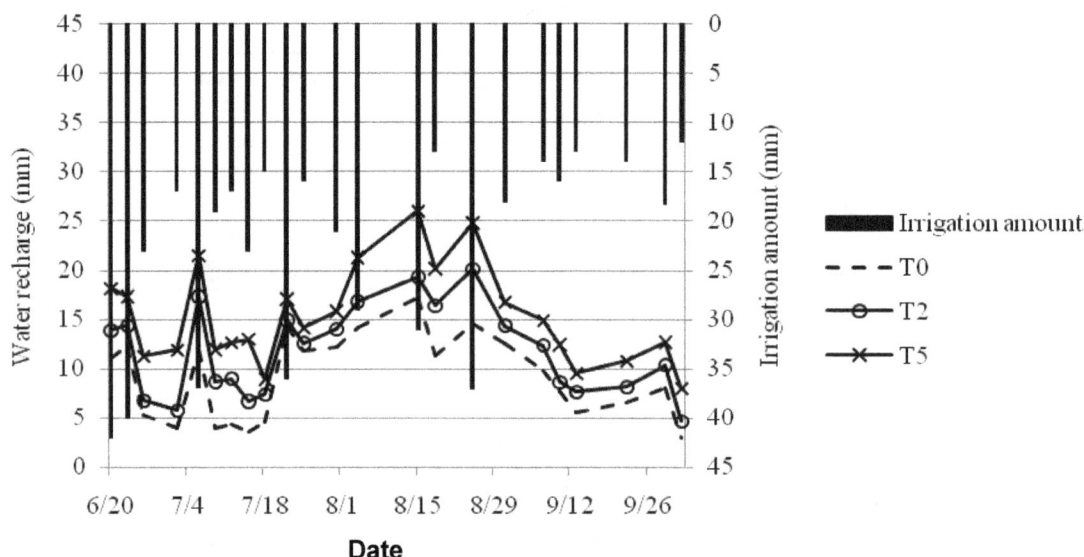

Figure 5. Amount of irrigation water and water recharge under T0, T2 and T5 for 1.0Ep.

Table 3. Water recharge under T0, T2 and T5 for 0.6Ep and 1.0Ep.

Symbol of mulch	0.6Ep			1.0Ep		
	Total (mm)	Mean (mm)	SD (mm)	Total (mm)	Mean (mm)	SD (mm)
T0	136	5.9[a]	2.6	210	9.1[a]	4.3
T2	170	7.4[b]	2.8	270	11.7[b]	4.5
T5	212	9.2[c]	3.4	351	15.2[c]	4.9

Means followed by different letters are significantly different at P ≤ 0.05; SD is standard deviation.

Table 4. Average plant height under T0, T2 and T5 for 0.6Ep and 1.0Ep.

Symbol of mulch	0.6Ep			1.0Ep		
	ES(cm)	MS(cm)	LS (cm)	ES (cm)	MS (cm)	LS (cm)
T0	38.9[a]	117.1[a]	192.9[a]	38.8[a]	129.9[a]	185.0[a]
T2	41.4[a]	109.4[a]	197.3[a]	41.4[a]	137.3[a]	197.2[a]
T5	46.0[b]	104.9[a]	200.5[a]	46.0[b]	164.2[b]	219.2[b]

Means followed by different letters are significantly different at P ≤ 0.05; ES is early growth stage; MS is middle growth stage; LS is late growth stage.

growth stage and late growth stage under T0, T2 and T5 for 0.6Ep and 1.0Ep are shown in Table 4.

For 0.6Ep, the plant heights under T0, T2 and T5 varied little for the three growth stages. Covering the drip lines with either a 2 cm thickness or a 5 cm thickness of dune sand had no effect on vegetative growth of sorghum. The drip lines were observed to expand and protrude to the surface under 2 cm sandy mulch thus causing non-uniform water distribution along the drip lines. The thin layer of dune sand resulted in higher temperature variations on the drip lines which caused expansion during daytime and contraction during night time. For 1.0Ep, the plant heights under T5 were greater than and significantly different (p = 0.05) from those under T0 and T2. Covering the drip lines with a 5 cm thickness of the dune sand significantly supported the vegetative growth of sorghum. The possible explanation for the support of vegetative growth of sorghum could be that the 5 cm dune sand mulch improved the amount of water recharged into the root zone.

At the early growth stage, the plant heights of 0.6Ep and 1.0Ep did not vary significantly. At the middle growth stage, the plant heights of 1.0Ep were significantly greater (p = 0.05) than those of 0.6Ep. At the late growth

Table 5. Dry matter yield under T0, T2 and T5 for 0.6Ep and 1.0Ep.

Symbol of mulch	0.6Ep			1.0Ep		
	Yield (kg/m^2)	Mean weight (kg/plant)	SD (kg)	Yield (kg/m^2)	Mean weight (kg/plant)	SD (kg)
T0	1.8	0.19[a]	0.02	2.5	0.19[b]	
T2	1.9	0.22[a]	0.03	2.2	0.24[a]	0.08
T5	2.1	0.32[b]	0.07	3.0	0.34[a]	0.07

Values followed by different letters are significantly different at P ≤ 0.05; SD is standard deviation.

stage, the plant heights of T0, T2 and T5 showed little variation.

An irrigation level of 0.6Ep did not supply sufficient amount of irrigation water to support sorghum growth. The combination of T5 and 1.0Ep resulted in the highest amount of water recharged into the root zone which supported sorghum growth. This combination supplied sufficient amount of irrigation water and improved water recharge to the root zone.

Dry matter yields under T0, T2 and T5 for 0.6Ep and 1.0Ep are shown in Table 5. For 0.6Ep, the dry matter yield under T5 was marginally higher than those of T0 and T2. For 1.0Ep, the dry matter yield of T5 was higher than those of T0 and T2. As shown in Table 5, for 1.0Ep, dry matter yield under T2 was higher than under T0. The dry matter yields under T0, T2 and T5 for 1.0Ep were higher than those of 0.6Ep. The combination of T5 and 1.0Ep resulted in the highest dry matter yield.

Conclusion

Though the 2 cm dune sand mulch also increased water recharged into the root zone, it was not practical as dune sand mulch on drip lines due to protrusion. Based on these results, the minimum thickness of the dune sand mulch should be 5 cm.

Water recharged into the root zone was highest under T5 and 1.0Ep. The combinations of the 5 cm dune sand mulch and a sufficient amount of irrigation water resulted in increased plant height and dry matter yield of the sorghum.

The 5 cm dune sand mulch is a better alternative to subsurface drip irrigation since no special machine is necessary to cover the drip lines by 5 cm of the sand soil. It is easy to remove the drip lines after the growing season.

REFERENCES

Agodzo SK, Nishio T, Yamamoto T (1997). Trickle irrigation of okra based on small pan evaporation schedule under glasshouse condition. Rural Environ. Eng. J. 33:19-36.

Allen RG, Pereira LS, Raes D, Smith M (1998). Crop evapotranspiration, FAO Irrigation and Drainage Paper 56. Food and Agriculture Organization of the United Nations, Rome, Italy.

Brouwer C, Goffeau A, Heibloem M (1985). Irrigation water management: Training manual No. 1- Introduction to irrigation. Food and Agriculture Organization of the United Nations, Rome Italy.

Cobos DR, Chambers C (2010). Calibrating ECH$_2$O soil moisture sensors. Decagon Devices Inc., Pullman, Washington.

Doorenbos J, Pruit WO (1977). Crop water requirements, FAO Irrigation and Drainage Paper 24. Food and Agriculture Organization of the United Nations, Rome, Italy.

Hillel D (1998). Environmental soil physics, Academic Press, San Diego, California. p. 534.

Klute D (1986). Methods of soil analysis. Part 1 Physical and mineralogical methods monograph 9ASA SSSA, Madison, WI.

Lamm FR, Trooien TP (2005). Dripline depth effects on corn production when crop establishment is non-limiting. Applied Eng. Agric. 21(5):835-840.

Longo FD, Spears TD (2003). Water scarcity and modern irrigation. Valmont water management group, Valmont Industries, Inc., Valley, Nebraska.

McIntyre BD, Speijer PR, Riha SJ, Kizito F (2000). Effects of mulching on biomass, nutrients, and soil water in banana inoculated with nematodes. Agron. J. 92:1081-1085.

Neelam P, Rajput TBS (2007).Effect of drip tape placement depth and irrigation level on yield of potato. Agric. Water Manage. 32(3):209-223.

Phene CJ, Blume MF, Hile MMS, Meek DW, Re JV (1983).Management of subsurface trickle irrigation systems. Transactions of the American Society of Agricultural. pp. 83-2598.

Qiu GY, Ben-Asher J, Yano T, Momii K (1999). Estimation of evaporation using differential temperature method. Soil Sci. Soc. Am. J. 63:1608-1614.

Richards LA (1948). Porous plate apparatus for measuring moisture retention and transmission by soil. Soil Sci. 66:105-110.

Solomon K (1993). Subsurface drip irrigation: product selection and performance, Subsurface drip irrigation: Theory, practices and applications. CATI Publication No. 9211001, California State University FRESNO, California.

Ssali H, McIntyre BD, Gold CS, Kashaija IN, Kizito F (2003). Effects of mulch and mineral fertilizer on crop, weevil and soil quality parameters in highland banana. Nutrient cycling Agroecosyst. 65:141-150.

Wang D, Shannon MC, Grieve CM, Yates SR (2000). Soil water and temperature regimes in drip and sprinkler irrigation, and implications to soybean emergence. Agric. Water Manage. 43:15-28.

Yamamoto T, Cho T (1978). Soil moisture content distribution in main root-zone and water application efficiency of crop-studies on trickle irrigation method in sand field. Trans. Japanese Soc. Irrig. Drain. Reclam. Eng. 75: 33-40.

Yamanaka T, Yonetani T (1999). Dynamics of the evaporation zone in dry sandy soils. J. Hydrol. 217:135-148.

2

Elimination of lime causing clogging in emitters by chemical methods in drip irrigation

Ömer Faruk KARACA[1] and Kenan UÇAN[2]

[1]Department of Biosystems Engineering, Faculty of Engineering and Architecture, University of Bozok, 66200 Yozgat, Turkey.
[2]Department of Biosystems Engineering, Faculty of Agriculture, University of Kahramanmaras Sutcu Imam, 46100 Kahramanmaras, Turkey.

Emitter clogging is one of the most serious problems facing users of drip irrigation systems using poor irrigation water and groundwater. It can affect negatively irrigation system performance and crop yield. The objectives of this study were to to determine the appropriate HCl amount (2.5, 5, 10, 500 and 1000 ppm) and whether these acid rates could open the clogged emitters (25, 50 and 100%) with additional $CaCO_3$ or not. Firstly, laterals were filled with water without using any pomp, then water-filled laterals were inversed and placed on the ground under the sun. But, it was realized that there were no clogging using that method. Therefore, based on calculated Langelier saturation index, lime-water was used. After the desired clogging, water with acid mixture were injected into the system and water samples were collected one in five meters for each lateral. Chemical analysis was carried out for each sample. The results of the study showed that 2.5, 5 and 10 ppm acid rates could not sufficiently open the clogged emitters. The highest percentages of opening were obtained in 1000 ppm with 40.91 for 25% clogged laterals.

Key words: Drip irrigation, HCl application, Langelier saturation index, water quality.

INTRODUCTION

Drip irrigation is the most efficient method for irrigation, especially arid and semi-arid regions. However, the clogging of emitters is one of the most serious problems facing users of drip irrigation systems using poor irrigation water and groundwater. This undesirable case can adversely affect the system performance and crop yield (Dehghanisanij et al., 2007; Gilbert et al., 1979). Bucks and Nakayama (1982) proposed an irrigation water quality classification for potential clogging hazard. They categorized the clogging into three essential categories. These are physical (by suspended solids), chemical (by precipitation of lime and scales) and biological (by bacterial and algal growth). Clogging mostly is caused by one or more of these factors. The most common physical causes of clogging of drip emitters are sand particles, which are usually found in surface and unstrained underground water. Biological clogging results from the growth of bacterial and algal growth within drip tapes and emitters. They combine with clay particles to block the emitters. Chemical precipitate is usually caused by exceeding precipitation of one or more of such minerals as calcium, magnesium, iron and manganese.

The most dangerous of these are sulphur and iron precipitations. Calcium carbonate or lime precipitation problem essentially depends on the pH of the water. Keller and Bliesner (1990) found that the water containing bicarbonate concentration greater than 2 me/l and a pH value greater than 7.5 likely could create calcium precipitation. Calcium carbonate precipitation can be observed in the form of a whitish deposit in emitters and micro-sprinklers. The relationship between bicarbonate and the pH of water showed that falling pH could prevent or reduce clogging of the carbonate in the system.

In order to predict what might cause chemical plugging of microirrigation system emitters, the process of mineral deposition must be understood. Carbon-dioxide gas (CO_2) is of particular importance in the dissolution and deposition of minerals. Water adsorbs some CO_2 from the air, but larger quantities are adsorbed from decaying organic matter as water passes through the soil. Under pressure, as is groundwater, the concentration of CO_2 increases to form carbonic acid. This weak acid can readily dissolve mineral compounds such as calcium carbonate to form calcium bicarbonate which is soluble in water. This process allows calcium carbonate to be dissolved, transported, and under certain conditions, again redeposited as calcium carbonate (Pitts et al., 1990).

There are some methods used to prevent clogging or partially clogging of emitters. The most usable and easy methods of these is acid injection. As a rule of thumb, the pH of water in the system should be lowered to near 2.0 to achieve maximum effectiveness of the acidification (Rible and Meyers, 1986). Acidification treatment lowers the pH so that the compounds and precipitates that are strongly bonded and insoluble under normal conditions can dissociate. The acid supplies an excessive amount of highly reactive positive hydrogen ions to the irrigation water. These hydrogen ions react with the dissociated anions to form soluble compounds that can be flushed from the system (Boman and Ontermaa, 1994).

Acids have been used to lower pH and reduce the potential for chemical precipitation (Pitts et al., 1990). Sulfiric (H_2SO_4), hydrochloric (HCl), phosphoric (H_3PO_4) and nitric (HNO_3) acid are used for this purpose. Coelho (2001) found that for all types of emitters, excepting one, it was seen that applied chlorine (Cl) rate enhanced the flow rate. To ensure continuity of free Cl in closed pipe system, the initial concentration of 5-6 ppm was required (Granberry et al., 2009). Stephen (1985) Suggested injecting Cl level with 30-60 min day^{-1} and 10-20 ppm concentration in order to remove the consisted deposits. The objectives of this study were to determine the appropriate HCl amount (2.5, 5, 10, 500 and 1000 ppm) and whether these amount could open the clogged emitters (25, 50 and 100%) with additional $CaCO_3$ or not. In addition, the study also aims to determine when the natural clogging of emitters with sun effect and to determine the effect of clogging along lateral line.

MATERIALS AND METHODS

Experimental location

Due to its location, Kahramanmaraş has a mild Mediterranean climate with a twist of South-Eastern Anatolian climate. The average temperature is 16.5 °C all year round in the city centre. Summers are hot and dry with an average of 35°C (95°F) but can go higher than +40°C (104°F). Winters are cold and damp with temperatures ranging from 0-5°C (32-41°F). The average annual rainfall is 709.8 mm. Rain falls mostly during the winter. Kahramanmaraş has 14.237 km^2 area and is the 13th big province of Turkey.

Experimental layout

The study was conducted in a research field in Agricultural Faculty of Avsar Campus of Kahramanmaras Sutcu Imam University in 2009. Research area is located at 37° 36' N latitude, 36° 55' E longitude and altitude is 600 m. For the experiment, a flat topography was selected (Figure 1).

In this study, the percentages of clogging were selected as 25, 50 and 100% and the concentrations were selected as 2.5, 5, 10, 500 and 1000 ppm. The main reason for high concentration selections as 500 and 1000 ppm were to determine which concentration opens the emitters completely. The study was designed in a randomized complete plot design with three replications.

The schematic view of system is given in Figure 2. The system had 4 plastic tank with 1000 L volume, main pipe with 40 and 63 mm diameters. Laterals had 20 cm emitter space, 3 l/h emitter flow rate and 16 mm diameter in-line emitters. Each lateral had 50 emitters and each of them was 10 m long. To prevent the accumulation of chemicals along the flow path in the emitter, operating pressure must not be less than 1 atm (Güngör et al., 2004). Therefore, the system was operated under 1.5 atm constant pressure.

Properties of water source

Irrigation water used in the study was provided from tap water based in Avsar Campus of the Kahramanmaras Sutcu Imam University. The irrigation water was analysed by methods as specified by Tüzüner (1990).

Determination of Langelier saturation index value

Langelier saturation index (LSI) was used to provide desired clogging levels in emitters. According to this index, lime water constituted by adding lime in specific rates were injected into the emitters. LSI is an index that determines the carbonate deposit. This helps determine the scaling potential of the water. A positive LSI indicates that $CaCO_3$ can precipitate. A negative LSI indicates that the water is corrosive to steel. Metcalf and Eddy (2003) used the index of LSI or Ryzner for estimation of calcium carbonate concentration in water so as to determine the potential of $CaCO_3$ precipitation formation of waste waters. LSI classification and factors used in calculation is shown in Table 1. According to data analysis, LSI values were determined with the following equation by the help of actual factors:

$$LSI = pH + AF + CF + TF - 12.1$$

Where, pH, Irrigation water acidic-alkaline value (measured by pH meter); AF, alkalinity factor (according to Na ion concentration); CF, calcium hardness factor (according to Ca ion concentration); TF,

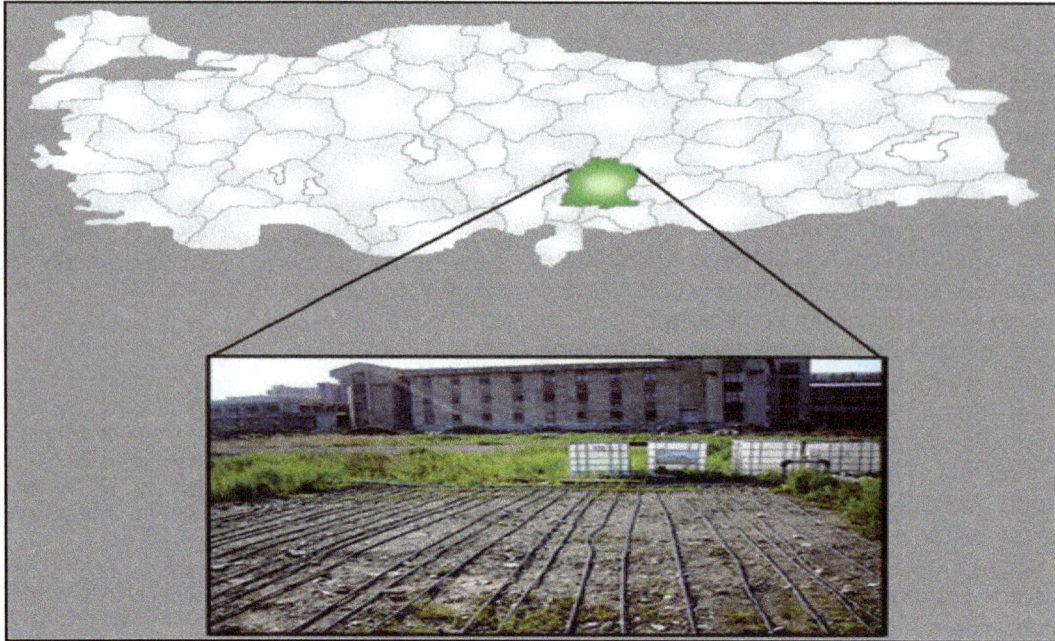

Figure 1. Experimental Layout.

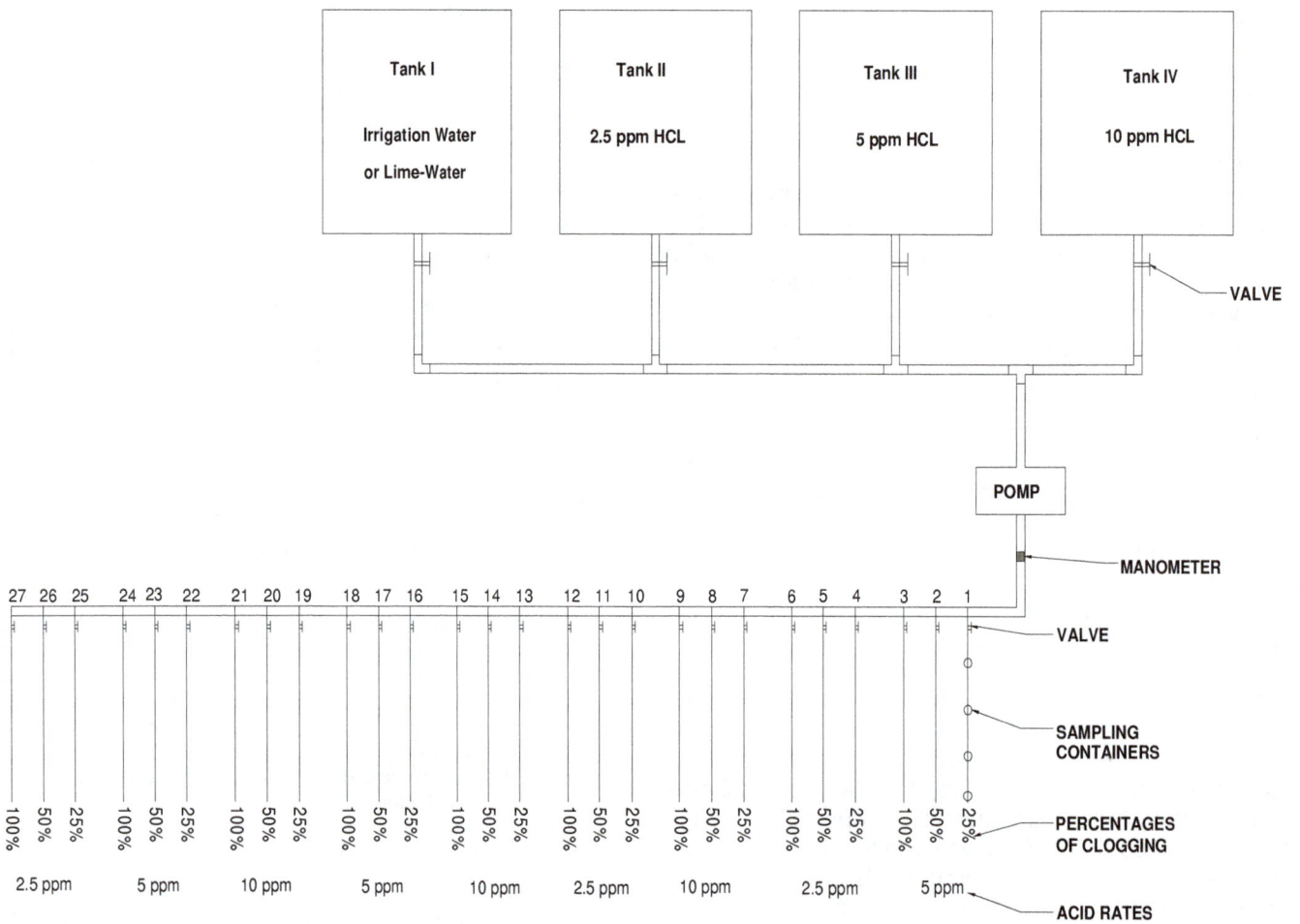

Figure 2. The schematic view of system.

Table 1. Langelier saturation index factors.

Temperature (°C)	Factor	Alkalinity (ppm)	Factor	Calcium hardness (ppm)	Factor
0	0.0	5	0.7	5	0.3
3	0.1	25	1.4	25	1.0
8	0.2	50	1.7	50	1.3
12	0.3	75	1.9	75	1.5
16	0.4	100	2.0	100	1.6
19	0.5	150	2.2	150	1.8
24	0.6	200	2.3	200	1.9
29	0.7	300	2.5	300	2.1
34	0.8	400	2.6	400	2.2
41	0.9	800	2.9	800	2.5

Figure 3. Scale and deposit on laterals during lime-water application.

temperature factor (according to measured irrigation water temperatures)

According to obtained LSI values, if, LSI > +2.0, Highly scale occurs; +0.5 < LSI < +2.0, scale occurs; LSI = 0, water is neutral; -2.0 < LSI < -0.5, water is slightly corrosive; LSI < -2.0, water is highly corrosive.

Distribution of water into the system

Firstly, laterals were filled with water without using any pomp, water-filled laterals were inversed and placed on the ground under the sun. Until the water in the laterals had evaporated (approximately 2 h), water samples were not taken. After the water in the laterals had evaporated, water samples were taken by measuring the flow rate. So, percentages of clogging were determined. These procedures were repeated throughout the day and totally, it was continued for a month. Finally, it was realized that there were no clogging using these procedures.

Hence, until the desired percentages of clogging were reached (25, 50 and 100%), lime-water mixtures was determined by LSI values. At the end of each lime-water application, the clogged emitters were determined by injecting clean water and measuring the flow rates. When the desired percentages of clogging were reached, the mini-valve of this lateral was closed (Figure 3). For the other laterals, it was continued in the same way until it reached the desired rates. In the determination of percentages of clogged

emitter, similar to that done by Smajstrla et al. (1983), the method of control of clogging emitters in total emitters were used. For this reason, the systems were installed in order that each lateral has 50 emitters.

Properties of acid and usage

Muriatic (HCl) acid is probably the most effective in dissolving majority of the scales. Muriatic acid generally does not produce harmful precipitation reactions and also it is often chemically the most effective in dissolving the precipitates. Therefore, HCl acid was used to open the clogged emitters. By injecting HCl acid to the system, it was aimed to removing the clogging by dissolving the lime and removing it from the emitter.

In this study, acid evenly was mixed in the tank and it was injected throughout the system and allowed to stay for 30 min. intermittent chlorination method was used for injection as described by Netafim (2000). When the water was seen at the end of laterals, valve of the mixture of acid-water were turned off and waited for occuring reaction between acid and lime. Application was maintained until the mixture of acid-water in tank was run out. Meanwhile, as the acid application proceeds, water samples were taken from the tank and the laterals for chemical analysis. As described in Stephen (1985) and Evans (2001), water samples were taken from emitters in every two meters with 30 min intervals. The main mixture was obtained by mixing all subsamples for one

Table 2. Irrigation water analysis results and water quality criteria.

Ion	me/L	ppm	Other water quality criteria	
Ca^{+2}	1.34	26.75	EC (dS/m)	0.42
Mg^{+2}	2.53	30.39	pH	7.60
Na^+	0.21	4.80	SAR (me/l)	0.15
K^+	0.03	1.20	RSC (me/l)	-3.72
CO_3^{-2}	0.00	0.00	% Na	5.08
HCO_3^-	0.15	9.15	TDS (ppm)	256.70
Cl^-	0.43	15.27	Hardness ($CaCO_3$) (ppm)	57.14
SO_4^{-2}	3.53	169.42		

Table 3. Important Quality Parameters of Irrigation Water (Çolakoğlu, 2010).

Quality class of water	EC (dS/m)	TDS (ppm)	Na (%)	Cl (me/l)
Excellent	<0.25	<175	<20	<0.5
Good	0.25 - 0.75	175 - 525	20 - 40	0.5 - 1.25
Middle	0.75 - 2.0	525 - 1400	40 - 60	1.25 - 2.5
Poor	2.0 - 3.0	1400 - 2100	60 - 80	2.5 - 5.0
Very Poor	>3.0	>2100	>80	>5.0

EC, Electrical conductivity; TDS, total dissolved solids, SAR, sodium adsorption ratio.

lateral. As described in Tüzüner (1990), electrical conductivity (EC) - pH measurements and anion (CO_3^-, HCO_3^-, Cl^- and SO_4^-) and cation (Ca^{++}, Mg^{++}, Na^+ and K^+) analyses were performed. Then the changes on total dissolved solids (TDS), sodium adsorption ratio (SAR), Na% and Cl were determined. Consequently, the relationship between acid rates injected into the system and percentages of openings were compared. Data was analyzed by using Statistical Package for Social Sciences (SPSS) program for the randomized complete plot design with three replications. Chemical analyses data were compared with lower and upper limit values of current existing literature.

RESULTS AND DISCUSSION

Properties of water source

The irrigation water was analysed by methods as described by Tüzüner (1990) and the results obtained are given in Table 2. According to chemical analysis results, the hardness leading to the formation of $CaCO_3$ wads calculated as 57.14 ppm. pH values of irrigation water were determined as 7.6. TDS were determined as 256.70 ppm. If hardness is between 150 and 300, it causes medium sized emitter clogging (James, 1988; Hills et al., 1989; Kuslu et al., 2005). On the other hand, the amounts of TDS less than 550 ppm, causes a small sized clogging, while higher values causes medium sized plugging (Pitts et al., 1990). So, water source used had no hardness value.

Table 2 indicates that current irrigation water source belong to C_2S_1 class. US Salinity Laboratory Staff (USSL) (1954) method were used to determine class type. This

method considered two factors such as EC and SAR. As a result, current irrigation water was determined as 2nd class and had no negative effect on soil and plants. Since Cl value is less than 0.5 and TDS value between 175-525 ppm, irrigation water had no negative effect on soil and plants. It can be used safely (Table 3) (Çolakoğlu, 2010).

Determination of Langelier saturation index value

As a result of analysis, the optimum LSI value was determined as 2.48. The first value created highly scale was selected. This mixture was made up to 1000 milliliters with water, 0.30 g of lime and 0.2 g of sodium hydroxid (NaOH). The NaOH in this mixture were added to increase dissolution amount of lime (Table 4).

Distribution of water into the system

In order to achieve the desired percentages of clogging, in total, 12 tests were performed. As soon as the desired percentages of clogging were reached, the existing lateral was closed, but injection of the mixture into the other laterals continued.

The process was terminated when the desired cloggings were ensured for all laterals. In total, 9 HCl acid applications were performed in different dates. These applications were continued approximately for 39 days. During each application, the concentration of 2.5, 5, 10, 500 and 1000 ppm acid were diluted with 1 ton of

Table 4. Determination of appropriate LSI value.

No	Water (mlt)	Lime (g)	NaOH (g)	pH	EC (dS/m)	T (°C)	TF	Ca (ppm)	CF	Na (ppm)	AF	LSI	LSI criteria
1	-	-	-	7.60	0.41	26.5	0.650	16.6	0.706	3.5	0.490	-2.65	Highly corrosive
2	500	10.00	1.00	11.69	12.86	28.3	0.686	104	1.616	42	1.604	3.50	Highly scale
3	500	7.50	1.00	11.98	13.55	27.1	0.662	211	1.922	177	2.254	4.72	Highly scale
4	500	5.00	1.00	11.89	12.70	27.4	0.668	239	1.978	172	2.244	4.68	Highly scale
5	1000	2.00	0.70	11.89	8.63	26.9	0.658	305.8	2.106	502	2.677	5.23	Highly scale
6	1000	0.50	0.58	11.53	3.84	28.1	0.682	49	1.288	435	2.626	4.03	Highly scale
7	1000	0.30	0.20	11.00	1,03	27.3	0.666	15.2	0.657	178.6	2.257	2.48	Highly scale

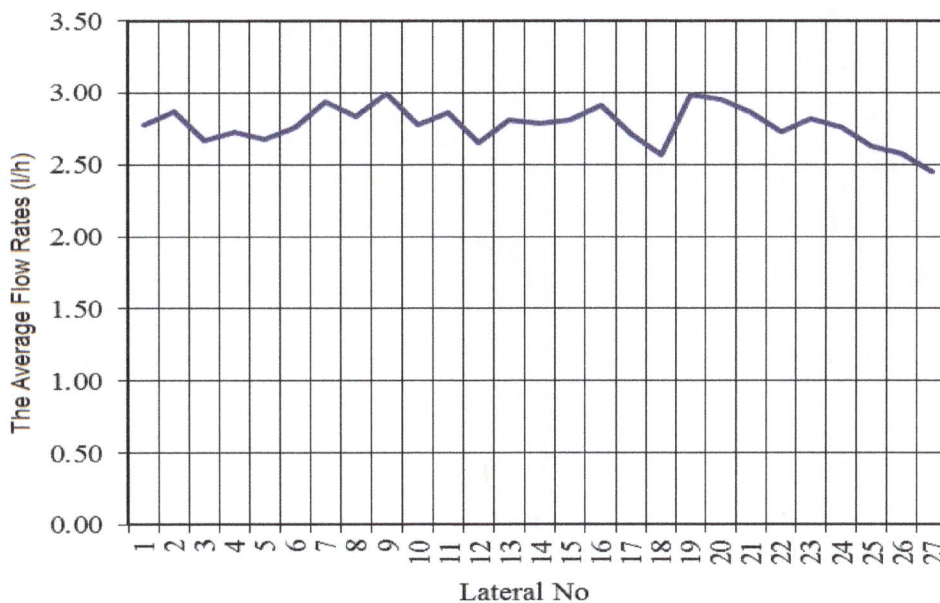

Figure 4. The average flow rate changes for 27 lateral lines.

irrigation water.

Statistical analysis

Analysis of variance of data showed that there was a significant difference between acid applications. But differences between the percentages of clogging were found not significant (Table 5). In the study, it was determined that the percentage of opening depended on the percentage of clogging and the application amounts of acid shown in Table 6. In the other acid applications excluding 1000 ppm, the percentages of opening increased with increased percentage of clogging. The lowest opening was at laterals of 2.5 ppm with 3.25% and the highest opening was at lateral of 1000 ppm with 40.91%. At the concentrations of 500 and 1000 ppm, higher percentages of opening were not reached. The reason of this could be that the calcium carbonate reacted

with HCl acid and the Cl remained free. Then, free Cl ions were precipitated by forming composite with ions such as Ca, Mg, Na in irrigation water.

Effect of acid on clogging and system

According to water quality classification, water analysis results were classified as excellent, in terms of Cl values. They had no effect. However, the concentration of 500 and 1000 ppm was classified as severe (Ayers and Westcot, 1989). In terms of toxic effect in the soil, the concetrations of 500 (55.86) ppm and 1000 (94.25) ppm tended to be slightly harmful, but the concentration of 2.5, 5 and 10 ppm were found to be harmless (Table 7). In the study, it was determined that flow rates along lateral lines differ by percentages of clogging. An average value was determined by using 27 lateral lines (Figure 4). The number one represented the nearest lateral to the pump,

Table 5. Result of variance analysis.

Parameter	Error sum of squares	Degree of freedom	Mean square errors	F	P
Application amounts of acid (V_1)	83.780	2	41.890	7.365	0.004**
Percentage of clogging (V_2)	21.811	2	10.906	1.917	0.171[ns]
Error	125.125	22	5.688		
General	1244.068	27			

**, Significant at $p < 0.01$; ns, not significant.

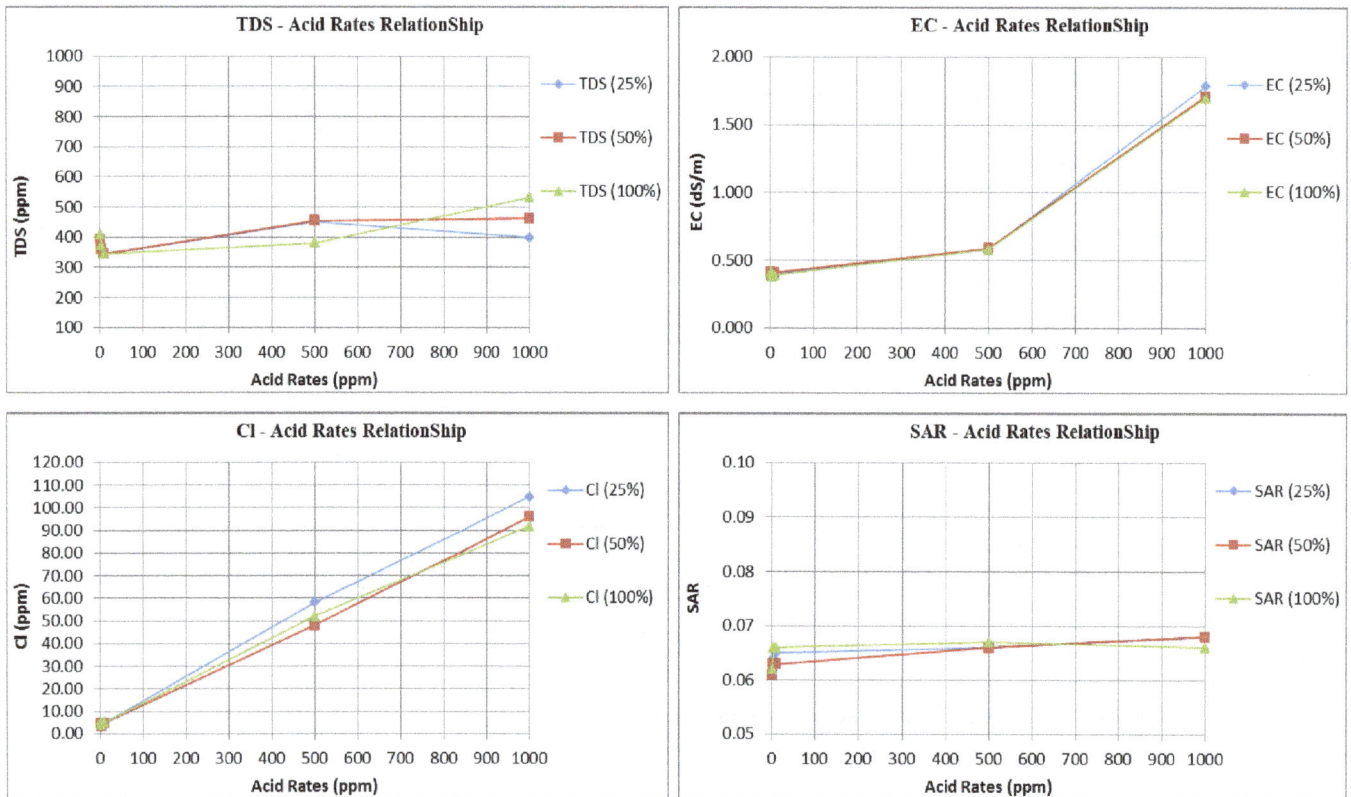

Figure 5. Changes in TDS, EC, Cl and SAR with different acid rates.

the number twenty-seven represented the furthest lateral to the pump. The furthest lateral had the most clogged emitters, in other words, the lowest flow rates.

pH, EC, SAR and TDS changes were determined that depended on amounts of acid. As acid rates increased, EC and Cl increased. But TDS and SAR had no changes (Figure 5). As specified in Evans (2001), Cl should be between pH 5.5-6.0 for optimum effectiveness. In this study, pH values (for all application) ranged from 7.0-8.0. This situation prevented the elimination of clogging. Besides, 500 and 1000 ppm had lower pH values. This meant that more clogged emitters were opened. But, this high amount of acid also led to deterioration of quality parameters of irrigation water and showed a negative

effect. These results indicated that 500 and 1000 ppm were economically not a good choice.

Conclusion

In general, it was seen that acid rates (2.5, 5 and 10 ppm) had no enough effect on clogged emitters. Besides, acid rates with 500 and 1000 ppm had almost enough percentages of opening (15.38% and 40.91%). The highest percentages of opening were seen at application rate of 1000 ppm with 40.91%, and the lowest were seen at application of 2.5 ppm with 3.25%. Nevertheless, the percentages of clogging used in this study were greater

Table 6. Changes between percentages of opening depends on percentages of clogging and application amounts of acid.

Application amounts of acid (ppm)	Percentages of clogging (%)			Mean percentage of opening (%)
	25	50	100	
	Percentage of opening (%)			
2.5	3.25^c	3.80^c	5.15^{bc}	4.07d
5	4.76^{bc}	6.99^c	7.59^{bc}	6.45^cd
10	8.20^b	8.53^b	8.38^b	8.37^{bc}
500	5.56^{bc}	9.09^b	15.38^b	10.01^b
1000	40.91^a	38.46^a	28.95^a	36.11^a

[a, ab, b, bc, c, cd, d]: Duncan groups of irrigation methods ($P > 0.05$).

Table 7. Chlorine toxic effect on soils (Çolakoğlu, 2010).

Cl (me/l)	Cl (ppm)	Effect class
<2	<70	None
2-4	70-140	Slightly harmful
4-10	140-350	Middle harmful
>10	>350	Severe

than those percentages of opening. This indicated that acid solutions with 500 and 1000 ppm did not reached the desired percentages of opening compared with 2.5, 5 and 10 ppm acid solutions. These results revealed that 500 and 1000 ppm is economically not desired. Consequently;

1. Percentages of clogging should not be more than 9% and acid applications should be at least 10 ppm for elimination of this clogging in the system.
2. At first glance, although the increase in the amount of acid is seen as a good method, it can create harmful effect on plant, soil and system with time.
3. Plants that are very sensitive to Cl must not be grown, especially in high acid rate.
4. If irrigation water has low quality and high lime, a filter system must be used before acid injection.

ACKNOWLEDGEMENT

This study was supported by Academic Research Projects of Kahramanmaras Sutcu Imam University.

ABBREVIATIONS

SAR, Sodium adsorption rate; **LSI,** Langelier saturation index; **EC,** electrical conductivity; **TDS,** total dissolved solids; **SAR,** sodium adsorption ratio.

REFERENCES

Ayers RS, Westcot DW (1989). Water Quality for Agriculture. FAO Irrigation and Drainage, Rome. p. 29

Boman B, Ontermaa E (1994). Citrus Microsprinkler Clogging: Costs, Causes, and Cures. University of Florida, IFAS Ft. Pierce Agricultural Research and Education Center 2199 South Rock Road Fort Pierce, FL 34945-3138 2A. Duda & Sons, Inc. P. O. Box 788 LaBelle, FL 33935.
Bucks DA, Nakayama FS (1982). Principles and Potentialities of Trickle (Drip) Irrigation. Adv. Irrigation 1: 219-298.
Coelho RD (2001). Biological Clogging of Netafim's Drippers and Recovering Process Through Chlorination Impact Treatment. ASAE Annual International Meeting, 012231, California, USA.
Çolakoğlu H (2010). Sera Yetiştiriciliğinde Gübreleme. Toros Agricultural Industry, Trading Joint Stock Company. Agricultural Adviser. (http://www.toros.com.tr/dosyalar/ciftci-dostu/gruplar/1_18801401_tarim_serayetistiriciligi.pdf)
Dehghanisanij H, Anioji H, Riahi H, Abou Elhassan, W (2007). Effect of Emitter Characteristics and Irrigation Schemes on Emitter Clogging Under Saline Water Use. J. Arid Land Stud. 16(4):225-233.
Evans RG (2001). Microirrigation, Irrigated Agriculture and Extension Center, Department of Biological Systems Engineering, Washington State University, Prosser, WA.
Gilbert RG, Nakayama FS, Bucks DA (1979). Trickle Irrigation: Prevention of Clogging. Trans. ASAE. 22(3):514–519.
Granberry DM, Harrison, KA, Kelley WT (2009). Drip Chemigation: Injecting Fertilizer, Acid and Chlorine. University of Georgia.
Güngör Y, Erözel AZ, Yildirim O (2004). Sulama. Ankara Üniversitesi Ziraat Fakültesi Yayınları, 1540/493, Ankara.
Hills DJ, Navar FM, Waller PM (1989). Effects of Chemical Clogging on Drip-Tape Irrigation Uniformity. Trans. ASAE 32(4):1202–1206.
James LG (1988). Principles of Farm Irrigation System Design. John Willey and Sons Inc., New York. p. 543.
Keller J, Bliesner RD (1990). Sprinkle and Trickle Irrigation. New York: Van Nostrand Reinhold.
Kuslu Y, Sahin U, Anapali O, Kiziloglu FM (2005). Damla Sulama Sistemlerinde Tikanma ve Giderilmesi ile Farkli Damlatici Tiplerinin Özellikleri. GAP IV. Tarim Kongresi Bildiri Kitabı, 2.Cilt, 1094–1101. 21–23 Eylul, Sanliurfa.
Netafim Irrigation, Inc. (2000). Chlorination of Drip Irrigation Systems. www.netafim.com, Fresno, CA.
Pitts DJ, Haman DZ, Smajstrla AG (1990). Causes and Prevention of Emitter Plugging in Microirrigation Systems. Bulletin 258, Fla. Coop. Ext. Serv., IFAS, Univ. of Florida (EDIS document no. AE032).
Rible JM, Meyer JL (1986). Cleaning Drip Irrigation Systems. p. 189-191. in: D. F. Zoldoske MY. Miyaski (eds.). Micro-Irrigation Methods and Materials Update. Center for Irrigation Technology, Fresno, CA.
Smajstrla AG, Koo RC, Weldon JH, Harrison DS, Zazueta FS (1983). Clogging of Trickle Irrigation Emitters Under Field Conditions. Proc. Fla. State Hort. Soc. 96:13–17.
Stephen DE (1985). Filtration and Water Treatment for Microirrigation. Drip/Trickle Irrigation in Action, The Third International Drip /Trickle Irrig. Cong., ASAE publ, p.(1).
Tüzüner A (1990). Toprak ve Su Analiz Labaratuvarları El Kitabı. T.C. Tarım Orman ve Köyişleri Bakanlığı Köy Hiz. Genel Müd., Ankara, p.375s.
USSL (1954). Diagnosis and Improvement of Saline Alkali Soils. USDA Agr. Handbook N6o0. , Washington, D.C.

3

Adoption and impacts of an irrigation technology: Evidence from household level data in Tigray, Northern Ethiopia

Tadesse Getacher[1], Amenay Mesfin[2] and Gebrehaweria Gebre-Egziabher[2]

[1]College of Business and Economics Mekelle University, Ethiopia.
[2]Research Economist, International Water Management Institute (IWMI), Ethiopia.

The occurrence of frequent droughts due to low and erratic rainfall poses high uncertainty and agricultural production risks leading to wide spread poverty and food insecurity. Hence, for agricultural intensification, water is an entry point implying that irrigation development especially small holder irrigation and adoption of water saving technologies is very important. This paper aims at indentifying factors that affect farm households' decision on whether to adopt pumps for lifting irrigation water and at making comparisons between the income of adopters and non-adopters. A logit model and an ordinary least-squares (OLS) regression are estimated using a total sample of 301 farm households to identify the factors that influence adoption of the technology and to determine impacts. Consistent with the findings of other studies, the regression results indicated that the most important determinants of the adoption of the small-scale irrigation technology include access to ground and surface water, yearly availability of water, sex of household head, level of education, access to credit and number of adult family members. The results also showed that the adoption of irrigation technologies has significant positive effect on agricultural income relative to the non- adopters.

Key words: Small-scale, household, irrigation technology, pumps, adoption, logit.

INTRODUCTION

Ethiopia in general and Tigray region in particular faces a major challenge in poverty reduction and food security mainly due to climatic factors that severely affect its agricultural performance. According to the Ministry of Finance and Economic Development (MoFED, 2010), the major challenges encountered in implementing the past Five Year Plan (2005/06-2009/10), known as PASDEP, were mainly due to irregular rainfall patterns. On the other hand, land scarcity driven by population pressure puts agricultural intensification high on the agenda of food security and poverty reduction.

Although Ethiopia has immense surface and groundwater potential (Makombe et al., 2007; Awulachew, 2010; Awulachew et al., 2006), the country fails to use this potential to produce enough food to feed its population. While irrigation in Ethiopia has the potential to increase cereal yields by up to 40% (Diao et al., 2010), agricultural producers have used only about 5 to 6% of the country's irrigation potential mainly through large and small-scale community irrigation schemes. The reality is similar in Tigray region. Tigray has a potential of about 350 thousand hectares of irrigable land of which, only about 83 thousand hectares or 6% of the potential is currently under irrigation (Awulachew et al., 2007).

The government and people of Tigray have been working intensively on agricultural development as a result of which, crop production has increased continuously in the past 5 years. Although most of it is through scheme/community irrigation systems, irrigated land has increased from 4000 ha in 2003 to 83000 ha in 2010. Moreover, the general direction of the region is to increase the use of irrigation with a target goal of irrigating about a quarter of the cultivated area in the region (Tigray Agriculture and Irrigation development Bureau, 2010). Scheme-level irrigation is capital intensive and limited to reach many farmers. Consequently, there is a need to give more emphasis to small household-level irrigation technologies. Small household irrigation technologies, including small pumps, are smallholder-friendly that can be used in diverse water sources. Based on its technical and socioeconomic attributes, small pump irrigation is categorized as smallholder private irrigation technology.

In the context of this paper, motor pumps are small size if they are between 1 and 10 horse power (HP) and less expensive if they cost between US$ 200 and 1000. They are initiated and financed by small farmers themselves mainly to irrigate small plots of cash crops. Unlike the community level irrigation infrastructures, they are owned and managed individually or by informal group of farmers to pump water from diversified sources such as rivers, reservoirs and shallow aquifer. Furthermore, investment in small-scale household level irrigation plays crucial role to reduce the dependence of agricultural production on rainfall and climate risk and improve crop production (MoFED, 2006; Diao et al., 2010).

The use of small scale pump irrigation schemes has proven successful in various parts of Asia, Africa and Latin America over the last fifteen years. Similarly, based on information from the Bureaus of Agriculture and Water Resources of the Tigray region, motorized small pumps are among the emerging small private irrigation technologies in the rural areas. The spread of this technology is made through two channels: motor pumps distributed through the regional bureau of agriculture mainly on credit and farmers' direct purchase motor pumps using own resource which happens spontaneously and unregulated. Accordingly, data from the bureau of agriculture shows that between 2004 and 2010, the Bureau of Agriculture and Rural Development has distributed about 9675 and 20184 of motor pumps and treadle pumps, respectively (Tigray Agriculture and Rural Development Bureau, 2010).

The occurrence of frequent drought due to low and erratic rainfall poses high uncertainty and agricultural production risks leading to widespread poverty and food insecurity. This is a serious problem in the drought prone areas of Tigray region. Population pressure has pushed land holdings to its low level. On the other hand, cultivable agricultural land has almost reached its limit implying that agricultural intensification is the best available option to increase agricultural production and food security.

For agricultural intensification (such as improved input use), water is an entry point implying that irrigation development, especially smallholder private irrigation and adoption of smallholder private irrigation technologies is very important. On the other hand, lack of simple and affordable irrigation technology that fits the production conditions of smallholder is a serious limiting factor to achieve food security. While there is evidence that there is high demand from smallholder farmers' for different types of water lifting technologies (such as motorized and human powered technologies), the level of adoption of the technologies is very low. Therefore, the general objective of this paper is to investigate factors that affect farm households' decision to adopt motor pump and its impact on household income. More specifically, it is intended to identify factors that affect decision to adopt irrigation technologies and examine if adopters attain significantly higher income levels.

METHODOLOGY

Description of the study area

The total area of Kilite Awulaelo Wereda[1] is about 105,758 ha. Out of the total area, about 21,620 ha are currently being cultivated and 84,138 ha are forests and sloppy areas. From the total cultivable areas 5,235 ha (5%) are currently irrigated. Up to now only 105 water pumps, 480 treadle pumps and 26 drip irrigation are being used in the wereda – all of which were distributed by the wereda adminstration. From this figures, we can see that most farmers of the wereda are still applying traditional methods of irrigation while applying irrigation technologies has strategic importance in improving the productivity of irrigation efforts in the wereda (Wereda Kilte Awlaelo Rural and Agriculture Office, 2010).

Sampling methods and data collection

The data used in this study comes from a household survey carried out in Abraha-Atsebeha and Adikesindad tabias (communities) of Kelete-Awlaelo wereda (district) in Tigray region. Primary data was collected from 301 randomly selected farm households of which 200 households were from Adikesindad and the rest from Abraha-Atsebeha. The research used a multi stage stratified random sampling method. In the first stage; we used information from the regional bureau of agriculture and Rural Development to identify wereda (district) with a high concentration of smallholder irrigation technologies, such as bucket, treadle pump and motorized pump. In the second stage, we used information from wereda agricultural offices to select Tabias (communities) that have high adoption rates of these technologies where we found that Adikesindad and Abraha-Atsebeha have high adoption rate of motor and treadle pumps, respectively. In the third stage, we used the list of farm households in selected communities and disaggregated them into adopters and non-adopters. Out of the total households of the two tabia (2373), every 8th from the list of the households (n+8) was

---

[1]Wereda (which means district) is the third smallest administrative unit next to kebele (meaning peasant association) and tabia (community).

Table 1. Total population and sample household heads by Tabia (community).

Tabia	Total number of household heads	Number and percentage of households in the Tabia		Sample size		
		Adopters	Non-adopters	Adopters	Non-adopters	Total
Adikisindad	1427	206 (14%)	1221 (86%)	54 (4%)	146 (10%)	200 (13%)
Abraha- Atsbeha	946	510 (54%)	436 (46%)	48 (5%)	53 (6%)	101 (12%)
Total	**2373**	**716**	**1657**	**102**	**199**	**301 (12.5%)**

considered for the study. Finally, a proportional random sampling technique was used to select sample households. Out of the total household 716 are adopters whereas 1757 are non-adopters. The 8th of 716 are more or less 102 and that of 1757 were 199 households. Of the total 301 sample households, 102 were using motor and/or treadle pumps (Table 1).

A comprehensive structured questionnaire was used to capture both qualitative and quantitative information. Secondary data from bureau of agriculture, wereda office of agriculture and tabia administration offices were also used.

Methods of analysis

In this paper, both descriptive and econometric data analysis methods are used. The production system in the study areas represents a multi-crop agricultural production where land holdings is fixed, the allocation of land into crop type and the adoption of irrigation technology is possibly endogenous (Negri et al., 1990). The adoption decision of irrigation technology is discrete where a farmer can decide to adopt or not to adopt in which case the farmer faces a dichotomous decision problem to adopt or not to adopt smallholder micro irrigation technologies. In this context, smallholder irrigation technology adopters are those who were using motor pump (treadle pump) to irrigate part of their land during the survey, while the rest are non-adopters.

Empirical model

The binomial logit model is used to estimate the probability of water lifting technology adoption that is, $\Pr(y_i = 1 | x)$ where the model is transformed into the odds ratio specified as follows (Long, 1997):

$$\frac{P(y_i = 1 | x)}{P(y_i = 0 | x)} = \frac{\Pr(y_i = 1 | x)}{1 - \Pr(y_i = 1 | x)} \tag{1}$$

The odds indicate to what extent farmers have adopted smallholder irrigation technology $(y = 1)$ relative to those who didn't adopt $(y = 0)$. The log of the odds specified in Equation (2) suggests that it is linear in the logit.

$$\ln\left[\frac{P(y_i = 1 | x)}{1 - P(y_i = 1 | x)}\right] = x\beta_i \tag{2}$$

Which is equivalent to the logit model derived as:

$$P(y_i = 1 | x) = \frac{\exp(x\beta_i)}{1 + \exp(x\beta_i)} \tag{3}$$

where P denotes the probability that the ith farmer has adopted one or more type of the smallholder irrigation technologies, x_i captures household and farm level characteristics that affect household's adoption of smallholder irrigation technology, while β_i s are parameters to be estimated. A binomial logit model is useful for investigating the influences of household and farm level attributes on household's technology adoption relating the probability of smallholder irrigation technology adoption to the underlying characteristics. The dependent variable (y) is the logarithm of the odds in favor of motor/treadle pump adoption, and the parameters are interpreted as derivatives of this logarithm with respect to the independent variables. The estimated coefficients can be used to predict the adoption probability of motor/treadle pump. In the logit model, like in any nonlinear regression model, the parameters are not necessarily the marginal effects (Greene, 2000; Kennedy, 2001), but represent changes in the natural log of odds ratio for a unit change in the explanatory variables.

The logit model specified above estimates the probability of adoption of smallholder irrigation technology.

On the other hand, to estimate the effect of irrigation technology adoption on agricultural income, we used a simple Ordinary Least Squares (OLS) regression model specified as follows:

$$Y_i = \beta_0 + \beta_1 W_i + \beta_2 P_i + \varepsilon_i$$

$$y_i = \beta_0 + x_{1i}\beta_1 + x_{2i}\beta_2 \dots x_{ki}\beta_k + p_i\beta_{k+1} + \varepsilon_i$$

Where; y_i = agricultural income; x_{ki} = is a vector of household's asset endowments, and household characteristics such as household size, education level, adult hour, oxen, fertilizer usage, access to loan, contact to local development agents (workers), age of household head, TLU (Tropical Livestock Unit); p_i = is irrigation technology adoption status, a dummy variable which is 1 for adopters and 0 for non-adopters; ε_i is the error term, which is assumed to be normally distributed (zero mean and unit variance).

To predict the probability of water lifting technology adoption, family size, gender, age, river source availability, access to ground water source, education, contact to development agents (DA), distance to market, access to loan and TLU were used as explanatory variables. Variables type, unit of measures and expected signs are shown on Table 2.

RESULTS AND DISCUSSION

Analysis of the survey data showed that about 68.6% of the irrigation technology adopters were producing at least two crops per year. Furthermore, variables such as livestock and oxen ownership show that adopters of the

Table 2. Variables included in the regression equation and their expected signs.

Variable type	Type of variable	Unit of measurement	Expected direction of effect on adoption
Adult labour	Continuous	Number	+
Household sex	Binary	1 if Female, for male	-
Household age	Continuous	Years	+
River source	Binary	1 if Available, 0 for otherwise	+
River yearly availability	Binary	1 if Yes, 0 for otherwise	+
Ground source	Binary	1 if Yes, 0 for otherwise	+
Yearly ground water availability	Binary	1 if Yes, 0 for otherwise	+
Irrigation season	Binary	1 if Yes, 0 for otherwise	+
Education	Binary	1 if Literate, 0 for otherwise	+
Contact to DAs in wet season	Binary	1 if Yes, 0 for otherwise	+
Access to loan	Binary	1 if , 0 for otherwise	+
Distance to market	Continuous	Kilometres	+
Motor pump awareness	Binary	1 if Aware, 0 otherwise	+
Farm size	Continuous	*Timad*= 1/4 of hectar	+
Tropical Livestock Unit (TLU)	Continuous	Number	+

+, Positive effect; -, negative effect.

Table 3. Descriptive statistics of the major variables.

Variable description	Non-Adopters (N=199)	Adopters (N=102)	Significance Test (T-test and Chi^2)
	Mean	Mean	
Number of adult family members	4.317 (1.037)	4.873(1.912)	0.001***
Household head age	46.201 (13.957)	43.971(11.692)	0.167
Household head sex (1=male)	0.688 (0.464)	0.922(0.270)	0.000***
River source accessibility (1=available)	0.101(0.301)	0.363(0.483)	0.000***
River availability yearly (1=yes)	0.065(0.248)	0.118(0.324)	0.119
Ground source accessibility (1=yes)	0.050(0.219)	0.510 (0.502)	0.000***
Ground water availability yearly (1=yes)	0.040 (0.197)	0.245(0.432)	0.000***
Irrigation season (1=throughout the year)	0.131(0.338)	0.686 (0.466)	0.000***
Motor pump awareness (1=aware)	0.357(0.480)	0.814 (0.391)	0.000***
Farm size (Timed)	4.364(2.542)	4.437 (2.555)	0.8127
Household's livestock holding in TLU	2.078 (1.669)	2.636 (1.729)	0.007***
Male adult labour (Number)	2.106 (0.806)	2.441 (1.215)	0.004***
Total harvest (birr)	3597.623(5106.762)	6715.558(5505.570)	0.001***
Adult female labour (No)	2.211(0.616)	2.431(1.239)	0.040***
Fertilizer value (birr)	398.472 (388.142)	734.480(825.257)	0.000***
Education (1=literate)	0.276 (0.448)	0.569 (0.498)	0.000***
Distance to market (Km)	6.699 (4.570)	6.130 (4.955)	0.304
Access to loan (1=yes)	0.543 (0.499)	0.637 (0.483)	0.117
Contact to DAs in wet season	0.653 (0.477)	0.676 (0.470)	0.162

***$p < 0.01$; **$p < 0.05$; *$p < 0.1$.

new irrigation technology are wealthier than the non-adopters (Table 3). Estimates from the logit and Ordinary Least Square (OLS) models are presented in Tables 4 and 5, respectively.

The regression results show that access to water sources (ground and surface water) have positive relationship with adoption of irrigation technology implying that farm households which have better access to water sources are more likely to adopt the technology. As an irrigation technology does not stand alone, the result is not unexpected. The implication is that motor pump as well as other technologies are suitable in groundwater and surface water potential areas and that enables farm households to produce during the dry season. Evidence

Table 4. Determinants of adopting irrigation technology - regression results from binary logit model.

Variable	Coef.	SE
Adult labour	0.350**	0.170
Household sex	-2.140***	0.730
Household age	-0.031	0.020
River source	2.033***	0.516
River yearly availability	-1.097	0.729
Ground source	3.662***	0.695
Ground yearly availability	-1.164	1.032
Irrigation season	2.586***	0.560
Education	0.788*	0.466
Contact to DAs in wet season	0.319	0.471
Access to loan	0.746*	0.438
Distance to market	-0.006	0.043
Motor pump awareness	0.783	0.638
Farm size	0.017	0.077
Tropical Livestock Unit (TLU)	0.132	0.128

***$p < 0.01$; **$p < 0.05$; *$p < 0.1$.

Table 5. The impact of motor pump adoption on agricultural income: OLD result.

Variable	Coef.	SE
Irrigation status	1525.438**	672.79
Adult labour	681.951***	203.66
Household sex	-827.522	711.84
Household age	25.041	22.92
Oxen quantity	1354.876***	501.13
Fertilizer value	1.957***	0.49
Contact to Das	-910.944	605.49
Education	2656.810***	614.76
Access to loan	192.944	574.49
Tropical Livestock Unit (TLU)	.-660.717***	245.38

***$p < 0.01$; **$p < 0.05$; *$p < 0.1$.

from Debrekerbe watershed, Tigray (Tadesse et al., 2008) supports the argument. Hence, as watershed development and other type of water harvesting activities are likely to improve groundwater recharge, investment in watershed development and environmental rehabilitation activities has paramount importance.

Literate household heads are also found to be more likely to adopt motor pumps as compared to illiterate household heads implying the positive role of education on technology adoption. Tjornhom (1995) and Feder and Slade (1984) argue that educated people develop positive attitude towards new innovations and are relatively ready to take risk of new farming practices. As educated people are better informed, they can easily respond to unforeseen events.

Furthermore, access to credit can reduce capital

constraints. Results indicated that access to credit positively affects the adoption of irrigation technology. In this case, it is likely that liquidity constrained households are rationed out in the adoption process.

Male headed households are found to have higher likelihood of adopting the technology than female headed households. Given the finding above where liquidity is a major constraint for adoption, a possible explanation for this result is that female headed households are often poorer and hence are less likely to adopt the technologies.

Smallholder irrigation technologies like small pumps are labour intensive and smallholder-friendly because farmers can use family labour and diverse water sources. In line with this, our result confirms that households who have more adult family labour are more likely to adopt motor pump. In ordinary least-squares (OLS) -based impact analysis, the dependent variable is households' total agricultural income. Similar to its effect on adoption of motor pump, level of education of household head captured by years of schooling has positive effect on household's agricultural production (Table 5). This is not surprising, because relatively educated farm households are more likely to easily understand the benefits of extension packages and adopt other production enhancing inputs. This has been demonstrated in the probability of adoption of motor pump where literate farm household heads were found to be more likely to adopt smallholder irrigation technologies.

We hypothesized that adoption of irrigation technologies positively contributes to agricultural production. Both the descriptive data and regression results show that agricultural production of irrigation technology adopters was significantly higher as compared to non-adopters providing evidence to validate the hypothesis. This result is consistent with our expectation because, farm households who adopt an irrigation technology can produce more than one crop per year - usually high value crops for the market, that enable them to generate additional income. Small scale irrigation is generally considered as labour intensive compared to scheme level irrigation. In imperfect labour market conditions, households' family labour is an important factor for adoption of labour intensive smallholder technologies. Accordingly, the results show that number of adult family members of the farm households positively and significantly affects agricultural income. Furthermore, households' oxen and other livestock (in TLU) ownership has significant effect on agricultural income. As expected, the value of fertilizer that farm households have used has positive and significant influence on agricultural income.

CONCLUSIONS AND POLICY IMPLICATIONS

This paper examines factors influencing farmers' decision on whether or not to adopt smallholder irrigation technologies in the Kilete-Awlaelo district of Tigray region, in Ethiopia. Regression results showed that the availability of family labour positively affects the adoption of smallholder irrigation technologies. This is because the availability of higher number of family members working on the farm reduces the farm's external labour requirements. Moreover, in a situation where the opportunity cost of family labour is low, farm households with higher number of adult labour are likely to adopt labour intensive technologies. This is in line with the policy direction that gives due consideration for the use of agricultural technologies that can intensively use farm household labour and land.

Male headed households are found to be more likely to adopt irrigation technologies as compared to female headed counterparts. This indicates that women have not benefited much from innovations in micro-irrigation technologies. To change this gender imbalance, programs that target both gender groups will be necessary to ensure equitable adoption of practice between male and female headed households. Level of education increases the likelihood of adopting small scale technology. This indicated the fact that small scale irrigation technologies need special technical and managerial skills for their proper utilization. Hence, special training programs (on both operation and maintenance of the technologies) need to be instituted to manage irrigation technologies. Furthermore, access to credit positively influences the adoption of pumps. It is important to stress that due to capital requirement for acquisition of irrigation technologies, targeted credit programs will ameliorate the financial constraints of farmers. At the moment, the training for credit handling to small scale household farmers is poor. Access to ground and surface water is found to positively and significantly influence the adoption of smallholder irrigation technologies. This implies that the use of such micro irrigation technologies is suitable and appropriate in surface and/or shallow ground water potential areas and an extension approach of 'one fits all' is not appropriate. Hence, even though, the use of smallholder irrigation technology needs to be promoted, it is suggested that areas where these technologies are suitable need to be identified. Furthermore, the effect of wider application of smallholder irrigation technologies on the environment needed to be considered.

In general, there are important and significant differences between farm households who did and did not adopt smallholder irrigation technologies. In terms of output, in 2010 harvest season, the average per season value of production of adopters was significantly higher (about two fold) compared to the value of production of non-adopters. Furthermore, smallholder irrigation technology adopters are more likely to employ more labour as compared to non-adopters, which might be an indicator of the multi-dimensional role of smallholder

irrigation in generating employment opportunities and then highlighting the broader community benefits from adopting smallholder irrigation technologies.

The policy implications of the above findings are that improved access to financial services, educating and raising farmers awareness through extension and provision of other complementary services would enhance the adoption of these technologies. Particularly, affirmative action, in the form of targeted interventions, is needed to help female headed households benefit from the new technologies.

REFERENCES

Awulachew SB (2010). Irrigation potential in Ethiopia: Constraints and opportunities for enhancing the system. International Water Management Institute.

Awulachew SB, Menker M, Abesha, D, Atnafe T, Wondimkun Y (2006). Background: About the Symposium and Exhibition. Best practices and technologies for small-scale agricultural water management in Ethiopia: MoARD/MoWR/USAID/IWMI symposium and exhibition, held at Ghion Hotel, Addis Ababa 7-9 March: International Water Management Institute (IWMI).

Awulachew SB, Yilma AD, Loulseged M, Loiskandl W, Ayana M, Alamirew T (2007). Water Resources and Irrigation Development in Ethiopia. Working. International Water Management Institute. P. 123.

Diao X, Hazell P, Thurlow J (2010). The Role of Agriculture in African Development. World Develop. 38(10):1375-1383.

Feder G, Slade R (1984). The Acquisition of Information and the Adoption of New Technology. Am. J. Agric. Econ. 66:312-20.

Greene WH (2000). *Econometeric analyses*. New York: Macmillan.

Kennedy P (2001). *A guide to econometrics*. Fourth Edition. Cambridge, Massachusetts: The MIT Press.

Makombe G, Kelemework D, Aredo D (2007). A comparative analysis of rainfed and irrigated agriculture production in Ethiopia. Irrig. Drainage Syst. 21:31-44.

Ministry of Finance and Economic Development (MoFED) (2010). Growth and Transformation Plan 2010/11-2014-15, 1: Main Text.

MoFED (Ministry of Finance and Economic Development) (2006). Ethiopia: Building on Progress: A Plan for Accelerated and Sustained Development to End Poverty 2005/6 –2009/10, September 2006.

Negri, Donald H, Douglas H, Brooks (1990). Determinants of Irrigation Technology Choice." Western J. Agric. Econ. 15(2):213-23.

Tadesse N, Asmelash B Bheemalingeswara K (2008)."Initiatives, Opportunities and Challenges in Shallow Groundwater Utilization: a Case Study from Debrekidane Watershed, Hawzien Woreda, Tigray Region, Northern Ethiopia". Agricultural Engineering International: the CIGR Ejournal. Manuscript LW 08 008.

Tigray Rural Agriculture and irrigation development bureau, (2010). Rural and irrigation development report.

Tjornhom JD (1995). Assessment of policies and socioeconomic factors affecting pesticide use in the Philippines. MS. Thesis, Virginia Polytechnic Institute and State University, Virginia.

Wereda Kilite Awlaelo Rural and Agriculture Office (2010). Rural and irrigation development report.

4

A computer model for designing of permanent gully control structures

D. Khalkho[1], N. S. Raghuwanshi[2], S. Khalkho[1] and R. Singh[2]

[1]Indira Gandhi Krishi Vishwavidyalaya, Raipur (Chhattisgarh) – 492012, India.
[2]Department of Agricultural and Food Engineering, Indian Institute of Technology, Kharagpur (W.B.) – 721302, India.

Engineering measures of soil and water conservation are an integral part of watershed management or development, and thus they must be designed accurately and economically. Designing of various gully control structures plays a vital role in the success of any watershed development programme. Analytical designing of these structures in terms of hydrologic, hydraulic and structural stability is very important. The study aims at developing comprehensive module for software namely SCS_Designer model for designing of different permanent gully control structures (drop spillway, drop inlet spillway and chute spillway) and other structures along with testing and validation of the developed module of the model. SCS_Designer model also has the provision to check the step-wise procedure followed by the model for the designing of various structures. These help the users especially students to analyze the differences in design parameters by changing the input data. Also these saved step-wise procedure files make SCS_Designer an effective teaching tool. SCS_Designer was tested with the help of standard examples for different permanent gully control structures. The design results obtained using the SCS_Designer were compared with the corresponding design results from different sources and error analysis was performed. In most cases, the design results obtained using the model matched exactly with the respective design results given in the source. Small deviation in design parameters was due to truncation and round off errors. Thus, it was concluded that SCS_Designer is validated and can be used as a standard designing tool for different permanent gully control structures *viz* drop spillway, drop inlet spillway and chute spillway. Furthermore, it can be used as an effective teaching tool.

Key words: Gully control, structures, validation, computer model.

INTRODUCTION

Erosion encompasses a series of interrelated natural processes that have the effect of loosening and moving away soil and rock material under the action of water, wind and other geologic agents. The happening of soil removal creates a very serious problem to perform agricultural activities and thereby causes reduction in crop yield, too. The implications of soil erosion extend further when sheet erosion converts to rill erosion. Once rills are large enough to restrict vehicular access they are referred to as gullies or gully erosion. Major concentrations of high-velocity run-off water in these larger rills remove vast amounts of soil. Removal of topsoil and subsoil by fast-flowing surface water creates abrupt deep and wide gullies, of two different kinds: Scour gullies and headward erosion. In scour gullies, run-off water concentrated in rills or depressions removes soil particles through sluicing - the washing effect of running water on loose grains. Singh et al. (1981), shows that the problems of soil erosion in India, their extent, severity and nature vary greatly in different parts of the country depending upon climate, topography, soil, land use and also with the pattern of agricultural economy. According to Dhruva and Ram (1983), it was estimated that about 5334 m ton (16.35 ton/ha) of soil is detached annually due to agricultural and associated activities alone. The country's rivers carry about 2052 m tons (6.26 ton/ha). Of this, nearly 1572 m ton (29% of the total eroded soil) is carried away by the rivers to sea every year and 480 m ton is being deposited in various reservoirs, resulting in the loss of 1 to 2% of storage capacity. Das (1985) estimated that out of a total reported geographical area of 329 Mha, about 167 Mha (51% of the total) is affected by serious water and wind erosion. Out of which, about 127 Mha is subjected to serious soil erosion and 40 Mha degraded through gully, ravines, shift cultivation, water logging, salinity, alkalinity, shifting of river courses and desert area etc. The objectives of every watershed development programmes is to increase infiltration into the soil, to control damaging excess runoff, to manage and utilize runoff and to reduce soil erosion to protect land. Altogether, soil and water conservation programmes are called as watershed development programmes. Economic and safe design of structures requires sound knowledge of the subject and engineering skills along with a good field experience. It is also important to keep in mind the local design consideration, depending upon the soil texture, rainfall pattern and farming culture of the local people. Many people involved in watershed management do not posses sufficient design skills and also the available textbooks on soil and water conservation lack information on comprehensive designs specifically for particular regions. Therefore, there is a need to develop comprehensive software for soil and water conservation structures design for professional uses as well as to be used as an educational tool. Therefore, keeping these points in consideration, study was undertaken to develop software with easy and friendly GUI for the designing of various soil and water conservation structure. The model was developed using Visual Basic 6.0. The model was conceptualized to work as soil and water conservation structures designer for field worker as well as an educational tool for students to analyses the changes in the design output with changes made in the input parameters.

Overview of the permanent gully control structures module of SCS_Designer

The module for permanent gully control structures designing is an important module of the model. For permanent gully control structures designing, the user has to enter the information about the drop and the storage requirement of the structure. The software will then decide the optimal structure namely drop spillway, drop inlet spillway and chute spillway, as per the information provided by the user data. Anyhow the user can change the selection and can select the desired structure manually. The design and result displays separately by clicking on different tabs of the model. Finally, the model has provision for displaying the detailed design in text format. The user has to save the input data after the design in order to save the procedure file. This procedure file can be viewed separately from the main menu. Permanent structures, built of masonry, reinforced concrete or earth, are efficient supplemental control measures in soil and water conservation. Permanent structures are generally used in medium to large gullies with medium to large drainage area. Three basic permanent structures employed in stabilizing gullies are: (1) Drop spillway, (2) Drop inlet spillway and (3) Chute spillway. Each of these is adaptable to specific site conditions. The basic components of permanent gully control structures are inlet, conduit and outlet. These are of various types and the structures are classified and named in accordance with the form these components parts take. The working is explained by flow chart as shown in Figure 1.

Drop spillway

The drop structure is a weir structure, is limited to a maximum drop of 3 m and it is not a favorable structure where temporary spillway storage is desired to obtain a large reduction in the discharge at or d/s from the structure. The three designs in consideration were:

(1) Hydrologic design, involving the estimation of design runoff rate and flood volume, which the structures have to handle safely. This is done by computing the runoff rate for 25 to 30 years return period, which indicates that the heaviest rain occurred once in 25 to 30 years period. Runoff rate is calculated by rational method.

(2) Hydraulic design, assuming that the total drop, F should be known to the user. The design consists of determining the length of crest, L, and depth, h, of the weir to provide required capacity and to maintain an adequate freeboard under free flow condition. (3) Structural design, involving the determination of strength and stability of different parts of the structure. The various forces which act on the structure are mainly, water pressure, force due to overflow and effect of water flow below the structure (that is, seepage and sub-surface flow). The structural design comprises of estimation of horizontal pressure (equivalent fluid pressure), uplift pressure, contact pressure and factor of safety.

The discharge capacity of an aerated, rectangular weir under free flow condition is

$$Q = C\left(H + \frac{V_a^2}{2g}\right)^{3/2}$$

(1)

where, Q = discharge, cumec; H = head on weir, m; V_a =

```
                          ┌──────────┐
                          │  Start   │
                          └────┬─────┘
                               ▼
              ┌────────────────────────────────────┐
              │ Selection of Conservation Structure │
              └────────────────┬───────────────────┘
                               ▼
              ┌────────────────────────────────────┐
              │  Permanent Gully Control Structures │
              └────────────────┬───────────────────┘
                               ▼
              ┌────────────────────────────────────┐
              │     Enter fall and site condition   │
              └────────────────┬───────────────────┘
                               ▼
                    ┌─────────────────────┐
                    │   Select structure  │
                    └──────────┬──────────┘
```

| Drop Spillway | Drop Inlet Spillway | Chute Spillway |

Calculation of Discharge

Design Options
1. Hydrologic design
2. Hydraulic Design
3. Structural Design

Calculations of Dimensions of spillway

Data Input — Correction calculation — Designing crest profile — Discharge carrier design — Results — Check procedure

Data Input — Determination of R_s — Structure Design — Discharge Curve — Results — Check procedure

Equivalent pressure | Uplift pressure | Contact pressure | Stability pressure

Data input for structural design

Field conditions

Results of design

Check procedure

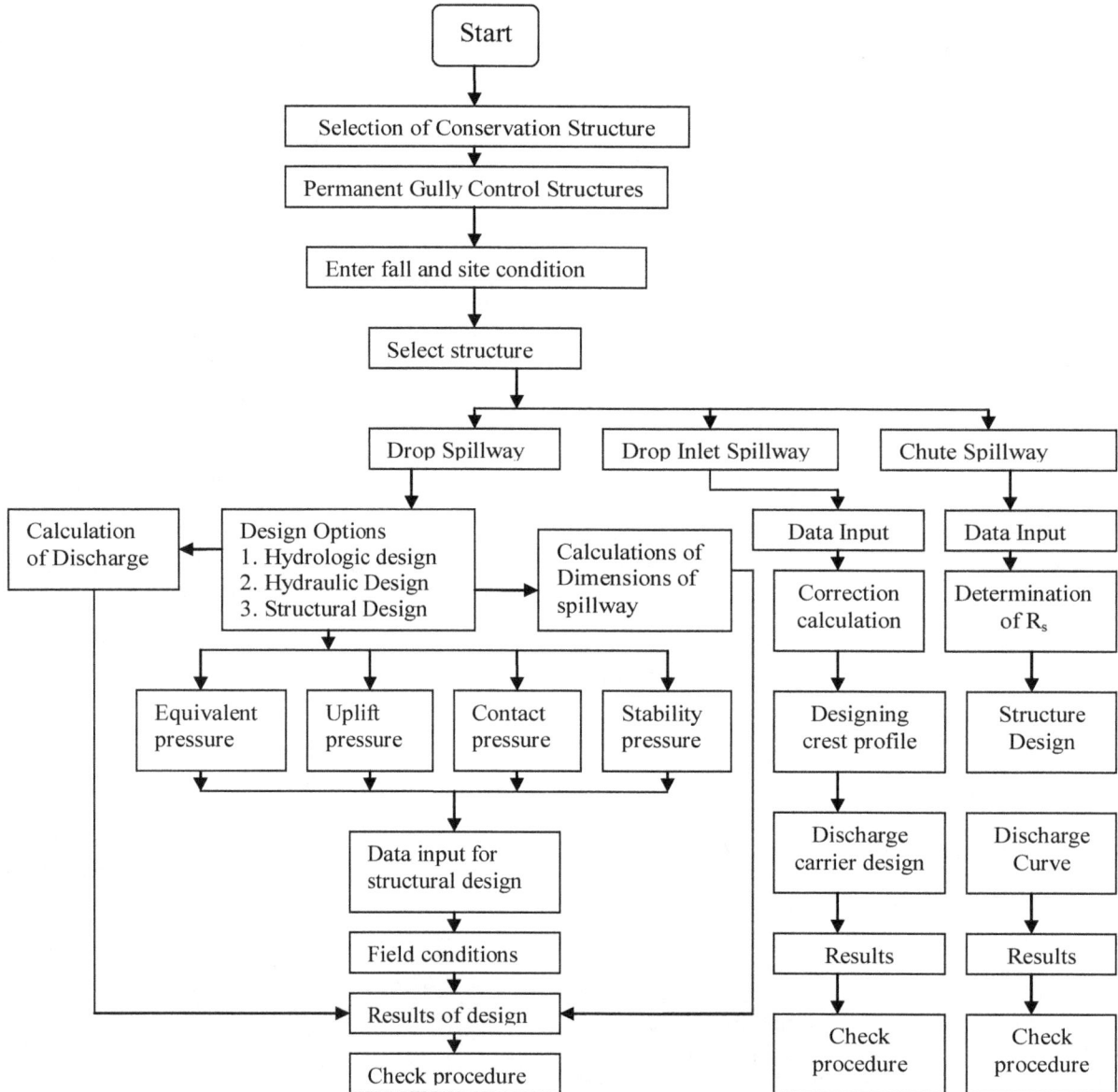

mean velocity of approach, m/sec; and C = discharge coefficient (1.77). From functional studies, it is found that

$$Q = \frac{CLh^{3/2}}{(1.10 + 0.03F)} \qquad (2)$$

$$h = \left[\frac{Q(1.10 + 0.03F)}{CL}\right]^{3/2} \qquad (3)$$

$$L = \frac{Q(1.10 + 0.03F)}{Ch^{3/2}} \qquad (4)$$

In order to determine the dimensions of the weir for a given discharge and fall condition, different combinations of L and h can be obtained for a range of assumed h values using Equations (3) and (4). Out of these combinations a suitable combination can be selected based on local site conditions. The design equations for component parts are given as follows:

Height of transverse sill (S),   S = h/3   (5)

Height of headwall (H_B), H_B = F + S   (6)

Height of headwall extension (H_E), H_E = F + S + h   (7)

Minimum length of headwall ext., E = 3h + 0.6 or 1.5 F, whichever is greater   (8)

Minimum length of apron,

$$L_B = F\left[2.3\frac{h}{f} + 0.05\right] \text{ or } 2[F + S + 0.3 - J],$$

whichever is greater   (9)

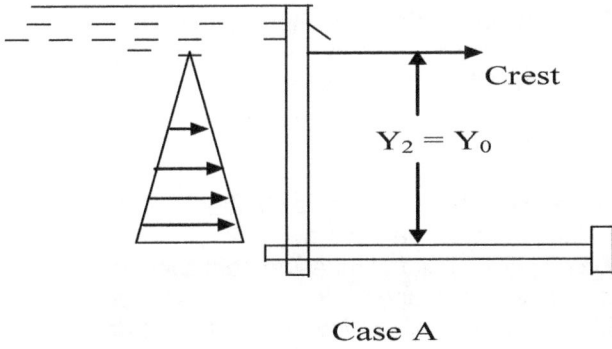

Figure 2. No fill against head wall. Y0 = elevation of crest, Y2 = elevation of saturation line above the top of apron.

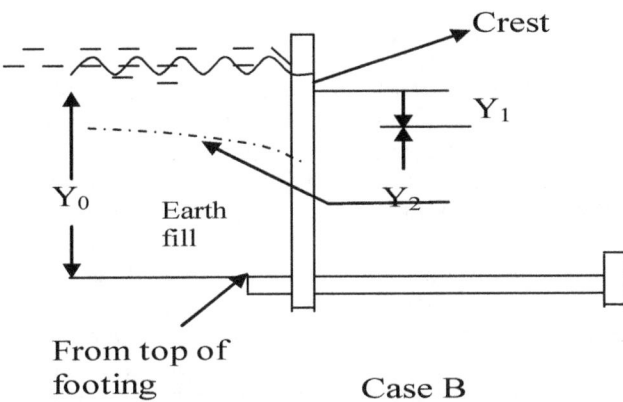

Figure 3. Sectional view of drop spillway. Y0 = elevation of crest, Y2 = elevation of saturation line above the top of apron.

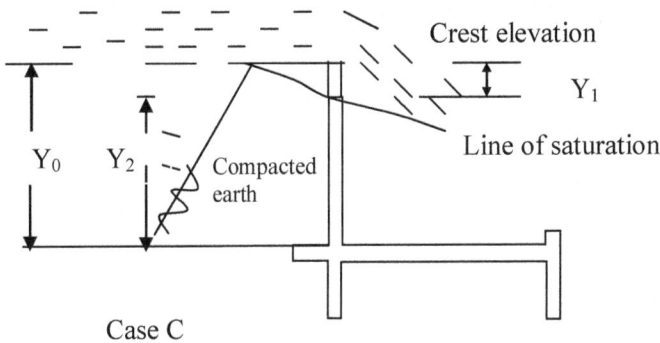

Figure 4. A compacted earth fill berm constructed to crest elevation.

Height of sidewall and wing wall at the junction, J:

$$J = 2h \text{ or } \{F + h + S - \frac{L_B + 0.13}{2}\}, \qquad (10)$$

whichever is greater

Depth of cutoff wall (C) and toe wall (T):

$$C = T = \frac{1.65(S + 0.4F + 0.75)}{4} \qquad (11)$$

Horizontal pressures (equivalent fluid pressure): It is calculated on the basis of relative permeability of foundation and backfill material, different conditions of backfill, water table conditions, flow conditions and drainage conditions. Case A (no back fill against head wall), Case B (gully graded full to crest elevation) and Case C (compacted earth fill berm constructed to crest elevation) are the three conditions of back fill that must be considered as shown in Figures 2 to 4.

Piping: The design against failure by piping is done by the line of creep theory because the majority due to piping occurred along the line of creep. Lane (USDA, 1957) defined the weighted line of creep as the sum of all steep contacts and one third of all those contacts which are flatter than 46°, between the head and tail water along the contact surface of the structure. The line of creep is the contact surface between the structure and the soil at the base of the foundation. To prevent failure of structure due to piping the weighted creep ratio (C_w) as estimated using Equation (12) must be greater than the recommended C_w values. The formula for weighted creep ratio is:

$$C_w = \frac{\sum L_v + \frac{1}{3}\sum L_H}{H} \qquad (12)$$

where, L_H = horizontal or flat contact distance, L_v = vertical or step contact distances, and H = head between head water and tail-water (difference in head).

Uplift: Due to the pressure transmitted through the water in the saturated foundation material, upward hydrostatic pressures may exist on the base of the spillway.

Contact pressure: Contact pressure should be computed for three loading conditions, namely before any backfill has been placed around the spillway, after all the backfill has been placed but without flow over spillway and with the spillway operating at design discharge capacity. Contact pressure can be estimated for rectangular base as follows:

$$P_1 = \frac{V}{A}\left(1 \pm \frac{6e}{d}\right) \qquad (13)$$

where, P1 = contact pressure at u/s or d/s edge of base, V = algebraic sum of all the vertical loads and weights that act on the structure, A = base area, e = eccentricity and d = base length.

Overturning and sliding: The structure is safe against overturning if positive contact pressures exist over the entire base area. The force resisting sliding is,

$$\mu \sum V + CA \qquad (14)$$

where $\mu = \tan\phi$, ϕ = angle of internal friction of foundation material, $\sum V$ = total vertical load, C = cohesion resistance of foundation material, and A = area of plane of sliding. The structure is safe against overturning and sliding for the ratio greater than 1.5.

Anchor design: If $\mu \sum V + CA$ is not greater than or equal to 1.5 H then it will be necessary to provide an anchor whose pull or resistance to sliding, T, will satisfy the equation:

$$T = 1.5\,H - (\mu \sum V + CA) \qquad (15)$$

Drop inlet spillway design

The drop inlet spillway is suitable in the location with fall greater than 3 m that provides an appreciable amount of temporary storage above the inlet. If the inlet is funnel shaped, this type of structures is often called as "Morning glory" or "Glory Hole" spillway. Discharge characteristics of the drop inlet spillway may vary with the range of head. As the head increase on a glory hole spillway, the control will shift from weir flow over the crest to tube flow in the transition and then to full pipe flow in the downstream portion. Equations (16) and (17) represent discharge through box inlet and pipe respectively.

$$Q = \frac{2}{3} C_d \sqrt{2g}\; L \; H^{3/2} \qquad (16)$$

where, Q = discharge (cumec) , L = crest length of a straight weir or sum of the lengths of 3 sides of box inlet (m), H = energy head of water over crest and C_d = coefficient of discharge (fraction).

$$Q = \frac{A\sqrt{2gh}}{\sqrt{1 + ke + k_c l}} \qquad (17)$$

where, l = length of the pipe, K_e = coefficient due to head loss at entrance, K_c = coefficient due to friction head loss in the pipe and depends upon pipe diameter and resistance coefficient. The K_c values can be obtained from standard graphs. The neutral slope (S_n) governs the condition for riser or pipe flow. The neutral slope for a conduit is given by, $S_n = \tan\theta =$

$$\frac{K\dfrac{V^2}{2g}}{\sqrt{1 - \left(K\dfrac{V^2}{2g}\right)^2}} \qquad (18)$$

Crest discharge: For small heads, flow over the drop inlet spillway is governed by the characteristics of crest discharge, which can be expressed as:

$$Q = C\,L\,H^{3/2} \qquad (19)$$

where, H = head measured either to the apex of the under nappe of the overflow, and length 'L' is related to some specific point of measurement such as the length of the circle at the apex, along some other reference line. The value of C will change with different definitions of L and H. If L is taken at the outside periphery of the overflow crest and if the head is measured to the apex of the overflow shape, the above equation can be written as,

$$Q = C_o(2\pi R_s)\,H_o^{3/2} \qquad (20)$$

It will be apparent that the coefficient of discharge for a circular crest differs from that for a straight crest because of the effects of submergence and back pressure incident to the joining of the converging flows. Thus, C_o must be related to both H_o and R_s and can be expressed in terms of H_o/R_s. These coefficients are valid only if the crest profile and transition shape conform to that of the jet flowing over a sharp crested circular weir at H_o head and if aeration is provided so that sub-atmospheric pressures do not exist along the lower nappe surface contact. The weir formula (Equation 19) is used as the measure of flow through the drop inlet entrance regardless of entrance, by using a coefficient which reflects the flow condition through the various H_o / R_s ranges.

Crest profiles: Values of co-ordinate to define the shape of the lower surface of the nappe flowing over an aerated sharp-crested circular weir for various conditions of p/R_s and H_s/R_s are available in tables developed by U.S.B.R.

Typical upper and lower nappe profiles for various values of H_s/R_s are plotted in terms of $\dfrac{x}{H_s}$ and $\dfrac{y}{H_s}$ for $\dfrac{p}{R_s}$.

Transition design: If friction and other losses are neglected. For a circular jet, the area is equal to πR^2. Therefore

$$Q = A.V = \pi R^2 \sqrt{2gh_v}\;\text{ and }\; R = \frac{Q}{\pi\sqrt{2gh_v}} = \frac{Q^{1/2}}{KHa^{1/4}} \qquad (21)$$

where H_a = different between the water surface and the elevation under consideration, and K = unit conversion factor (5 for FPS and 3.73 for MKS). Thus, the diameter of jet decreases with the distance of the free vertical fall for normal design applications. If an assumed total loss (to allow for jet contraction losses, friction losses, velocity losses due to direction change etc.) is taken as 0.1 H_a,

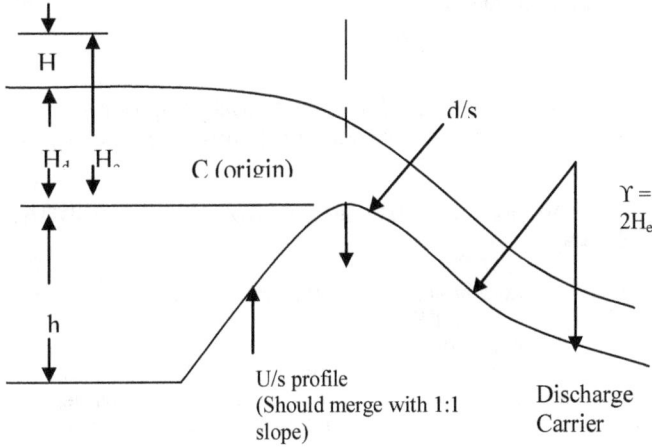

Figure 5. Control structure of the chute spillway.

the equation for determining the approximate required shaft radius can be written as:

$$R = k \frac{Q^{1/2}}{H_a^{1/4}} \qquad (22)$$

where, k = unit conversion factor (k=0.204 FPS, k=0.275 MKS).

Design of chute spillway

Chute spillway consist of four components, namely, entrance channel or approach channel, control structure, chute channel or discharge carrier, and energy dissipation arrangement in the form of a stilling basin. Flow is generally super – critical at energy dissipater. It is generally used when the drop exceeds 3.0 m. It is superior to drop–inlet spillway when large discharges are required.

Approach channel and control structure: Approach channel is constructed to lead the reservoir water upto the control structure. The friction head $h_f = S_f . l$, lost in the channel upto the spillway crest, may be calculated by Manning's equation, as follows:

$$h_f = \frac{n^2 \, V^2}{R^{4/3}} . \ell \qquad (23)$$

where, S_f = energy gradient between two points, n = Manning's roughness coefficient, V = velocity in the channel, R = hydraulic radius and L = length of the channel. Ogee shape spillway is generally provided as control structure (Figure 5). The important feature of the

ogee shape spillway is that the profile of this spillway is made in accordance with the shape of the lower nappe of the free falling jet.

Design of crest profile: The crest profiles (U/S and D/S) are drawn considering the known heads, namely, head due to velocity of approach (H_a), design head (H_d), total head ($H_e = H_d + H_a$) and height of weir crest from bed level(h). Standard table is used to determine the u/s profile coordinates of the ogee weir. The u/s profile is joined at an angle of 45° to the bottom.
Discharge equation for ogee spillway:

$$Q = C \, L \, H_e^{3/2} \qquad (24)$$

where, L = length of the spillway crest, H_e = total head over the crest including velocity head and C = coefficient of discharge, depends on many factors such as depth of approach, downstream side submergence and u/s slope. The effect of individual component is described following.

Depth of approach: Coefficient of discharge (C), depends on the depth of approach or in other words on ratio $\dfrac{h}{H_d}$; e.g., If $\dfrac{h}{H_d} > 1.33$, velocity of approach has negligible effect on Q. In such a case $H_a = 0$ (that is, $H_e = H_d$) and C = 2.2. However C is evaluated considering $\dfrac{H_e}{H_h}$ and $\dfrac{h}{H_d}$ ratios in standard graph for $\dfrac{h}{H_d} < 1.33$.

Submergence: There is no effect on C, if $\dfrac{h_d + d}{H_e} \geq 1.7$.

Thus we get the final value of C.

Chute channel or discharge carrier: A concave vertical curve is provided whenever the slope of the chute changes from steeper to milder. The curvature is expressed by,

$$y = -x . \tan \phi - \frac{x^2}{K\left[4(d + h_v)\cos^2 \phi\right]} \qquad (25)$$

where, ϕ = slope angle of the floor u/s , (d + h_v) is the specific energy of flow at junction point; K = constant $\geq$ 1.5 and x, y are the coordinates of the point under consideration.

Stilling basin design: Hydraulic jump phenomenon is generally used for designing these basins:

$$\frac{y_2}{y_1} = \frac{1}{2}\left(\sqrt{1 + 8 \, F_1^2} - 1\right) \qquad (26)$$

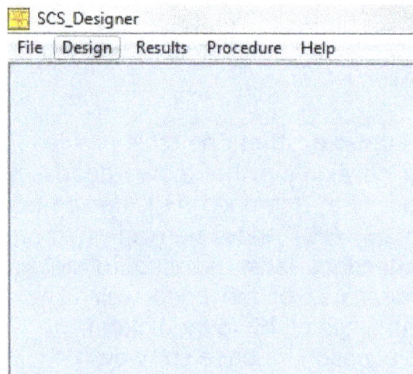

Figure 6. Model main window.

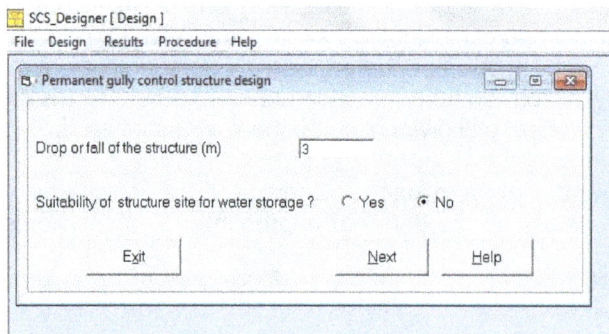

Figure 7. Input window for drop.

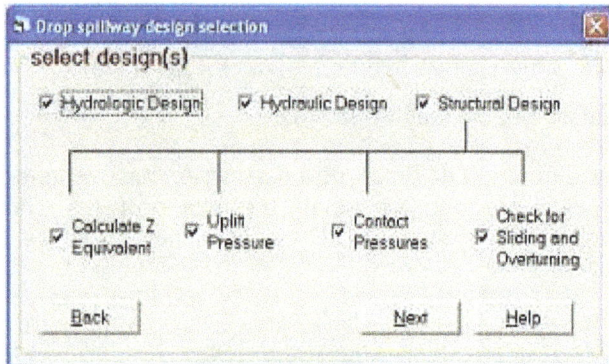

Figure 8. Structure selection window.

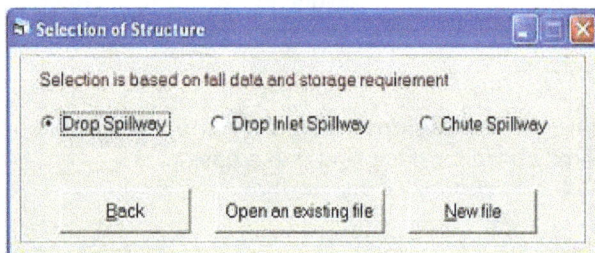

Figure 9. Drop spillway design(s) selection.

$$F_1 = \text{Froude No.} = \frac{V_1}{\sqrt{gy_1}} \qquad (27)$$

where, y_1 = pre jump depth, y_2 = post jump depth and F1 = Froude number corresponding to the jump velocity (V_1).

DSS FOR PERMANENT GULLY CONTROL STRUCTURES DESIGN

To design a permanent gully control structure, the user has to click on Design menu and Permanent gully control structures submenu (Figure 6). A window is displayed in which the user has to enter information on drop and has to select either YES or NO option for site suitability for storage (Figure 7). After providing all necessary data, clicking on Next button leads to a structure selection window (Figure 8). Depending upon the information provided in the previous form, one of the three options (drop spillway, drop inlet spillway and chute spillway) is selected. The drop spillway is selected if the fall is less than or equal to 3 m. The drop inlet spillway is selected for fall greater than 3 m, and site suitable for storage (YES option). The chute spillway is selected for higher discharge, fall greater than 3 m and site unsuitable for storage (NO option). However, the user has been provided with a flexibility to change the selection to design a specific structure. Once selection is finalized, the design process can be initiated by clicking on the New file button.

DSS for drop spillway

If drop spillway is to be designed, a selection window (Figure 9) is displayed with the three designs of drop spillway, that is, hydrologic, hydraulic and structural design. The user has to select whether he wants to design only one or two or all of them at the same time. If structural design is selected then the user also has to select one of the four check boxes of Z equivalent, uplift pressure, contact pressure and check for sliding and overturning. Thereafter clicking on the Next button will display the window as per the selection. If only hydrologic design is selected in window of Figure 8, the design discharge can be calculated using the rational method by clicking on the Calculate Discharge button (Figure 10). Thereafter clicking on OK button will finalize the design.

In case if only hydraulic design is selected (Figure 9) then also the hydrologic design window will be displayed at first. Entering the fall and discharge and then clicking on Next will display the window for hydraulic design (Figure 11), showing several combinations of practical length and depth of the drop spillway. The user can select any one of the combinations according to the site constraints for length of spillway. Clicking on Design button will finalize the dimensions. When the design is completed a completion message appears on the screen. The input data and procedure can be saved by Save button and can be retrieved in future by clicking on Open an existing button.

In case if only structural design is selected, then the data entry window for structural design will appear (Figure 12) after clicking on the Next button on the design selection window (Figure 8). The user has to enter all the information on material properties and the dimensions of the structure as well. When the user clicks on the Next button, next window is displayed (Figure 13), which shows different conditions about the site. The site conditions include three conditions of backfill namely, no backfill against headwall (Case A), gully graded full to crest elevation (Case B), and compacted earth fill berm constructed to crest elevation (Case C), three conditions of relative permeability, namely the permeability of foundation material greater than, equal to or less than the permeability of backfill, two conditions of drainage (Type a and type b) and flow conditions. The description on these conditions is provided in the previous section.

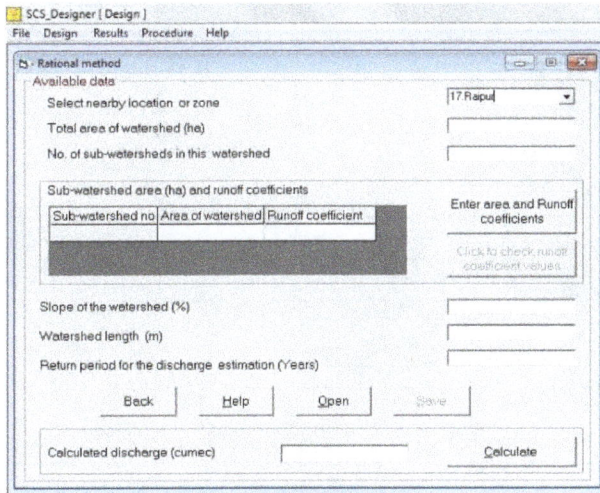

Figure 10. Discharge by rational method.

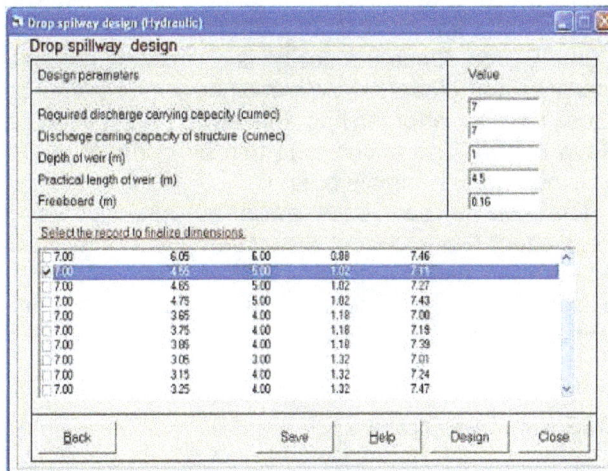

Figure 11. Drop spillway design window.

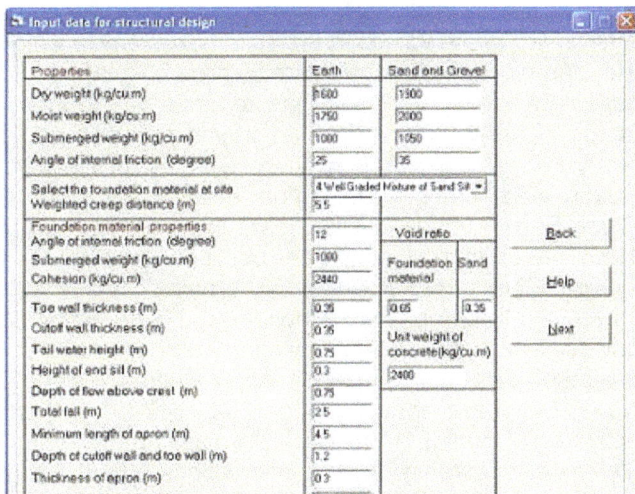

Figure 12. Data input for structural design.

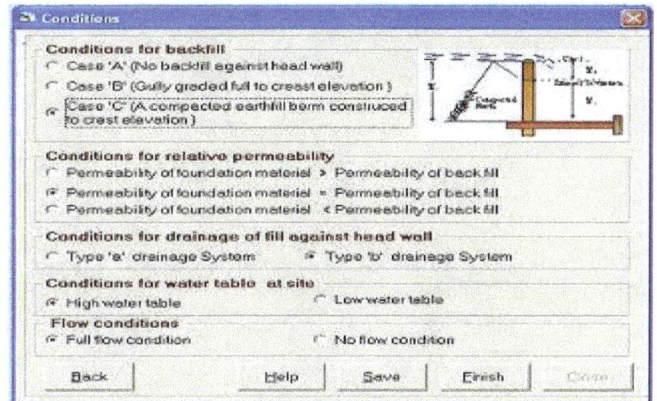

Figure 13. Site selection condition window.

The user has to select appropriate conditions according to the site and click on the Finish button for completing the design. When the design is completed, a completion message appears on the screen. The input data and procedure can be saved by clicking on the Save button and can be retrieved in future by clicking on Open an existing file button.

If all the three viz. hydrologic, hydraulic and structural design is selected then the procedure will be the same with the only difference that on hydrologic design window the user has to click on Next button (Figure 8) after entering the fall and discharge values. Thereafter, in hydraulic design window, again the user has to click on Next button after selecting one combination according to the site constraints for the length of spillway. Then the structural design window will appear with all the dimensions of the structure entered into their respective boxes. After entering all the information on material properties, click on the Next button. The user has to select appropriate conditions, as explained above, according to the site and click on the Finish Button for completing the design. When the design is completed a completion message appears on the screen. The input data and procedure can be saved by clicking on the Save button and can be retrieved in future by clicking on Open an existing button.

The results can be viewed from the Results menu by clicking on Drop spillway under Permanent Gully Control Structures sub menu. The results window (Figure 14) includes tabs for hydraulic design, uplift pressure calculations, contact pressure calculations and stability analysis with the tabs of their figures. The input data and procedure can be saved with Save button.

DSS for drop inlet spillway

The Drop Inlet Spillway design consists of two options, namely, Full Drop Inlet Design and Hydrographs of Drop Inlet, and any one-design option can be selected (Figure 15). Clicking on the Next Button leads to the data input window for the chosen design. Figure 16 shows the data input window for the Hydrographs of Drop Inlet option. After entering all the required data, clicking on OK button completes the design and the results can be seen from Results | Permanent Gully Control Structures | Drop Inlet Spillway.

On the other hand when the user selects the Full Drop Inlet Design, a window as shown in Figure 17 is displayed. The user has to select the design option and

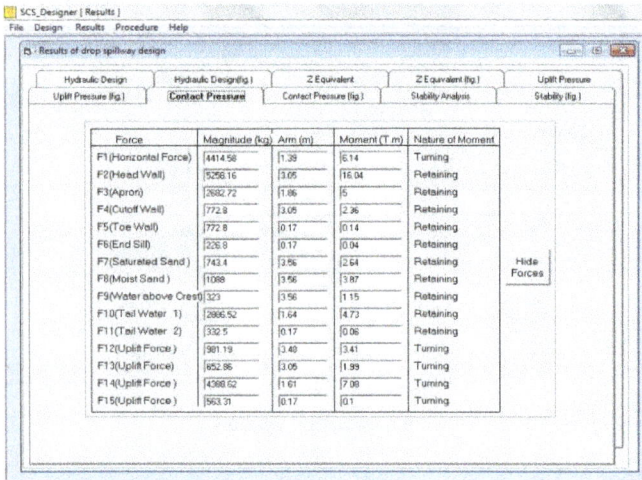

Figure 14. Results window of drop spillway.

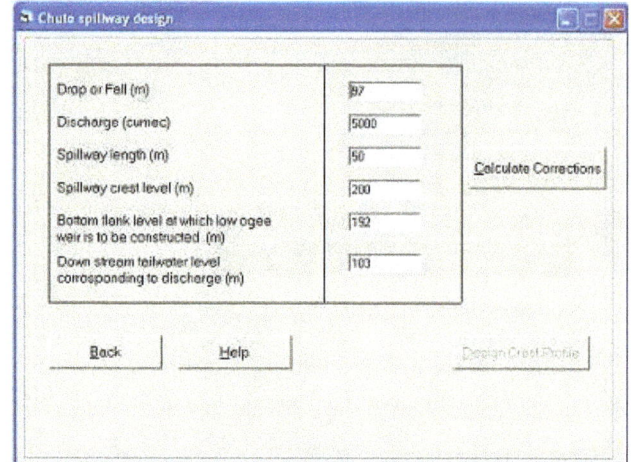

Figure 17. Drop inlet spillway design.

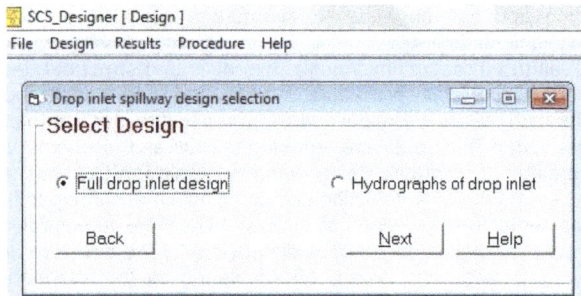

Figure 15. Drop inlet spillway design.

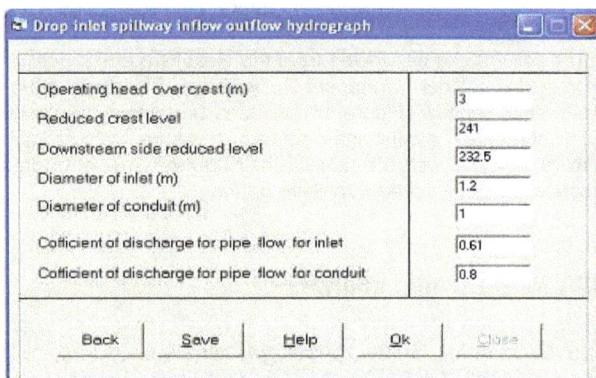

Figure 16. Data input for discharge hydrographs

P/R_s ratio, enter all input data with an approximate value of R_s. Then the user has to click on the Determination of R_s Button to get exact value of the radius of conduit. Thereafter, clicking on the Design button will open a new window with the calculated ratios displayed in it. The user is provided with an option to verify the calculated ratios using the Show Graphs for Reference. Thereafter, the

user has to enter different discharge coefficients to get hydrographs for the design and has to click on Discharge Curve Button. Once the design is completed successfully, the user can close the active window by clicking on the Close Button after saving the data by clicking on the Save button. The saved data can be opened by clicking on Open an existing file button.

The results can be viewed by clicking Results | Permanent Gully Control Structures | Drop Inlet Spillway. Again the result window contains four tabs, namely, design parameter (Figure 18), discharge curve, transition curve and drop inlet figure.

DSS for Chute spillway

Figure 19 shows the data input window for the design of chute spillway. The user has to enter input data and has to click on the Calculate Corrections Button. Two input box will appear for entering the length of clear width of spillway length and the thickness of each spillway pier. After entering the values in the input boxes a new window will appear with the calculated ratios displayed in it, based on which the user has to enter the correction factors. The correction factors graphs can be seen for ready reference by clicking Show Graph's for Reference Button. After entering correction factors, clicking on the Close Button brings back to data input window (Figure 19). The crest profile can be obtained by clicking the Design Crest Profile Button. Thereafter, user has to select an appropriate equation for the design of downstream side profile of low ogee weir depending upon the ratios and has to click on Apply Button (Figure 20). Now, the data input window for discharge carrier activates (Figure 21), in which the user has to enter all the data according to the instructions given on the window and click on the Design Discharge Carrier Button. This is the last stage in the chute spillway design. The

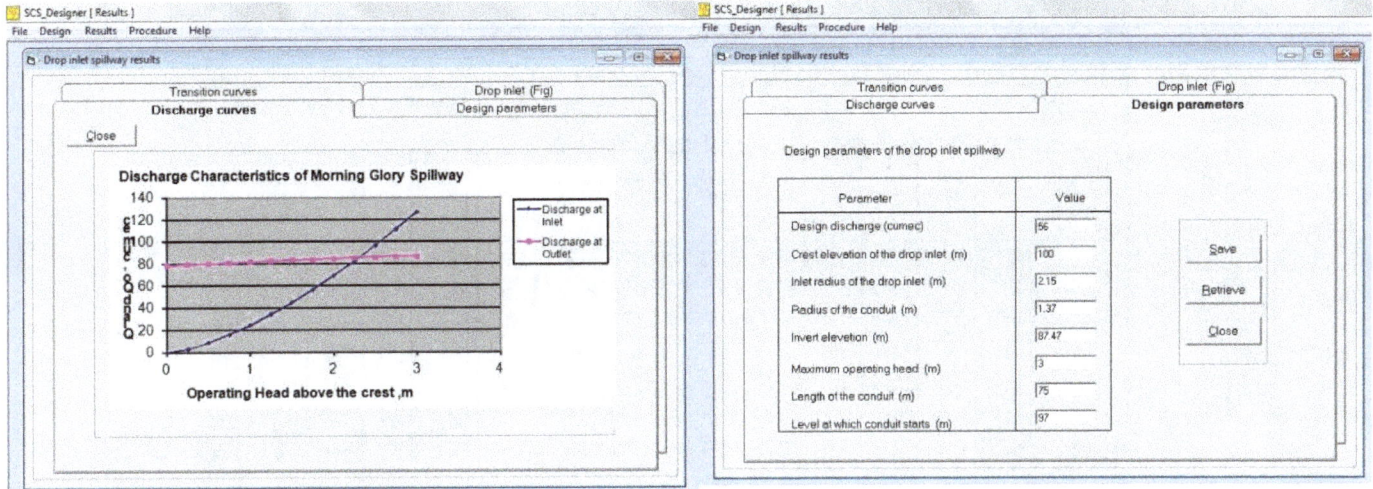

Figure 18. Results window for design of drop inlet spillway.

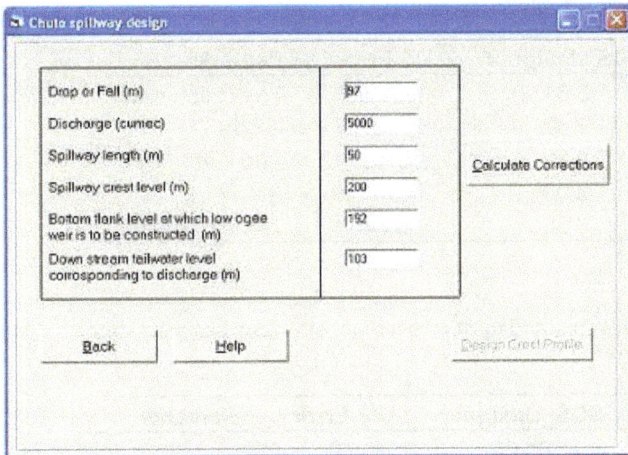

Figure 19. Data input for chute spillway.

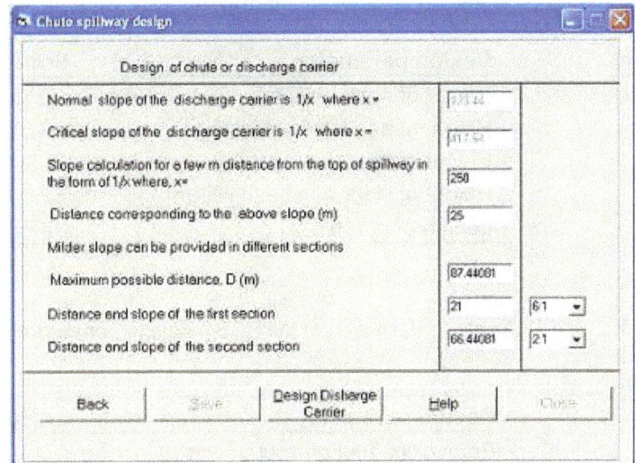

Figure 21. Design of discharge carrier.

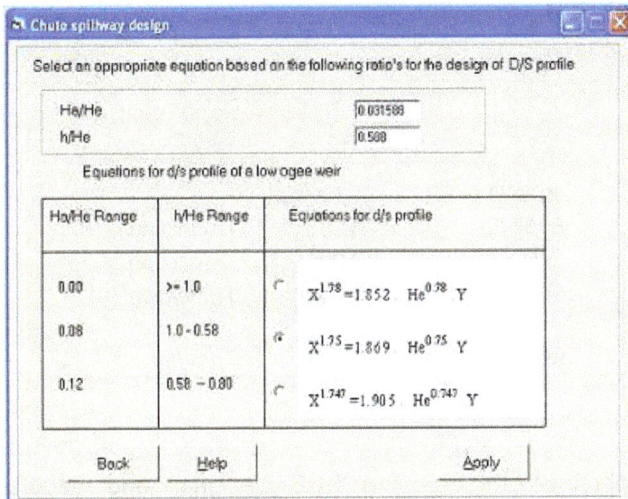

Figure 20. Equation for profile of low ogee weir.

input data can be saved by clicking on the Save button. The data can be retrieved in future by clicking on the Open an existing file button. The results can be viewed by clicking Results | Permanent Gully Control Structures | Chute Spillway. Again the result window contains six tabs, namely, design of discharge carrier (Figure 22), co-ordinates of u/s and d/s profiles, design of curve No. 1 and 2, curve No. 1 (graph), curve No. 2 (graph) and stilling basin design. The user has to click on Click here to draw graph button to view graphs of curve no. 1 and 2.

RESULTS AND DISCUSSION

The model was tested against the solved numerical from different sources. Wherever, solved numerical were not available, the model was tested against the hand calculations. For validation and testing of drop spillway

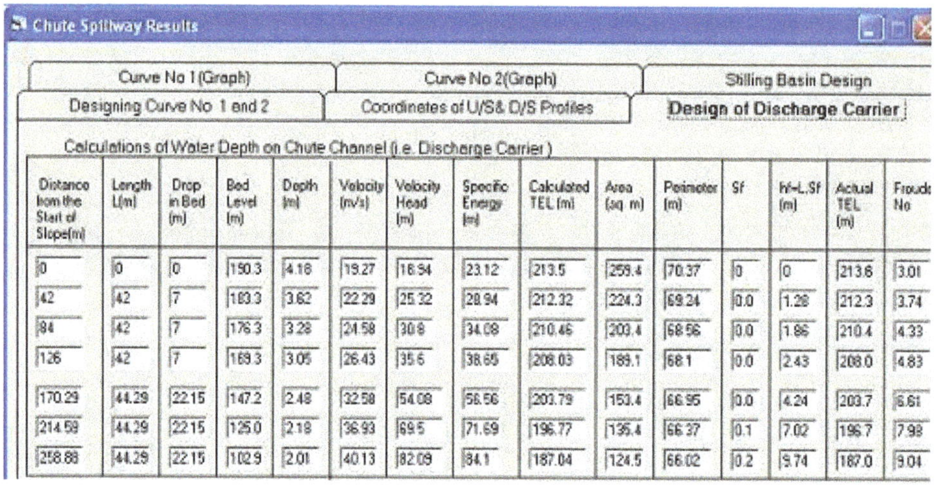

Distance from the Start of Slope(m)	Length L(m)	Drop in Bed (m)	Bed Level (m)	Depth (m)	Velocity (m/s)	Velocity Head (m)	Specific Energy (m)	Calculated TEL (m)	Area (sq. m)	Perimeter (m)	Sf	hf=L.Sf (m)	Actual TEL (m)	Froude No
0	0	0	190.3	4.18	19.27	18.94	23.12	213.5	259.4	70.37	0	0	213.6	3.01
42	42	7	183.3	3.62	22.29	25.32	28.94	212.32	224.3	69.24	0.0	1.28	212.3	3.74
84	42	7	176.3	3.28	24.58	30.8	34.08	210.46	203.4	68.56	0.0	1.86	210.4	4.33
126	42	7	169.3	3.05	26.43	35.6	38.65	208.03	189.1	68.1	0.0	2.43	208.0	4.83
170.29	44.29	22.15	147.2	2.48	32.58	54.08	56.56	203.79	153.4	66.95	0.0	4.24	203.7	6.61
214.59	44.29	22.15	125.0	2.18	36.93	69.5	71.69	196.77	135.4	66.37	0.1	7.02	196.7	7.98
258.88	44.29	22.15	102.9	2.01	40.13	82.09	84.1	187.04	124.5	66.02	0.2	9.74	187.0	9.04

Figure 22. Results of chute spillway discharge carrier design.

Table 1. Validation of the SCS_designer for drop spillway design (hydraulic).

Design parameter	Source	SCS_designer	% Error	Remarks
Height of head wall (m)	2.82	2.81	-0.355	Truncation error
Height of headwall extension	3.82	3.82	0	---do--
Height of transverse sill (m)	0.315	0.32	1.587	--do--
Height of sidewall at end sill (m)	2.00	2.00	0	--
Length of apron (m)	3.55	3.58	0.845	Truncation error

Table 2. Validation of the SCS_designer for equivalent fluid pressure and uplift pressure for structural design of drop spillway.

Design parameter		Source	SCS_Designer	% Error	Remarks
Equivalent fluid pressure (T/m^3)		0.805	0.805	0	
Weighted creep distance between points (m)	A, B	1.2	1.2	0	-
	B, C	0.117	0.117	0	-
	C, D	1.2	1.2	0	-
	D, E	1.1	1.1	0	-
	E, F	1.2	1.2	0	-
	F, G	0.117	0.117	0	-
	G, H	1.2	1.2	0	-
	H, I	0.167	0.167	0	-
Uplift pressure between points (kg)	B,C	498.71	498.71	0	-
	D, E	5057.22	5057.38	0.003	Truncation
	F, G	574.03	574.06	0.005	Truncation
	H, I	860	860.02	0.002	Truncation
Total uplift (kg)		6989.96	6990.17	0.003	Truncation

(Rao, 1986), data were entered in the data input window (Figure 11) and corresponding design results are shown in Figure 14. Comparison between the design parameters obtained through the SCS_Designer and source is presented in Table 1 for hydraulic design and in Table 2 for structural design.

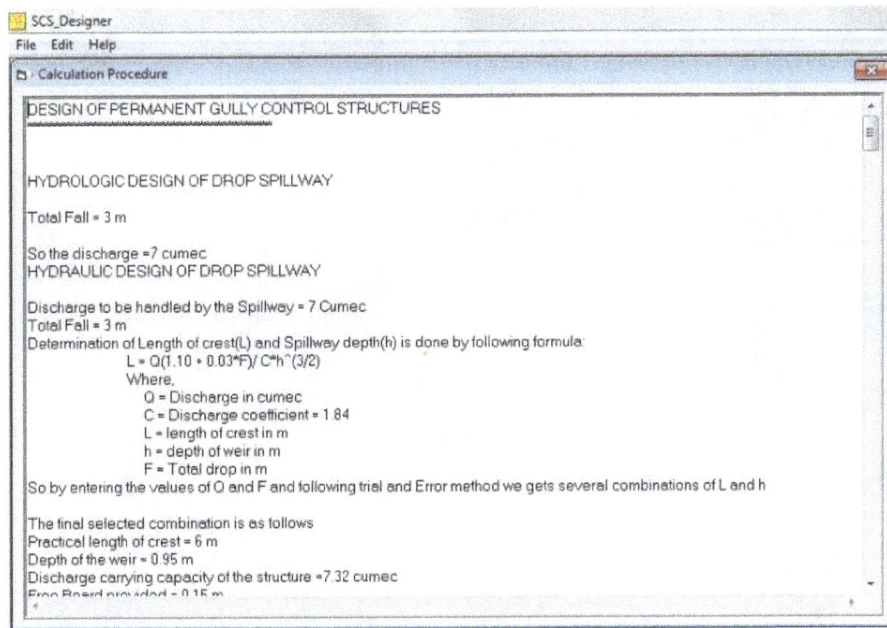

Figure 23. Procedure file window for SCS_designer.

The design results for permanent gully control structures obtained using the SCS_Designer matched well with the respective designs taken from various sources. Slight variation in values is due to truncation and round off errors. This suggests that the model provides accurate design of the selected permanent gully control structures and can be used for their designs.

Conclusions

The comprehensive package for the designing of permanent gully control structure module of SCS_Designer was developed, tested and validated as user-friendly software with advanced graphical user interface (GUI) using Visual Basic as the programming language. The important property of GUI is that, it allows user to select and edit data, determine the design parameters and coefficients, implement the simulation and display the results in tabular form. It also informs the user of any wrong activity and even the way to rectify it by its message boxes. In drop spillway design, user has the flexibility to choose hydrologic, hydraulic and structural designs separately. The design of drop inlet spillway and chute spillway are made easier and simpler for the users.

SCS_Designer has the provision for step wise design procedure to be displayed. These procedure files (Figure 23) are generated when the input data are saved, after the successful designing of a structure by the model. The procedure file has the same file name as that of the saved input data file but with *.txt, as its filename

extension. The SCS_Designer is equipped with help file for user's reference. The help file contains the detailed information about the theoretical consideration of the structures as well as the step-by-step procedure of the working of the model. These help files are window based easy to use reference files which will appear automatically on the screen by pressing the 'F1' key on the keyboard or by clicking on the 'Help' buttons of the respective windows of SCS_Designer.

The other objective of this study was to validate the SCS_Designer for the portion which has been modified and included so far. The SCS_Designer was validated against the available design examples for each of the structures it supports (Raghuwanshi, 2003; Suresh, 2002; Singh et al., 1981). Mostly these design examples were taken from standard books and in case if a proper example was not available, then problem was formulated and solved by hand calculations. Sometimes solved problems from the class notes were also used to validate SCS_Designer. In some cases, the model was validated in stepwise manner due to changes in design parameters. Thus, it can be concluded that the SCS_Designer is validated successfully and can be used as a standard tool to design permanent gully control structures namely drop spillway, drop inlet spillway and chute spillway.

REFERENCES

Das DC (1985). Problems of soil and land degradation in India. National Seminar on Soil Conservation and Watershed Management, September 17-18, New Delhi.

Dhruva NVV, Ram B (1983). Estimation of soil erosion in India. J. Irrigat. Drain. Eng. 109(4):419-434.

Raghuwanshi NS (2003). Lecture notes on Soil and Water Conservation Structures. IIT Kharagpur. Dept. of Agric. Food Eng. IIT, Kharagpur.

Rao D (1986). Development of optimal stilling basin for drop structures. Ph.D thesis. Dept. of Agril. and Food Engg., IIT Kharagpur. pp. 50-55.

Singh G, Venkataraman C, Sastry G (1981). Manual of Soil and Water conservation practices in India. Central Soil and Water Conservation Research and Training Institute (I.C.A.R.) Dehradun, India.

Suresh R (2002). Soil and Water Conservation Engineering. Standard Publishers Distributors, New Delhi.

USDA (1957). Drop spillway, Section II, Engineering Hand Book, Soil Conservation Service Washington D.C.

5

Farmers' perception of water scarcity and components influencing on this challenge in Fars province

Yaser Mohammadi, Hussein Shabanali Fami and Ali Asadi

Department of Agricultural Development and Management, College of Agricultural Development and Economic, University of Tehran, Tehran, Iran.

As water scarcity is alarmingly on the increase in national and international level especially in agriculture and due to the importance of examination of components influencing how water scarcity is dealt with by farmers, this research was carried out to analyze the situation of water scarcity by farmers of Zarindasht County and to examine influencing components on this challenge. A questionnaire was used for the collection of data which its reliability was confirmed through computing Cronbach's Alpha coefficient which was above 0.80. About 150 farmers were selected as a sample population through calculating Cochran's formula among 4648 farmers in this County and were sampled using multi-stage sampling technique. Findings revealed that more than 70% of farmers were highly faced with water scarcity, and there was a significant relationship but negative between farmers' perception of water scarcity (FPWS) and variables such as "the depth of water in the well", "income", and a positive significant relationship between this challenge and variables such as "length of water transmission canal" and "the volume of water decrease". In addition, the older farmers, the ones without a second job, the ones using soil canals for water transmission, and the ones with salty water irrigation were faced with water scarcity challenge more often. Also, geographical situation was recognized significant as there was a significant difference among villages and states. Finally, the results showed that, among all the examined components, the decrease of used water volume, having a second job and, the length of water transmission canal were the most important factors, explaining 40% of the variance, influencing on FPWS in Zarindasht County. It is suggested that to decrease the rate of water scarcity, farmers should be supported financially to change their irrigation method or transmission canals as the efficiency of irrigation be improved.

Key words: Water scarcity, influencing factors, water transmission canals, Zarindasht County, Iran.

INTRODUCTION

Water has been the most important factor in the development of the world for a very long time (Khalilian and Zare-Mehrjardi, 2005; Azizi, 2001). Ninety seven percent (97%) of the world's water resources are salty, and a very limited amount of it is directly being used by human beings. Almost 1.76% of the water on the planet is crystallized or has changed into frozen rivers and whatever left is stored in the underground. Increasing demands for water by industrial and urban users will intensify the competition to get it. At the same time, water scarcity is increasing in several important agricultural areas (Fraiture and Wichelns, 2010). According to the scientists' prediction, in the following decades water scarcity in the global scale will be experienced more than before, and the necessity of this vital substance will become more obvious (Sayer and O'Riordan, 2000). Nowadays, water crisis has become one of the controversial issues among all countries' scientists, researchers and politicians. As Frank Rager Berman one of the managers of International Water Institute said, about one fourth of the world are faced with water scarcity because of physical factors including natural disasters, overuse of water resources and, poor management in agriculture which cause rivers and underground water resources to dry up early (Sistan and Baluchestan Regional Water Company, 2007). Iran is located in the arid and semi-arid area in the world, and as a result, water scarcity is a big problem (Forooghi et al., 2006). Generally, water scarcity happens when the rate of water users' pressure is more than water supply, and increasing demands of different parts of the environment like agriculture, industry and urban users are not answered completely (FAO, 2007). Water scarcity, can broadly be understood as the lack of access to adequate quantities of water for human and environmental uses. The term 'water scarcity' is regularly used by the media, government reports, non-governmental organizations (NGOs), international organizations such as the United Nations (UN) and Organisation for Economic Cooperation and Development (OECD), as well as in the academic literature, to highlight areas where water resources are under pressure (White, 2012).

Zarindasht County is one of the 13 counties of Fars province in Iran faced with water scarcity which is due to the persistence of drought and overuse of the underground resources (Ansarifar, 2006). This County with an area of 4626 km^2 and the population of about 65000 people is located in the Southeast of Fars province of Iran. With an average amount of 236 mm rainfall and the average temperature of 22.7°C, Zarindasht is a hot and semi-arid County of Iran and in recent years, water scarcity has been highly increasing due to continuum droughts which has made the farmers leave their unproductive farms behind and migrate to big cities (Jihad-E-Agriculture Management of Zarindasht County, 2010). While agriculture in this County is of great importance, the main source of supplying agricultural water is groundwater. However, due to overexploitation of groundwater, the annual decrease of the water level from this resource is considerable. Decreasing number of agricultural water wells from 915 in years 2004 to 2005 to 870 wells in years 2008 to 2009 and decreasing average of discharge from 12 to 9 L$_S$ prove that groundwater, as the main source of supplying agricultural water, signifies an alarming case of water supply (Jihad-E-Agriculture Management of Zarindasht County, 2010). In fact, both sequential droughts and lack of groundwater optimal use in Zarindasht County have caused water scarcity problem which has ended in agricultural yield loss in this County (Asadi et al., 2009). Considering the importance of water scarcity problem in this County, this research aimed to investigate how water scarcity is dealt with by farmers

and to find out factors influencing this challenge. Scientists believed that different factors can cause water scarcity. Pereira et al. (2002) stated that overexploitation of water resources and water quality degradation is associated with water shortage. Also, overexploitation and poor management of groundwater threaten the resources (Fraiture and Wichelns, 2010). Luquet et al. (2005) showed that traditional irrigation methods are a big challenge in countries facing water scarcity. Research carried out in Rafsanjan County by Abdollahi and Soltani (1998) revealed that the land ownership and topography are the factors causing waste of water which results in water scarcity. Kardovani (2000) also believed that the long length of irrigation canals as well as their turns and twists are the factors causing water loss and water scarcity. Zehtabiyan (2005) referred to three points that is the fact that irrigation canals are made of soil, the farms are not leveled, and the long length of water canals. He believes that these factors result in decreasing irrigation productivity and water scarcity. Davarpanah (2005) concluded that agricultural product insurance against water scarcity and government supportive policies are the managerial components to overcome water scarcity and drought. Furthermore, studying water scarcity in Darab County (located in Fars province, Iran) and determining components leading to water scarcity, Forooghi et al. (2006) stated that turning soil irrigation canals into polyethylene pipes, and improving water consumption in the farm by hydroflom pipes are the best managerial strategies to tackle water scarcity.

RESEARCH METHODS

This study was based on survey study. A number of 4648 households from Zarindasht County (with 2 divisions, 5 rural districts and 23 villages from Fars province) were selected as statistical population 150 of which were selected as sample population using Cochran's formula (Figure 1). This formula estimates the sample size using standard deviation of a main variable in pretest stage. Sampling was done in two phases, first by using stratified sampling method and in the second phase by random sampling. A questionnaire was used to gather data and information which was included of farmers' characteristics, agronomy information, water and land data, and farmers' perception from water scarcity. Pilot study revealed statistically acceptable reliability of the questionnaire by estimating the Cronbach's Alpha which was above 0.80. Validity of the questionnaire was also confirmed by expert opinions like Fars water organization experts, power ministry experts. As the dependent variable, that is farmers' perception of water scarcity (FPWS) with an ordinal scale, Mann-Whitney test was used to compare the average of FPWS of farmers between dummy variables and for more than two levels, Kruskal-Wallis test were exploited.

RESULTS

Personal and professional characteristics

The results of this research showed that most of the farmers were middle aged (30 to 60 years old) with the

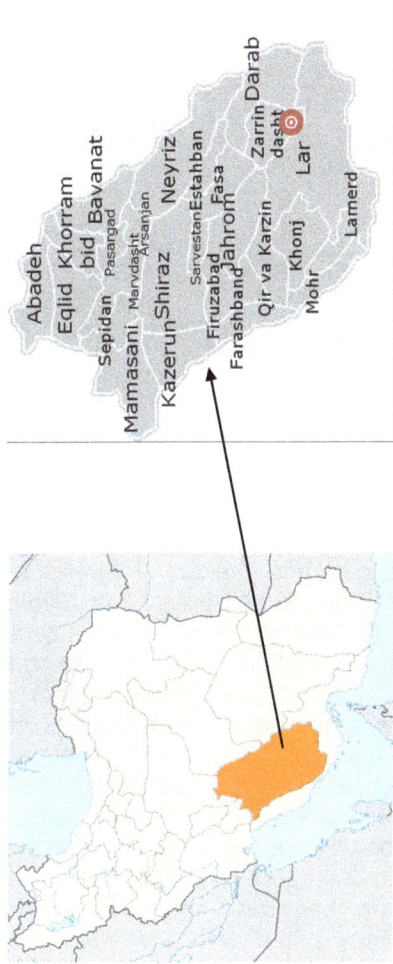

Figure 1. The position of Zarindasht County in Fars province and Iran Country.

Table 1. The average situation of farmers' information and their agricultural characteristics.

Variable	Age	Agri-exp-years	Literacy	Cultivated lands	Topography	Soil quality	Source of irrigation	Length of water canals	Water ownership	Irrigation method	Type of water canal	Water quality
Mean / mode	42	16	Read and write	6.5	Not flat (hilly)	Salty	Deep wells	>1 km	Joint	Traditional	Made by soil	Semi-salty

Source: Research findings (2011).

average age of 42, and the average years of their agricultural experience was 16. Most of them were able to read and write. Also, the average cultivated lands were 6.5 ha, and the average number of their lands' plots was three. About 50% of their farms were not flat (hilly) and the soil was salty with white spots on it. Furthermore, 95% of farmers use a deep well with pipes with an average diameter of 4 inches to irrigate their farms. In addition, 62% of the wells were located farther than 1 km away from farms. Moreover, most of the farmers had a joint ownership regarding the water resources, and more than 90% of them used the traditional deepwater irrigation method to irrigate the farms. Also, most

of them use soil canals to transmit water to the farms with an average length of 1.5 km. About 61% of the farmers used governmental credits to improve their irrigation systems. Water quality for most of the farms was semi-salty, and more than 97% of them did not have drainage system for their farms (Table 1).

Farmers' perception of water scarcity (FPWS)

It was established that 41.3% of the farmers believed that they faced very severe water scarcity, 32.7% of them faced severe scarcity, 21.3% faced average scarcity, 4% described it as

being low, and just 0.7% of them believed that they had faced water scarcity very low (Table 2).

As it is observed in Table 2, water scarcity is a very serious problem in Zarindasht because farmers faced water scarcity of above average were more than 70%, then the necessity to find out and pay attention to the factors influencing water scarcity is completely inevitable.

Relationship between FPWS, and personal and agricultural characteristics

Research findings in Table 2 show that there is a positive and significant relationship between how

Table 2. Farmers' perception of water scarcity (FPWS).

Intensity of FPWS	Frequently	Percent	Cumulative percent
Very low	1	0.7	0.7
Low	6	4.0	4.7
Average	32	21.3	26.0
Severe	49	32.7	58.7
Very severe	62	41.3	100
Total	150	100	

Source: Research findings (2011).

Table 3. The relationship between facing water scarcity with independent variables.

Variable 1	Variable 2	r	Significance
	Age	0.176*	0.032
	Literacy level	-0.021	0.801
	Agricultural experience	-0.070	0.397
	Water discharge (liter/S)	-0.446**	0.000
Facing water scarcity	Water depth in the well	-0.362**	0.000
	Income	-0.161*	0.050
	The length of irrigation canal	0.218**	0.007
	Amount of water decreasing	0.569**	0.000
	Amount of water consumption	0.006	0.946

**, $P \leq 0.01$. *, $P \leq 0.05$.

often farmers are faced with water scarcity, the length of irrigation canal in the level of 1% and the variable of age in the level of 5%. Also, there is a negative and significant relationship between the variable of the amount of water resource, water depth in the well, and income with facing water scarcity in the levels of 1 and 5% (Table 3).

According to the results in Table 3, as the age increases, the rate of FPWS increases which could be due to the fact that farmers pay more attention to their old principles and do not use new irrigation methods. In addition, the rate of water scarcity increases with increasing the length of irrigation canal. This seems to be due to the fact that increasing the length of the route causes over evaporation considering high penetration of canals. Decreasing the depth of water in the well and considered water also cause the problem of water scarcity. But increasing income causes a decrease in water scarcity which could be the result of purchase of new irrigation systems like polyethylene pipes, electro pumps. This is what farmers agreed with that when they have been asked. These help farmers to use water much better, and to face water scarcity less often.

Comparing the ranked mean of the FPWS between dummy variables

The results of comparing ranked mean of the FPWS

between two different levels of Dummy Variables could be observed using Mann-Whitney Test (Table 4).

As shown in Table 4, farmers with a second job face water scarcity less often. This could be to the fact that farmers with a second job have a higher income, and consequently are more financially empowered to buy water than the other group. Also, farmers who use product insurance have fewer problems than those without because, they are not worried about draught and water scarcity and use water better. However, those who do not have insurance try to exploit water more to compensate for the problems caused by water scarcity. As a result, they face a decrease in the level of water in the well.

There is a significant difference between the farmers who use modern irrigation method and the ones who use conventional irrigation. Therefore, the second group faces water scarcity more often than the first group. This is understandable considering how more water in deep water irrigation method is wasted compared with sprinkler irrigation methods.

Also, the results showed that the rate of facing water scarcity is high among people who use diesel pumps compared with the ones who use electro pumps, and people who have soil canals as compared to the ones with polyethylene or cement pipes. This could be because of the fact that soil canals are penetrable, and weeds grow along these canals.

Table 4. Ranked mean of the FPWS between virtual parameters.

Dependent variable	Grouped variable	Levels	Ranked mean	U	Significance
The rate of facing water scarcity	Second job	Have Don't have	70.48 83.93	2160.00**	0.050
	Insurance	Use Don't use	72.17 90.82	1179.50*	0.040
	Irrigation method	Modern Conventional	34.00 76.06	65.00*	0.201
	Water extracting tool	Electro pump Diesel pumps	66.01 83.80	2136.00**	0.008
	Soil canals	Use Don't use	83.29 65.59	2118.00**	0.008
	Drainage	Yes No	74.67 105.63	171.500	0.135
	Location of wells	<1 km >1 km	65.24 81.68	2076.00*	0.018
	Water quality	Salty Sweet	19.60 13.39	75.50*	0.096
	Land topography	Not-flat Flat	54.53 42.65	758.00*	0.061

** Difference is significant at the 0.01 level. * Difference is significant at the 0.05 level.

Others results also showed that the farmers whose pumps were more than 1 km away from their farms compared to the ones with a distance less than 1 km, and farmers who had salty water and not flat agricultural farms compared with the ones with sweet water and flat farm face water scarcity more often.

Comparing the ranked mean of FPWS among variables with more than two levels

In order to compare the ranked mean of FPWS among different levels of non-virtual parameters, Kruskal-Wallis test was used. The results are observable in the Table 5.

Table 5 results show that there is a significant difference among different districts and villages of Zarindasht County regarding the rate of facing water scarcity. The least and the highest rate of water scarcity were related to Khossuyeh and Izad Khast districts, respectively. This could be because of salty water in West Izad Khast district. Among villages, the least and the highest rate of those facing water scarcity were observed in Miandeh and Darreshoor villages. This could also be because of salty water in Darreshoor farms and

the shortage of rain and low level of underground water. In addition, there is a significant difference among farmers with different ownerships of water resource regarding facing with water scarcity which means that the least and the highest rate of water scarcity belong to personal and rental ownerships, respectively. The farmers who use polyethylene pipes to transmit water and the ones, who use soil canals, face with the least and the most rate of water scarcity respectively.

Also, the results showed that there is a significant difference among different topographies of the farms regarding the rate of facing water scarcity. The highest rate of facing water scarcity was observed in hilly farms and the least of it in flat farms. It can be related to amount of water wasted in hilly farms in comparison to flat farms.

The discriminant analysis of the components influencing the challenge of FPWS

There were definitely certain characteristics that could separate farmers who faced water scarcity more often compared to farmers who did less often. These

Table 5. Ranked mean of the FPWS among non-virtual parameters.

Dependent variable	Grouped variables	Levels	Ranked mean	Chi-square	Significance
		East Izad Khast	95.17		
		West Izad Khast	98.05		
	Districts	Khossuyeh	63.24	17.557**	0.002
		Dabiran	73.86		
		Zirab	73.00		
		Mazijan	95.17		
		Darreshoor	98.05		
		Khossuyeh	71.73		
		Sachoon	74.85		
	Villages	Tajabad	60.36	26.518**	0.001
		Miandeh	36.40		
		Dehno	73.86		
		Chahsabz	74.69		
		Galugah	71.77		
The rate of facing water scarcity		Joint	79.12		
	Water ownerships	Personal	80.99	6.567*	0.037
		Rental	58.82		
		Polyethylene pipes (1)	63.07		
		Cement canals (2)	69.00		
	Water transmission canals	Soil canals (3)	83.59	12.342*	0.030
		(1) + (3)	81.52		
		(2) + (1)	119.50		
		(3) + (2)	119.51		
		With up and down (1)	79.45		
		Smooth and flat (2)	61.52		
	Land topography	Smooth and gradient (3)	71.66	11.407*	0.044
		(1) + (2)	119.50		
		(1) + (3)	68.50		
		(2) + (3)	75.10		

** Difference is significant at the 0.01 level. * Difference is significant at the 0.05 level.

characteristics are the factors really influencing the challenge of facing water scarcity. To find out these factors, one has to find out what characteristics differentiate these groups regarding how they face water scarcity. Discriminant analysis is a technique that shows the discriminant characteristics of these two groups. Using estimated discriminant equation, we could identify the components affecting the challenge of water scarcity, and how important each factor is.

Considering the two groups of compared farmers, one discriminate equation with the Eigenvalue of 0.616 and canonical correlation of 0.617 was gained which explained about 40% of the discrimination between the two groups because the square root of canonical correlation coefficient indicate the percentage of explained discriminations by linear combination of independent variables (Table 6). Another criterion for the assessment of the function is referring to Eigenvalue which in this function also showed that the gained function was very powerful in discriminating the groups.

The results of Table 7 also showed a significant level for the discriminant function. Considering the value of Chi-square and Wilks' Lambda, the discriminating equation was significant and could discriminate groups well. The estimated equation from discriminant analysis can be written as:

$$D = -2.308 + 0.082x_1 + 0.264x_2 - 0.765x_3$$

According to Table 8 and discriminant equation, it was showed that between 13 components in discriminant analysis, during two stages, three came to be the

Table 6. Eigen values and canonical correlation of discriminant functions.

Function	Eigenvalue	% of variance	Cumulative %	Canonical correlation
1	.616[a]	100.0	100.0	.617

[a] First 1 canonical discriminant functions were used in the analysis.

Table 7. Wilks' Lambda.

Test of function (s)	Wilks' Lambda	Chi-square	df	Significance
1	0.619	68.863	3	.000

Table 8. Standardized canonical discriminant function coefficients.

Variable	Function
	1
Percent of water decrease	0.898
Length of irrigation canal	0.352
Having a second job	-0.364

significant components influencing the rate of facing water scarcity most by farmers. Of these three components, the percentage of water decrease was the most important component, having a second job comes next, and then the length of irrigation canal.

DISCUSSION

According to the results gained in this study, most of the studied farmers were older than 40 years, and had a low level of literacy. More than half of their farms were not flat and their soil was salty. In addition, the distance from their farms to their water resource was also a lot considering that most of them transmit water through soil canals, and that the canals were very long too. As a result, most of them face water scarcity a lot which is also observable in Table 1. Also, the method of conventional irrigation and lack of drainage system in farms worsen the situation as more than 90% of farmers had farms without drainage which explained why more than 74% of the farmers are faced water scarcity much or very much. The relationship between mentioned characteristics with the rate of facing the challenge of water scarcity was proved which means increasing age, increasing the length of irrigation canal and decreasing the depth of water in the well make farmers face water scarcity more often. The older a farmer is, the less able he is in maintenance and management of canals. The longer the canals are from the wells to the farm, and the fact that they use soil canals, the more water is wasted, and the productivity of water transmission decreases, too. Therefore, the farmer faces water scarcity more often.

Although, the more people's income increases, or the deeper their well water is, the less often they face water scarcity. Since they have more water, and they could also use their money to employ some workers, or, they could buy polyethylene pipes and increase the productivity of water transmission, so, the farmers who had a second job as compared with those without a second job faced water scarcity less often because having a second job means another source of income that the farmer can use to buy new equipment of irrigation, turn soil canals into polyethylene pipes, and change his pipes and irrigation pumps to increase the productivity of water transmission and consumption. Poor farmers who had to use the traditional irrigation method instead of sprinkler irrigation, use soil canals to transmit water, utilize diesel pumps instead of electro pumps definitely faced water scarcity more often as these farmers wasted more water, and they had low water productivity.

Other components for example, land topography, the quality of agricultural water and the place of water resource also decrease the productivity of water consumption. The farmers who did not have flat farms, with agricultural salty water and those whose farms were more than 1 km away from the pump, usually faced water scarcity more often because the farms that are not flat use lots of water, while the amount of water penetrating the earth is also less than flat farms. Salty water is also more penetrating and gets into the soil very quickly. Many weeds preferring salt grow there, and use most of the water. This results in less water productivity and more water scarcity. The longer the distance is between water pump and the farm, the more evaporation into the air, and more penetration into the land is. This decreases the transmission productivity dramatically.

Moreover, the kind of water resource ownership influenced how the farmers felt toward water scarcity. Those who owned water resources personally felt water scarcity more than the ones who had rented the resources. It seems that those who rented the resources considered the condition of water just at the moment, while the owners were worried about the water condition now and in the future.

The place where farmers lived was also a component that influenced the rate of facing water scarcity. The rate of facing the challenge of water scarcity between different districts and villages of Zarindasht County had significant differences. It can be concluded that different districts and villages are different considering the amount of rain, land topography, the amount of water in the wells, the quality of agricultural water, the amount of the farmers' income, and so on and all these components had a significant impacts on facing water scarcity.

All in all, of all these components, the rate of water decrease in wells (which is caused by draught or overuse), having or not having a second job (which could provide a financial support for farmers to increase the productivity of water transmission), the length of the irrigation canals were the most influencing components considering the results of discriminant analysis. This means that enough attention should be paid to these components to tackle water scarcity.

Conclusion

To sum up, the results showed that water scarcity in Zarindasht County is a serious problem and more than 70% of the farmers faced water scarcity. This challenge of facing water scarcity was influenced by farmers' age, their income, having a second job, water depth in well, decrease in the amount of consumed water, the length of irrigation canal, the kind of irrigation canal being made of soil, irrigation method, the quality of irrigation water, the place of settling water pump, and the farms of improper topography. Of these components, the decrease of the amount of the consumed water, having or not having a second job and the length of irrigation canal were known as the most important ones. This means that to tackle this challenge, and improve the condition of irrigation water in Zarindasht County, it is imperative to pay due attention to these components. Operational strategies should be taken to remove these obstacles. For instance, considering the fact that long length of water canal and soil canals increase the rate of facing water scarcity, it is suggested that we use new methods of water transmission like polyethylene pipes or concrete canals in order to prevent waste of water (Jin and Yong, 2001; Berim-nejad and Paykani, 2004; Foroghi et al., 2006). Considering the positive effect of a second job and high income in decreasing the rate of facing water scarcity, it is suggested that the government provide financial or credit support in order to improve farmers' financial situation (Alizadeh, 2001; Farzampour, 2001; Zehtabiyan, 2005; Mahdavi, 2005; Assare et al., 2005; Arjomandi et al., 2000).

ACKNOWLEDGEMENTS

This research is done using financial credit of the Deputy of Research and Planning of Natural Resources and Agriculture Faculty of Tehran University in the sixth type of research plan. Here, we would like to thank them all.

REFERENCES

Abdollahi EM, Soltani GH (1998). Optimal Allocation of Groundwater in agriculture (case study of Rafsanjan County), J. Agric. Sci. Technol. (In Persian) 12(2):26-35.

Alizadeh A (2001). Water scarcity and necessity of increasing water productivity. J. Dry Water Scarcity Agric. (In Persian) 2:3-8.

Ansarifar S (2006). Thirteen cities of Fars province have water guarantee problem. Sanitation and Water Corporation of Fars province, Iran. http://www.qudsdaily.com/archive/1385/html/5/1385-05-19/page1.html#5.

Arjomandi R, GHaysardehi F, Nagafi A (2000). The effect of agricultural exploitation systems changes on water management. Proceeding of the Article Collection of Technical Workshop on Farmers Participation in Irrigation Network Management, Proceeding of the 27th Aban Season,   pp.   1-4. http://www.irncid.org/workshop/pdf/W15pdf/3arjmandi.pdf.

Asadi A, Mohammadi Y, Shabanali FH (2009). Investigation of the Agricultural Water Management Mechanisms in Zarindasht County, Fars Province, Iran. Am. J. Agric. Biol. Sci. 4(2):110-117.

Azizi J (2001). Sustainability of Agricultural Water. J. Dev. Agric. Econ. (In Persian). 9(36):113-136.

Berim-nejad V, Paykani GH (2004). The Effects of Irrigation Efficiency Improvement in Agricultural Section on Increasing level of Ground water. Dev. Agric. Econ. (In Persian) 12(47):69-95.

Davarpanah GH (2005). Investigating the most important of economical and social effects of drought and government assistant mechanisms to reduce these effects. Proceeding of the Conference on Investigating Mechanisms to Dominate with Water Scarcity, Zabol University, Sistan and Baluchestan province, Iran. http://www.esnips.com/SharedFolderAction.ns;jsessionid=69E29FC1 C1482CBE83D8CED1FCC4914C.

Farzampour A (2001). Investigating Challenges of Iran Water Recourses Management. Budget Program J. (In Persian) 6:7-12.

Food and Agriculture Organization (FAO) (2007). Coping with Water Scarcity.   On   line   available   on: www.fao.org/nr/water/docs/escarcity.pdf.

Forooghi F, Mohsenkhani A, Karimi M (2006). Investigation the Circumstance of Fassarud (Darab county, Fars province) region water recourses in resent drought, payam-e-ab Pub. (In Persian) 4(26):65-68.

Fraiture, CD Wichelns, (2010).Scenarios for meeting future water challenges in food production. Agric. Water Manage. 97(4):502-511.

Jihad-E-Agriculture Management of Zarindasht County (2010). Fars province, Iran. The report of drought in 2006. On line available on: http://www.jk-zarindasht.ir/izadkhast.phtml.

Jin L, Yong W (2001). Water use in agriculture in china water policy. 3:215-228.

Kardovani P (2000). Resources and Problems of Water in Iran (Runoff and Ground Water and their Exploitation Problems), University of Tehran press. (In Persian) 1:457.

Khalilian S, Zare-Mehrjardi MR (2005). Groundwater Valuation in Agricultural Exploitation, Case study: Wheat Producers of Kerman province. J. Dev. Agric. Econ. (In Persian) 13(51):3-23.

Mahdavi M (2005). The Role of Management in Irrigation Water Optimum consumption, Proceedings of 10the conference of Iranian National Committee on Irrigation and Drainage, Tehran, Iran. From http://www.irncid.org/seminars/10.htm.

Pereira LS, Oweis T, Zairi A (2002). Irrigation Management under water scarcity. Agric. Water Manag. 57:175-206.

Sayer M, O'Riordan T (2000). Climate Change, water management and agriculture. Center for Social and Economic Research on the Global Environment, University of East Angelia and University College, London.

Sistan and Baluchestan Regional Water Company Report (2007). The water scarcity influence on one person from three people. http://www.sbrw.ir/default.asp?nw=news.

White CH (2012). Understanding water scarcity: Definitions and measurements, available online at: http://www.iwmi.cgiar.org/news_room/pdf/Understanding_water_scarcity.pdf

Zehtabiyan GH (2005). The Causes of Low Irrigation Efficiency in Varamin region, the 7th seminar of Iranian National Committee on Irrigation and Drainage. Tehran. Iran. (In persian). http://www.irncid.org/seminars/7.htm.

6

Mineral nutrition of mycorrhized tropical gum tree *A. senegal* (L.) under water deficiency conditions

Ndiaye Malick[1], Cavalli Eric[2], Leye El Hadji Malick[3], and Diop Tahir Abdoulaye[1]

[1]Laboratoire de Biotechnologies des Champignons, Département de Biologie Végétale, Faculté des Sciences et Techniques, Université Cheikh Anta Diop, BP. 5005 Dakar-Fann, Sénégal.
[2]Laboratoire de Nanomédecine, Imagerie et Thérapeutique, EA 4662, UFR Sciences Médicales et Pharmaceutiques, Université de Franche-Comté, 19 rue Ambroise Paré, 25030 Besançon cedex, France.
[3]Laboratoire National de Recherche sur les Productions Végétales/ Institut Sénégalais de Recherche Agricole, 'LNRPV/ISRA', Dakar, Sénégal.

A pot experiment was set to examine the effects of arbuscular mycorrhizal fungi (AMF) on mineral nutrition of a tropical legume tree (*Acacia senegal)* under three different water status. *Acacia senegal* seedlings were inoculated with three species of AMF, *Glomus intraradices, Glomus fasciculatum* or *Glomus mosseae*. Three water levels (field capacity, moderate water deficiency and severe water deficiency) were applied to the pots after transplantation. *A. senegal* seedlings were colonized by the three AM fungi. Twelve weeks after water stress imposition, uptake nutrient of *A. Senegal* was enhanced by mycorrhizal inoculation under moderate and severe water status. Root colonization varied from 30.4 to 62.5%. The lowest intensity (30.4%) was observed on field capacity associated with *G. intraradices* and the highest root colonization (62.5%) was observed on severe water deficiency associated with *G. fasciculatum*. Relative improvement was noted in the foliar nutrient content nitrogen (N), phosphorus (P), potassium (K) and shoot water content of the inoculated plants, whatever the water regime. Mycorrhizal inoculation has no significant effect on shoot calcium (Ca), magnesium (Mg) and sodium (Na) compared to uninoculated plants. *G. fasciculatum* was the most efficiency fungus in nutrient foliar of *A. senegal* plants under water defiency conditions. Inoculating *A. senegal* plant with the arbuscular mycorrhizal fungus *G. fasciculatum* increased ability to acquire N, P, and K under water deficiency conditions.

Key words: Arbuscular mycorrhizal fungi, *Acacia senegal*, mineral nutrient, water deficiency.

INTRODUCTION

The major factors affecting plant growth in Sahelian zone agrosystem are water and nutrient (Floret et al., 1993). Water availability has been recognized as the most critical determinant of plant survival rate after transplantation in Sahelian zones. Thus, it is necessary to improve the level of efficiency in the plant capture and use of nutrients which is important in plant growth. In recent years, there has been increasing evidence that the microbial communities of soil and plants have an important role in the development of sustainable agriculture. Among the microorganism living in plant rhizosphere, arbuscular mycorrhizal fungi (AMF) have

Table 1. Characteristics of the soil of study.

Component	Contents
Clay	3.6%
Silt	1.6%
Fine silt	2.9%
Fine sand	51%
Coarse sand	40.9%
Organic matter	1.06%
Total carbon	2.5
Total nitrogen	0.33
Conductivity (µS/cm)	658
Total phosphorus (ppm)	47
Available phosphorus (ppm)	3.1
pH (sol/water ratio 1:2)	6.7
pH (sol/KCl ratio 1:2)	4.5

been found to be essential components of sustainable soil-plant systems (Bucher, 2007). Therefore, mycorrhizal inoculation with suitable fungi has been proposed as a promising tool for improving restoration success in semi-arid degraded areas (Garbaye, 2000).

The symbiotic relationship between AMF and the roots of higher plants contributes significantly to plant nutrition and growth (Augé, 2001). These positive responses in productivity to AMF colonization have mainly been attributed to the enhanced uptake by AMF of relatively immobile soil ions such as phosphorus (P), potassium (K), calcium (Ca), magnesium (Mg), and manganese (Mn) (Smith and Read, 2008), but it involve the enhanced uptake and transport of more mobile nitrogen (N) ions, particularly under drought conditions (Liu et al., 2007). Since nutrient mobility is limited under drought conditions, AMF may have a larger impact on overall plant growth and development in water deficiency conditions (Sánchez-Díaz and Honrubia, 1994).

The objective of the present study was to examine the effects of AMF on foliar nutrient of *A. senegal* seedlings under water deficiency conditions.

MATERIALS AND METHODS

Planting of materials

The experiment was conducted at the Department of Plant Biology (University Cheikh Anta Diop/Senegal). The soil used in this study was collected from Sangalkam (50 km from Dakar, Senegal). Soil was sterilized by autoclaving at 12°C for 1h. Characteristics of the soil were given in Table 1. Seeds were scarified and surface-sterilized with concentrated sulfuric acid for 10 min then washed in sterile distilled water before germinated on sterile water agar at 0.8 at 30°C in the dark for 48 h. The germinated seeds were transferred to plastic bags used in nursery. Four germinated seeds were sown into each bag containing approximately 500 g sandy soil of Sangalkam and thinned to one seedling per pot, 1 week after emergence. All seedlings were watered daily for 1 week to allow proper establishment.

AMF Inoculation

The three species of AMF were used: *Glomus mosseae*, *Glomus fasciculatum* and *Glomus intraradices* obtained from the Laboratory of Fungal Biotechnology (LBC) of the Plant Biology Department (University Cheikh Anta Diop, Sénégal). Mycorrhizal inoculum for each endophyte consisted of mixed soil, spores, mycelium and infected root fragments obtained using maize as host plant. Inoculum from each AM fungus possessed similar infected characteristics (an average of 40 spores per gram and 85% of infected roots).

Experimental design and growth conditions

Plants were inoculated with one of three AMF species by placing 20 g of inoculum directly in the substrate at the position of the roots. The seedlings were subdivided into three blocks. Each block had a plant control and AM inoculation treatment with *G. mosseae*, *G. fasciculatum* or *G. intraradices* each in five replicates. The first block of control plants was kept to 100% of field capacity (FC). The second block of moderate water-deficiency (MWD) plants was maintained at 50% FC and the last block of severe water-deficiency (SWD) plants was maintained at 25% FC. Plants were raised from January to March in a greenhouse with the following conditions: average day/night temperature 30/25 ± 2°C, and relative humidity maintained between 55 and 65%. Plants were harvested after 3 months.

Measurements and chemical analysis

The percentage of mycorrhizal root infection was estimated by microscope observation of fungal colonization after clearing washed roots in 10% KOH and staining with 0.05% trypan blue in lactophenol (v/v), according to Phillips and Hayman (1970). The relative arbuscular richness (that is, percentage of arbuscules observations among the number of AMF in roots) were calculated using the grid-line intersect method (Giovannetti and Mosse, 1980). Chemical analysis of plants and soil were conducted at the Analysis and Characterization Service (SERAC) of the University of Franche-Comte in France. Plant mineral element compositions [Ca, Mg, sodium (Na), and K] were assessed after digestion with HNO_3 + H_2O_2. Elements were determined using flame atomic emission spectrometry (VARIAN spectra A 220FS) following the French AFNOR standards (FD T90-019, 1984). Kjeldahl Nitrogen was determined by volumetric determination using the French AFNOR standards (NF EN 25663, 1994). The content of the four mineral elements was expressed in mg. kg^{-1} of dry weight. The phosphorus was determined by atomic absorption (GANIMEDE P) using a molecular adaptation following the French AFNOR standards (NF EN 6878, 2005).

Statistical analysis

Statistical procedures were carried out with the software package R version 2.5. Two factors analysis of variance (ANOVA) was performed to partition the variance into the main effects and the interaction between inoculation and water status.

RESULTS

Symbiotic development

As shown in Table 2, roots of control seedlings were

Table 2. Effects of water status on mycorrhizal colonization (%) of *A. Senegal* seedlings.

Treatment	AM colonization (%)		
	FC	MWD	SWD
Control	0.00[d]	0.00[d]	0.00[d]
G. intrardices	30.40[c]	37.2 0[bc]	43.60[c]
G. fasciculatum	50.10[a]	55.60[a]	62.50[a]
G. mosseae	40.95[b]	50.35[a]	55.40[b]

Values in columns followed by different letters are significantly different (p ≤ 0.05). FC: Field capacity; MWD: Moderate water deficiency; SWD: Severe water deficiency.

Table 3. Foliar nutrient (mg/kg dry matter) of *A. Senegal* shoot under water status subjected to four different treatments.

Treatment	N	P	K	Na	Ca	Mg
FC						
Control	34.30[a]	1.58[a]	11.73[b]	2.75[a]	34.18[a]	6.15[a]
G. intraradices	32.60[a]	1.40[a]	13.22[b]	1.88[b]	32.22[a]	5.16[a]
G. fasciculatum	39.80[a]	1.64[a]	15.71[a]	2.44[a]	30.01[a]	6.10[a]
G. mosseae	37.40[a]	1.51[a]	12.97[b]	2.23[a]	29.99[a]	5.00[a]
MWD						
Control	36.30[c]	1.48[b]	9.63[c]	2.10[a]	31.74[a]	4.41[a]
G. intraradices	35.80[c]	1.65[b]	14.96[b]	1.70[b]	25.60[b]	4.88[a]
G. fasciculatum	57.00[a]	2.27[a]	20.54[a]	1.93[a]	30.02[b]	4.46[a]
G. mosseae	43.00[b]	1.70a	17.18[a]	1.87[b]	25.46[b]	4.47[a]
SWD						
Control	32.80[d]	1.18[b]	7.00[c]	0.89[a]	25.30[a]	3.99[b]
G. intraradices	60.00[c]	1.55[b]	10.33[b]	0.79[b]	23.50[a]	3.36[b]
G. fasciculatum	74.10[a]	2.02[a]	17.10[a]	0.90[a]	25.28[a]	5.07[a]
G. mosseae	67.00[b]	1.46[b]	10.38[b]	0.92[a]	24.19[a]	4.05[a]

Values in columns followed by different letters are significantly different (p ≤ 0.05). FC: Field capacity; MWD: Moderate water deficiency; SWD: Severe water deficiency.

observed after water treatments, indicating the absence of mycorrhiza. *A. senegal* seedlings were infected by any of AM fungi and the average root colonization varies from 30.4 to 62.5% (Table 2). Results showed that *G. fasciculatum* and *G. intraradices* had the highest and the lowest AM fungus colonization among the three AMF, regardless of water status. Water deficiency increased root colonization.

Foliar nutrient

Nutrient concentrations in the shoots of the seedlings after harvesting are shown in Table 3. *A. senegal* shoot N, P, and K contents showed that the nutrient uptake was a function of the applied treatment. Colonization by *G.*

fasciculatum improved the content of these nutrients more than the other two fungi. The greatest differences among inoculated plants were found for foliar K. This fungal effect was more obvious under water deficiency conditions where nutrient uptake by *A. senegal* plants is more limited. Moreover, the damaging effects of water deficiency conditions on nutrient acquisition of control plant were reduced in inoculated plant (Table 3). As observed for mycorrhizal colonization, the highest foliar concentrations of N, P, and K were seen in the plants inoculated with *G. fasciculatum* under water deficiency condition. Likewise, the field capacity did not have an effect on foliar nutrient concentrations in shoots.

The presence of water deficiency conditions reduced the contents of N, P and K by 13, 25.3 and 36.4%, respectively, in the control plants. However, the most

Table 4. Significance level (F-values) of effects of differents factors and factors interaction on foliar nutrient based on analysis of variance (ANOVA).

Variable	Inoculation	Water status	Inoculation x Water status
Df	3	2	6
N	1779.07***	9671.15***	181.10***
P	1636.975***	18.437***	32.294***
K	365.666***	23.892***	47.761***
Na	4.4865**	92.4347***	1.1571^{ns}
Ca	5.4989**	30.8064***	1.8789^{ns}
Mg	8.6854***	25.7115***	2.0110^{ns}

*, **, *** and ns indicate the level of significance at $P \leq 0.05$, 0.0, 0.001 and the absence of significance respectively.

active fungus in increasing water deficiency tolerance was *G. fasciculatum*. It increased N, P, and K concentrations by 125, 86.4 and 144%, respectively over those values recorded in control plants under severe water stress (Table 3). Shoot N, P, and K concentrations for water deficiency conditions were significantly higher than field capacity. Analysis of variance was seen to have significantly increased all foliar nutrient concentration by inoculation and water status. Statistical results also show that combined factors were significant for the N, P, and K contents (Table 4).

Na, Ca, Mg concentration declined with increasing water deficiency conditions (Table 3). Na concentrations of inoculated plant with *G. intraradices* decreased significantly with the water regime. Ca concentrations of inoculated plant decreased significantly under moderate deficiency condition, but not under FC and SWD conditions. Magnesium concentrations in different treatments did not change significantly in comparison with control except for Mg shoot content in severe water deficiency condition.

DISCUSSION

Human activity is one of the main factors reducing soil fertility in the vast Sahelian area. As it was observed in tropical gum trees, water deficiency conditions reduce plant mineral nutrition in areas distress from water limitation. In this study, *G. fasciculatum* was the most efficient fungus in terms of *A. Senegal* performance, and mostly in improving foliar nutrient concentration. Differences among AMF presenting foliar nutrient improvement under water deficiency conditions have already been reported (Qu et al., 2004; Miransari et al., 2007). When soil water is limited, K plays an important role in the control of water relations, helping to maintain a high tissue water level, even under osmotically impaired conditions. It is well documented that K, as the most prominent inorganic solute, plays a key role in

osmoregulation processes in photosynthesis (Smith et al., 2004; Li et al., 2006). It should also be noted that AM plants are able to absorb higher rates of P, which, among other important roles in the plant, can markedly enhance root growth (Miransari et al., 2008; Smith and Read, 2008). The increase of the uptake may help explain the important growth effects obtained in inoculated plants under water deficient conditions (Audet and Charest, 2006).

In our study N, P, K concentrations in inoculated plants were particularly increased under drought conditions. This is in agreement with the studies of other researchers comparing AM species under different stresses (Ruiz-Lozano et al., 2003). While water status significantly affected the concentration of N, P, and K, AMF species exerted a significant effect on all nutrient concentrations. Table 3 shows that the high efficiency of all AMF on the use of these macronutrients by tropical gum tree. Accordingly, previous researchers have found that the efficiency of AM increases with increased level of stress (Audet and Charest, 2006; Subramanian et al., 2006).

Significant interaction of water status and AMF on the concentrations of N, P, and K indicate differential nutrient uptake efficiencies for different AM-host plant combinations (Table 4). In agreement with these ideas, Subramanian and Charest (1997) contend that AM symbiosis improved drought tolerance of a tropical tree, primarily through the enhanced uptake of slowly diffusing nutrients. In fact, foliar nutrient have been found to increase in inoculated plants (Mortimer et al., 2008), but not under water deficiency condition. The opposite effect was observed by Huett et al. (1997). Na, Ca, Mg concentrations in inoculated plants were lower than those of the controls, however, improved Mg concentration under severe water condition was observed. The same results were found by Monzon and Azcon (1996) for inoculated plants.

The support of symbiotic associations between these fungi and tropical gum tree may be suggested in application of inoculation for successfully improve mineral nutrient uptake of seedlings in Sahelian zone agrosystem.

Conclusion

We may conclude that especially under water deficiency conditions, AMF enhances foliar essential nutrient uptake. This role is very much dependent on different species of arbuscular mycorrhiza. Accordingly, the symbiosis of A. Senegal with more efficient AM specie G. fasciculatum under such conditions can be very beneficial in water deficiency condition. The next step should be transposing this greenhouse experiment to the field.

ACKNOWLEDGEMENTS

The authors wish to thank OHM Téssékéré and Analysis and Characterization Service (SERAC) at University of Franche-Comté in France.

Abbreviations: AMF, Arbuscular mycorrhizal fungi; **FC**, field capacity; **MWD**, moderate water deficiency; **SWD**, severe water deficiency.

REFERENCES

Audet P, Charest C (2006). Effects of AM colonization on « wild tobacco » plants grown in zinc-contaminated soil. Mycorrhiza 16:277-83.

Augé RM (2001). Water relations drought and vesicular-arbuscular mycorrhizal symbiosis. Mycorrhiza 11:3-42.

Bucher M (2007). Functional biology of plant phosphate uptake at root and mycorrhiza interfaces. New Phytol. 173:11-26.

FD T90-019 (1984). Essais des eaux - Dosage du sodium et du potassium - Méthode par spectrométrie d'émission de flamme, Afnor.

Floret C, Pontanier R, Serpantié G (1993). La jachère en Afrique Tropicale. Man and Biosphere, Unesco. Paris. 16:86.

Garbaye J (2000). The role of ectomycorrhizal symbiosis in the resistance of forests to water stress. Agriculture 29:63-69.

Giovannetti M, Mosse B (1980). An evaluation of techniques for measuring vesicular-arbuscular mycorrhizal infection in roots. New Phytol. 84:489-500.

Huett DO, George AP, Slack JM, Morris SC (1997). Diagnostic leaf nutrient standards for low-chill peaches in subtropical Australia. Aust. J. Exp. Agric. 37:119-126.

Li H, Smith SE, Holloway RE, Zhu Y, Smith FA (2006). Arbuscular mycorrhizal fungi contribute to phosphorus-fixing soil even in the absence of positive growth responses. New Phytol. 172:536-543.

Liu A, Plenchette C, Hamel C (2007). Soil nutrient and water providers: how arbuscular mycorrhizal mycelia support plant performance in a resource limited world. In: Hamel C, Plenchette C (eds) Mycorrhizae in Crop Production. Haworth Food and Agricultural Products Press, Binghamton. NY. pp. 37-66.

Miransari M, Bahrami HA, Rejali F, Malakouti MJ (2008). Using arbuscular mycorrhiza to reduce the stressful effects of soil compaction on wheat (Triticumaestivum L.) growth. Soil Biol. Biochem. 40:1197-1206.

Miransari M, Bahrami HA, Rejali F, Malakouti MJ, Torabi H (2007). Using arbuscular mycorrhiza to reduce the stressful effects of soil compaction on corn (Zea mays L.) growth. Soil. Biol. Biochem. 39:2014-2026.

Monzon A, Azcon R (1996). Relevance of mycorrhizal fungal origin and host plant genotype to inducing growth and nutrient uptake in Medicago species. Agric. Ecosyst. Environ. 60:9-15.

Mortimer PE, Perez-Fernandez MA, Valentine AJ (2008). The role of arbuscular mycorrhizal colonization in the carbon and nitrogen economy of the tripartite symbiosis with nodulated Phaseolus vulgaris. Soil. Biol. Biochem. 40:1019-27.

NF EN 25663 (1994). Qualité de l'eau - Dosage de l'azote Kjeldahl - Méthode après minéralisation au sélénium, Afnor.

NF EN 6878 (2005). Qualité de l'eau - Dosage du phosphore - Méthode spectrométrique au molybdate d'ammonium, Afnor.

Phillips JM, Hayman DS (1970). Improved procedures for cleaning roots and staining parasitic and vesicular arbuscular mycorrhizal fungi for rapid assessment of infection. Trans. Br. Mycol. Soc. 55:93-130.

Qu L, Shinano T, Quoreshi AM, Tamai Y, Osaki M, Koike T (2004). Allocation of C14 carbon in two species of larch seedlings infected with ectomycorrhizal fungi. Tree Physiol. 24:69-76.

Ruiz-Lozano JM (2003). Arbuscular mycorrhizal symbiosis and alliviation of osmotic stress. New perspectives for molecular studies. Mycorrhiza 13:309-317.

Sánchez-Díaz M, Honrubia M (1994). Water relations and alleviation of drought stress in mycorrhizal plants In: S. Gianinazzi and H. Schüepp, Editors, Impact of Arbuscular Mycorrhizas on sustainable agriculture and natural ecosystems, Birkhäuser Verlag, Basel. Switzerland. pp. 167-178.

Smith SE, Smith FA, Jakobsen I (2004). Functional diversity in arbuscular mycorrhizal (AM) symbioses: the contribution of the mycorrhizal P uptake pathway is not correlated with mycorrhizal responses in growth and total P uptake. New Phytologist. 162:511-524.

Smith SS, Read DJ (2008). Mycorrhizal Symbiosis. Academic Press.

Subramanian KS, Charest C (1997). Nutritional, growth, and reproductive responses of maize (Zea mays L.) to arbuscular mycorrhizal inoculation during and after drought stress at tasselling. Mycorrhiza, 7:25-32.

7

Employment generation potential of watershed development programmes in semi-arid tropics of India

Biswajit Mondal and N. Loganandhan

Central Soil and Water Conservation Research and Training Institute, Research Centre, Bellary, Karnataka, India.

This paper examines the employment generation potential of watershed development programmes and identifies the factors that contribute to the shift in labour absorption in farming activities over control situation. Analysis of secondary data collected from the watershed implementing agencies revealed that on an average watershed programmes helped to generate one time employment ranged between 26 and 76 mandays per hectare, for soil conservation, forestry and other works. Primary survey revealed that utilization percentage out of available labour at average households increased due to watershed development programmes. Employment elasticity with respect to various factors of production in crop cultivation worked out and showed negative price elasticity of demand for labour whereas, employment elasticity with respect to all other factors were found to be positive indicating their positive influence on labour demand. A decomposition analysis revealed that about 61% of the employment growth in watersheds over control situation was attributable to technology effect. Labour efficiency was much higher in the watershed than the control villages as evident from higher labour income. The study establishes that the watershed development programme had the potential of creating huge employment opportunities at the farm level in semi-arid tropical region of India.

Key words: Decomposition analysis, employment elasticity, labour demand, profit function, watershed.

INTRODUCTION

Employment generation has come to occupy centre stage in research and development planning as well as implementation in many developing countries. Expanding productive employment was the principal aim of many developmental and relief programmes for sustained poverty reduction, as labour is the main asset for a majority of the poor. Underemployment or disguised unemployment is still a common phenomenon in the Indian agriculture. It results in out-migration of young and healthy rural workforce to sub-urban and urban areas. Thus agriculture is left on physically weaker workers of the village society, which leads to lower marginal productivity of labour and lower wages. Low levels of productivity and low input usage characterize the rainfed agriculture, which constitutes 60% of net sown area of the country. Bulk of the rural poor lives in rainfed regions. So, it is important to accord high priority to sustainable development of these areas through watershed development approach, which are having high potentials in terms of scope of different activities to be carried out as well as surplus labour to engage in such activities. Therefore, generation of gainful employment opportuneties in the rural areas was aimed through watershed management programme. Singh et al. (2010) reported that integrated watershed development led to increased employment opportunities for the community members with better wage earnings in construction work during the programme implementation phase and engagement in agricultural fields during the post-implementation periods, but no specific formal mechanisms were developed to enhance the opportunities. Several earlier studies (Dhyani et al., 1997; Arun, 1998; Arya and Yadav, 2006; Kalyan, 2007) also reported that watershed management programme had positive and significant impact in generating employment opportunities at the farm level.

The pre-requisite for proper planning of the available labour force in an area or region is the information about the technical coefficients with regard to labour absorption in various enterprises. An investigation of dynamics of labour employment would provide an increased understanding of the sources of employment growth and this understanding would indicate the directions in which efforts will have to be made for increasing employment opportunities. This research paper provided an account of employment generation potential of watershed development programmes and identified the factors that contributed to the shift in labour absorption in farming activities over control situation. The specific objectives of this paper is to estimate the availability and its utilization of labour for crop and non-crop activities, employment elasticities with respect to various inputs in agricultural production and decompose the employment growth into watershed technology effect and other factors.

MATERIALS AND METHODS

Data source

The study was carried out in semi-arid tropical region of India. A multistage stratified random sampling was employed for the selection of samples for the study. At the first stage, two states namely; Andhra Pradesh and Karnataka were selected purposefully, as major portion of these states comes under hot, semi-arid eco-region. At second stage, two watersheds from each state namely, Upparhalla (Watershed-I) and Kalvi (Watershed-II) from Karnataka state and Mallapuram (Watershed-III) and Chinnahothur (Watershed-IV) from Andhra Pradesh state were selected for detail investigation. One control village contiguous to each selected watershed was also chosen on the ground that such villages did not come under any watershed based activities or progarmmes. Thus, in all, a total of eight villages were selected to carry out the investigation. For selection of ultimate unit of sampling that is, sample households, a complete list of households of the selected villages was prepared. The third stage of sampling involved random selection of 200 households from four watershed villages and 202 households from control villages, in accordance with the probability proportional to number of households in each village for well representation.

The data for present investigation were collected from both primary as well as secondary sources. The primary data pertaining to socio-economic characteristics of respondents, cropping pattern, productivity and employment generation were collected by personal interview of respondents with the help of pre-tested comprehensive schedule. Information on various aspects of labour availability and utilization was collected using time allocation schedule particularly designed for this study. Secondary data like details of watershed works undertaken and temporary employment generated during

programme implementation were collected from Detailed Project Report (DPR) of the selected watersheds.

Analytical methods

To estimate various components of labour availability, initially, effective labour available per household were obtained by considering monthly labour availability and subtracting from it non-availability because of sickness, festivals and various unforeseen. To obtain estimates of labour utilization for crop and non-crop activities, the current magnitudes of labour use for all activities were calculated. These activities include crop production, animal husbandry and various non-farm activities. The returns to per unit of labour input have also been estimated for both watersheds and control villages by using budgeting technique.

For estimating partial employment elasticities with respect to various inputs, production function and labour demand function were fitted. To decompose total change in employment per hectare due to introduction of watershed technology an employment decomposition model based on Unit-output-price (UOP) profit function (Lau and Yotopoulos, 1972) were formulated and used. The UOP profit function formulation enables us to derive labour demand as a function of the normalized wage rate, variable input price and the quantities of fixed inputs. The Cobb-Douglas form of production function with usual neo-classical properties was used for this study with the following specification:

$$Y = AN^{\alpha}F^{\beta 1}K^{\beta 2}L^{\beta 3} \tag{1}$$

where, Y is output, N is the variable labour input and F, K and L are the fixed inputs of fertilizer and manures, flow of capital services and land, respectively. A is constant term of scale parameter, and α, β_1, β_2 and β_3 are the partial output elasticities of labour, fertilizer, capital and land, respectively. The output elasticity of land (β_3) was obtained by the maintained hypothesis of constant returns to scale as:

$$\beta_3 = 1 - (\alpha + \beta_1 + \beta_2).$$

Following Lau and Yotopoulos (1972), a UOP profit function and labour demand function were specified as below:

$$\pi^* = A^*W^{\alpha^*}F^{\beta 1^*}K^{\beta 2^*}L^{\beta 3^*} \tag{2}$$

$$N = A^{\psi}W^{\alpha^*-1}F^{\beta 1^*}K^{\beta 2^*}L^{\beta 3^*} \tag{3}$$

where, $\pi^* = \pi/P_y$ = normalized profit or unit-output-price (UOP) profit, π = profit defined as current revenues less current total variable input cost; $W = P_n/P_y$ = normalized labour wages, P_n = wage rate, and P_y = output price; $A^* = A^{\theta}(1-\alpha)\alpha^{\alpha\theta}$, $\theta = (1-\alpha)^{-1}$, $\alpha^* = -\alpha\theta < 0$, $\beta_1^* = \beta_1\theta > 0$, $\beta_2^* = \beta_2\theta > 0$, and $A^{\psi} = -\alpha^*A$.

Every concave production function has a dual which is a convex profit function; the proposition has been tested from the fact that parameters for profit function that are defined in production function (1) are closely related to the parameters of UOP profit function (2). On the basis of a-priori theoretical considerations, we know that the UOP profit function is decreasing and convex in normalized wage rate (W) and increasing in quantities of fertilizer and capital. It also follows that the function is increasing in the price of output (Bisaliah, 1978). The crucial assumption is that firms behave according to some decision rules which include profit maximization, given the price regime of output and labour, and given the quantities of F and K. In this study, the existence of these systematic decision rules is a maintained hypothesis.

In Equation (3) we have labour demand as a function of W, F and K. In this function, the total change in employment is brought about by shifts in parameters that define the function itself and by changes in W, F and K. It has been argued that ordinary last squares applied to UOP profit function (2) and labour demand function (3) separately are consistent (Lau and Yotopoulos, 1972). However, these estimates are argued to be inefficient because β's appears in both the equations. So, a more efficient approach is to estimate both the equations from production function elasticities imposing the restrictions that β's are equal in both the functions (Zellner, 1962). Our main concern was to decompose the total difference in employment between watershed and control areas farms, hence, the labour demand function in Equation (3) was differentiated totally and converted into per hectare terms as below:

$$dN/N = d\alpha^*/\alpha + dA^*/A^* + (\alpha^*-1)\, dW/W + \beta_1^*dF/F + \beta_2^*dK/K + \log(W)d\alpha^* + \log(F)\, d\beta_1^* + \log(K)\, d\beta_2^* \quad (4)$$

Following Bisaliah (1978) the decomposition equation was expressed in terms of elasticities of production and re-arranged as below:

$$dN/N = [\theta.dA/A] + [\theta.d\alpha/\alpha + \theta^2(\log A + \log \alpha)d\alpha - \theta^2(\log W)d\alpha + \theta^2\{(1-\alpha)d\beta_1 + \beta_1 d\alpha\}\log F + \theta^2\{(1-\alpha)d\beta_2 + \beta_2 d\alpha\}\log K] - [(\theta\alpha +1)\, dW/W] + [\theta\beta_1(dF/F) + \theta\,\beta_2(dK/K)] \quad (5)$$

Equation (5) allowed us to decompose per hectare change in employment (dN/N) into following three components:

(i) Technology effect: This includes the effects of shifts in scale parameter (A) and slope parameters (output elasticities) in production function (1), given W, F and K as under control. This effect was captured by adding the values of first and second bracketed expression of employment decomposition Equation (5).

(ii) Normalized wage rate effect: This effect was denoted by the third bracketed expression of Equation (5) which captured the effect of difference in normalized wage rates (Ws) confronting watershed and control areas farms, given the output elasticity of labour (α) in watershed areas.

(iii) Complementary inputs effect: This effect (further bracketed expressions) includes the employment effects of differences in quantities of fertilizers and capital, given the output elasticities of these inputs in watershed areas.

For the purpose of decomposition analysis, derivatives in Equation (5) were expressed in discrete form as below:

$$\Delta N/N = [\theta.\Delta A/A] + [\theta.\Delta \alpha/\alpha + \theta^2(\log A + \log \alpha)\, \Delta\alpha - \theta^2 (\log W)\, \Delta\alpha + \theta^2\{(1-\alpha)\Delta\alpha + \beta_1\Delta\alpha\}\log F + \theta^2\{(1-\alpha)\Delta\beta_2 + \beta_2\Delta\alpha\}\log K] - [(\theta\alpha+1)\Delta W/W] + [\theta\beta_1(\Delta F/F) + \theta\beta_2(\Delta K/K)] \quad (6)$$

where, ΔN = (N) Watershed technology - (N) Control. Likewise, ΔA, ΔF, ΔK and coefficients were computed and the base values of N, A, F, K were the values pertaining to control.

RESULTS AND DISCUSSION

Description of the study region and watershed development programmes

Despite the significant increase in irrigated area in India during planned development of over four decades, about 60% of net sown area in the country is still rainfed (Bhatia, 2005). Amongst different rainfed areas, the most vulnerable are semi-arid tropical regions. About 53.4% of India's land area comprises arid and semi-arid regions (GoI, 2004) characterized by low and erratic rainfall, periodic droughts and different associations of vegetative cover and soils. The states which are falling under semi-arid tropics (SAT) includes Andhra Pradesh (Prabhakar et al., 2011), Gujarat, Karnataka, Madhya Pradesh, including Chhattisgarh, Maharashtra, Rajasthan and Tamil Nadu. The share of rainfed agriculture in SAT states is about 73% of which the two selected states (Karnataka and Andhra Pradesh) contributes around 40% and the percentage of rainfed area to net area sown in these two states are 77 and 60, respectively (Bhatia, 2005). Major crops grown in this region include coarse cereals like jowar, bajra and ragi; pulses like bengal gram and horse gram and oilseed crop like groundnut and sunflower. Not only the yields per hectare are low in rainfed land in semi-arid region, the variability in both area and yield for most of the crops in semi-arid states is much higher than the all-India average. The semi-arid areas have been subject to large scale degradation of natural resources caused by the depletion of forests, soil erosion, declining common pool resources, etc. (Jodha, 1990) and around 300 million people depend for their sustenance on dryland agriculture, of which 30 to 40% can be classified as poor as per latest available Census (Ryan and Spencer, 2001). Seasonal migration could be seen as a form of spatial diversification, which is at the root of food and livelihood security strategies by the marginal and small farmers in the region (Mondal et al., 2012).

Watershed development programmes are being implemented in the region with the twin objectives of natural resource conservation and enhancing the livelihoods of the rural poor through enhancement of production levels. A total of 4.3 and 6.2 million hectare of land have been covered under the various projects (by different departments under Ministry of Rural Development, Ministry of Agriculture and Ministry of Environment and Forests, Government of India) in the State of Karnataka and Andhra Pradesh, respectively (Sharda et al., 2008). Different types of treatment activities that generally carried out in the watersheds includes soil and moisture conservation measures in agricultural lands (contour/ field bunding and summer ploughing), drainage line treatment measures (loose boulder check dam, minor check dam, major check dam, and retaining walls), water resource development/ management (percolation pond, farm pond, and drip and sprinkler irrigation), crop demonstration, horticulture plantation and afforestation. The aim of all these were to ensure the availability of drinking water, fuel wood and fodder and raise income and employment for farmers and landless labourers through improvement in agricultural production and productivity (Prabhakar, 2012).

Temporary employment generated by watershed programmes

In all the selected watersheds, soil and water

Table 1. Temporary employment generated at the study watersheds (in hundred mandays).

Components	Karnataka state		Andhra Pradesh state	
	Watershed-I	**Watershed-II**	**Watershed-III**	**Watershed-IV**
Soil and moisture conservation works	922.4	120.6	217.1	283.8
Forestry/horticulture	98.6	42.9	170.2	9.21
Total	1021.0	163.5	387.3	293.0
Average (per hectare)	0.74	0.33	0.45	0.26
Non-land-based activities	26	6.4	5.9	1.5

Table 2. Availability and utilization of labour in study villages.

Particulars	Watersheds	Control villages
Monthly labour available per household (mandays)	78	69
Monthly labour utilization per household (mandays)	53	38
Labour utilization percentage	67.95	55.07
Employment gap	32.05	44.93

Table 3. Labour absorption in farm and non-farm activities in the study domain (mandays per year per household).

Activities	Watersheds	Control villages	Difference
1. Crop enterprises	351 (55.10)	288 (62.34)	63[***]
2. Animal husbandry	142 (22.29)	120 (25.97)	22[**]
3. Non-farm employment	144 (22.61)	54 (11.69)	90[***]
Total	637 (100.00)	462 (100.00)	181[***]

Figures in parentheses indicate % of total; [***] and [**] indicates difference between two areas were significant at 1 and 5% level.

conservation as well as planting programmes were carried out by manual labour only. Analysis of secondary data reveals that the watershed programmes generated one time employment ranged between 26 and 74 mandays per hectare for soil conservation, forestry and other works (Table 1). The differences between the watersheds were mostly due to variations in the number and size of erosion control structures and in case of forestry and horticulture it was due to difference in intensity of programme. Further, the programme also created assets for livelihoods of poor and provided employment to the extent of 150 to 2600 mandays in different watersheds. The magnitude of employment is not in-significant; further these activities now providing recurring employment as the fruit trees are yielding as reported by the respondents.

Household availability and utilization of labour

The specific aspects covered in this objective are availability of family labour per household per month, contribution of labour for crop, livestock and non-farm activities under different land holding categories (Tables 2 and 3). On an average, 78 and 69 mandays of effective labour per household per month, respectively in watershed and control areas were available. Farm households in watershed could utilize only 68% of available labour and the utilization in control areas was only 55%. Employment gap was more in control areas to the extent of 13%.

Introduction of watershed technologies increased the cropping intensity, production levels and shifts the farming activities from less labour intensive (low value) to more labour intensive (high value) crops, livestock and other enterprises which in turn shift the labour absorption per hectare of cultivated area over control areas. Therefore, human labour use in farming as well as non-farm activities in both watershed as well as control areas were also analysed and the results are presented in Table 3. The regular labour employment in agriculture in watersheds (351 mandays per household per year) was significantly higher than the control villages (288 mandays per household per year). The results also indicated that the average number of days employed in animal husbandry activities were significantly higher in the watershed villages. Thus, it can be concluded that watershed management programme had positive and

Table 4. Input-output levels in crop production in study region (per farm per year).

Items	Watersheds (173)	Control villages (171)	Difference
Output (kg)	5795.69	3238.68	2557.01[***]
Labour input (mandays)	563.23	357.93	205.30[***]
Fertilizer input (INR)	10143.60	6460.09	3683.51[***]
Capital input (INR)	31780.02	20645.68	11134.34[***]
Farm size (cultivated land in hectare)	2.816	2.537	0.279[NS]

Figures in brackets indicate number of farms in each area; INR: Indian National Rupees, [***] indicates the significant difference between two areas at 1% level; NS: Not significant.

Table 5. Estimates of production function, UOP profit function and labour demand function.

Variables	Watersheds			Control villages		
	Production function (R^2=0.932)	UOP profit function	Labour demand function	Production function (R^2=0.899)	UOP profit function	Labour demand function
Constant	1.865(0.285)	1.132	2.969	1.342 (0.364)	0.347	0.821
Labour input	0.724[***] (0.080)	-	-	0.703[***] (0.103)	-	-
Labour wage (Normalized)	-	-2.623	-3.623	-	-2.367	-3.367
Fertilizer input	0.144[***] (0.063)	0.522	0.522	0.146[**] (0.066)	0.492	0.492
Capital input	0.212[***] (0.061)	0.768	0.768	0.243[***] (0.078)	0.818	0.818
Land	-0.08	-0.290	-0.290	-0.092	-0.310	-0.310

Figures in parentheses indicate standard errors; [***] and [**] indicates coefficients are significant at 1 and 5% level, respectively.

significant impact on generating employment opportunities at the farm level.

Input-output level in crop enterprises

Table 4 presents the pattern of input use and output level during a calendar year under predominantly groundnut growing farms at both watershed and control villages. It indicates that the level of output and labour usages was much higher in the watersheds which can be attributable directly to the watershed development programmes. Higher usages of other inputs also indicate the level of intensiveness in cultivation at watershed villages compared to control villages.

Elasticities of production function, UOP profit function and labour demand function

Table 5 presents the production function estimates as well as estimates of UOP profit function and labour demand function separately for watersheds and control areas. In production function estimation, the dependent variable (Y) considered was the output measured in kg per year per farm whereas the independent variable included were labour input measured in mandays per

year (N), expenditure on fertilizer and manures measured in INR per year (F) and expenses on capital inputs covering expenditure on bullock labour, value of seed, expenditure on irrigation, value of plant protection chemicals, expenditure on machinery use, land revenue, apportioned amount of interest and depreciation on fixed capital assets used in production, etc. measured in INR per year (K). It may be noted from Table 5 that about 93 and 90% of the total variation in the output, respectively at watershed and control areas were explained by the included variables which justify the use of production function coefficients for estimating profit function and labour demand function. The regression coefficients associated with all the independent variables were positive and statistically significant except farm size, indicating thereby their positive influence on the level of output at both the areas. Negative output elasticities of land emerged from the fact that small farms might received better attention and management and results more output.

The UOP profit function and labour demand function given in Table 5 were estimated using production function estimates. As expected, the profit function is decreasing in prices of labour and increases in fertilizer and fixed capital inputs. Employment elasticities with respect to different factors as envisaged in labour demand function to be different under watershed technology and control.

Table 6. Decomposition analysis of total change in employment (per hectare per year).

No.	Estimated sources of change in employment	Percentage attributable
A.	Technology effect	
	(i) Neutral technology effect	131.22
	(ii) Non-neutral technology effect	-70.70
Sub-total		60.52
B.	Normalized wage rate effect	-3.32
C.	Complementary inputs effect	
	(i) Effect of fertilizer and agro-chemicals	24.94
	(ii) Effect of capital inputs	21.81
Sub-total		46.75
D.	Total estimated change due to all effects	110.59

This implies that employment response to a given change in these key variables will be different on these two locations. Negative price elasticity of demand for labour implying withdrawal of a section of labour force consequent upon a rise in wage rate which indicated the presence of disguised unemployment in the production enterprises. Negative elasticities of demand for labour with respect to land might be due to mechanization of farm operations at larger sized farms. Employment elasticities with respect to all other factors viz. fertilizer and capital input are positive implying positive influence of these factors on labour demand.

Decomposition of employment growth

The aim of this objective was to decompose the total change in employment into watershed technology, wage rate and complementary inputs component for providing an empirical perspective on the sources of employment growth. For empirical implementation of this objective, an employment decomposition model based on the UOP profit function and labour demand function has been formulated. The decomposition analysis, assuming Hicks-non-neutral technical change between watershed and control areas yielded the following results (Table 6).

First, 131% of the increased employment is attributable to technology effect; we call it as watershed technology. A brief explanation on why technical change given rise to positive employment effect in the decomposition model based on the UOP profit function framework is in order. As every concave production function has a dual which is a convex profit function and vice-versa, technical change that shifts production function upwards, also shifts profit function upwards. This upward shift in profit function, other things being equal, shifts the demand for labour. The effect of non-neutral component of technical change on employment was negative to the extent of 71%. The negative effect was due to the normalized wage rate differentials which were slightly higher at control areas

than the watershed areas which, however, could not offset the total technology effect that was still positive to the extent of 61%.

Second, the employment effect of the difference in normalized wage rate is estimated to be -3%. This follows from the fact that the daily normalized wage rate in watersheds are slightly lower than one paid at control areas. Even though the money wage rate at watershed is higher than the control areas, lower normalized wage rate confronting watershed areas follows from the higher average output price which might be due to better quality of produce and/or higher market linkages. This result seems to indicate the importance of output price, given the money wage rate, in generating employment opportunities at the farm level. Several earlier studies (Mellor, 1976; Rao, 1977; Vyas and Mathai, 1978) also indicated that the output level and price to be closely linked to employment as evidenced from high employment elasticities.

Third, the complementary inputs effect on employment is estimated to be 25% with fertilizer and 22% with other capital expenses. This positive employment effect follows from the condition that an increase in quantities of these inputs shifts the marginal product curves of labour to the right. As a result, the profit maximizing farm tends to employ more labour in watersheds at a given wage rate. This result very well support the strategy of physical planning for the production and distribution of complementary inputs in any employment generation programme.

Rate of return per unit of labour

The returns to per unit of labour have been estimated for crop production at both watershed and control areas and presented in Table 7. The results showed that the labour rate was almost similar between watershed and control areas but due to higher gross return per hectare, corresponding difference in labour income was around

Table 7. Differences in rate of return per labour input between watersheds and control villages.

Particulars	Watersheds	Control villages	Difference
Gross return (INR in thousand per hectare per year)	32.99	19.75	13.24
Labour income (INR in thousand per hectare per year)	15.09	0.67	14.42
Labour unit used (number per hectare per year)	200	141	59
Labour income (INR per unit)	75.46	4.74	70.72
Cost of labour (INR per unit)	32.82	31.86	0.96
Net labour income (INR per unit)	42.64	(-) 27.12	69.76

INR 71 and net labour income in the control areas was negative to the extent of INR 27. It indicates that labour efficiency increased due to implementation of watershed development programmes in the study region.

CONCLUSION AND RECOMMENDATION

It is evident from above analysis that watershed development programmes have helped to create employment opportunities on temporary as well as permanent basis. If maintenance programmes are taken up as envisaged in the technical plan of the watershed development programme, it is possible to further improve the employment opportunities, which will help to stabilize annual incomes of the landless in particular and farming community in general. On the basis of various estimates of labour availability and utilization, the employment potentials of rural semi-arid region can be further boosted with the following specific points.

(i) Execution of watershed development programme should be done involving both land based and non-land based activities for creation of employment for the rural masses. Inclusion of horticulture/plantation works generates huge temporary as well as regular employment in the watershed programme.
(ii) Labour utilization per household was found to be higher in the watershed villages than the control villages, which might be due to higher cropping intensity and introduction of labour intensive remunerative crops.
(iii) Labour wages, product price and all the complementary inputs are important determinants of employment. Estimated employment elasticity and the labour demand model can be used to derive the output price adjustments and the use of complementary inputs to reach specific employment goals.
(iv) Positive technology effect in the decomposition analysis follows from the condition that technical change shifts profit function upwards. This upward shift in profit function, other things being equal, shifts the demand curve for labour to the right. As the output price is very important to bring the wage rate negative so that total effect increased, more emphasis should be given to high value crops. Various complementary inputs also shift the marginal product curves of labour to the right. As a result, the profit maximizing farm tends to employ more labour in the watershed at a given wage rate. This result very well support the strategy of physical planning for the production and distribution of complementary inputs in any employment generation programme.
(v) Higher net return per labour input in watershed indicates higher labour efficiency in the watershed. Hence, implementation of watershed development programme needs to be continued and extended to other villages for higher labour income and well being of the farmers by improving the productivity of land in the region.

ACKNOWLEDGEMENTS

Authors wish to acknowledge the support rendered by Mr. K.K. Reddy, Mr. K. Channabasappa, Mr. B. N. Sesadri and Mr. B. C. Eranna for field data collection and officials of the selected watersheds for their supports during survey.

REFERENCES

Arun KYS (1998). Economic Evaluation of Watershed Development: A Case Study of Kuthanagere Micro-watershed in Karnataka. Ph.D. Thesis Submitted to the University of Agricultural Sciences, Bangalore (India).

Arya SL, Yadav RP (2006). Economic Viability of Rainwater Harvesting by Renovating Village Ponds in Small Agricultural Watershed of Joharanpur. Agric. Econ. Res. Rev. 19:71-82.

Bhatia MS (2005). Viability of Rainfed Agriculture in Semi-arid Regions. Occasional Paper – 40, Department of Economic Analysis and Research, National Bank for Agriculture and Rural Development (NABARD), Mumbai (India).

Bisaliah S (1978). Decomposition Analysis of Employment Change under New Production Technology in Punjab Agriculture. Ind. J. Agric. Econ. 33(2):70-80.

Dhyani BL, Samra JS, Juyal GP, Ram Babu, Katiyar VS (1997). Socio-economic Analysis of a Participatory Watershed Management Programme in Garhwal Himalaya. Technical Bulletin No. T-35/D-24. Central Soil and Water Conservation Research and Training Institute, Dehradun (India). P. 113.

GoI (2004). India first National Communication to UNFCCC (2004). Ministry of Environment and Forests (MoEF), Government of India.

Jodha NS (1990). Rural Common Property Resources: Contributions and Crisis. Econ. Pol. Weekly, Quart. Rev. Agric. 25(26):65-78.

Kalyan KBM (2007). An Economic Evaluation of NWDPRA and Sujala

Watershed Programme in Northern Karnataka - A Comparative Study. M.Sc. Thesis Submitted to the University of Agricultural Sciences, Dharwad (India).

Lau LJ, Yotopoulos PA (1972). Profit, Supply and Factor Demand Functions. Am. J. Agric. Econ. 54(1):11-18.

Mellor JW (1976). The New Economics of Growth: A Strategy for India and the Developing World. Cornell University Press, New York. pp. 181-187.

Mondal B, Raizada, A, Loganandhan N (2012). Drought-induced Shocks and Coping Strategies by Small Farmers in the Dry Tropics. In: Abstract Proceedings of Conference on 'Livelihood and Environmental Security through Resource Conservation in Eastern Region of India (LESRC-2012)' held at Orissa University of Agriculture & Technology (OUAT), Bhubaneswar (India). P. 126.

Prabhakar K (2012). Watershed management: (ISBN 978-3-8443-8443-7), LAP LAMBERT Academic Publishing GmbH & Co. KG, Germany.

Prabhakar K, Lavanya LK, Rao AP (2011). Prakasam district farmer's participation in NGOs watershed programme, International NGO Journal, ISSN 1993–8225, Vol. 6(10), October 2011, pp. 219-223.

Rao CHH (1977). Agricultural Growth and Rural Poverty: Some Lessons from Past Experiences. Econ. Pol. Weekly 12 (33&34):1369-1374.

Ryan JG, Spencer DC (2001). Some Challenges, Trends and Opportunities Shaping the Future of the Semi-arid Tropics. In: Bantilan, M.C.S, Parthasarathy Rao, P. and Padmaja, R. (eds.), Future of Agriculture in Semi-arid Tropics. Proceedings of International Symposium on the Future of Agriculture in the Semiarid Tropics, 14 November 2000, ICRISAT, Hyderabad (India).

Sharda VN, Juyal GP, Naik BS (2008). Watershed Development in India – Status and Perspective. Central Soil and Water Conservation Research and Training Institute, Dehradun (India).

Singh P, Behera HC, Singh A (2010). Impact and Effectiveness of Watershed Development Programmes in India. Centre for Rural Studies, National Institute of Administrative Research, Lal Bahadur Shastri National Academy of Administration (LBSNAA), Mussoorie (India).

Vyas VS, Mathai G (1978). Farm and Non-farm Employment in Rural Areas: A Perspective for Planning. Econ. Pol. Weekly 13(6&7):333-347.

Zellner A (1962). An Efficient Method of Estimating Seemingly Unrelated Regression and Tests for Aggregation Bias. J. Am. Stat. Assoc. 57(2):348-375.

8

Allometry and condition factors of African pike "*Hepsetus odoe*" actinopterygii in a lake

ADEDOKUN, Mathew Adewale[1], **FAWOLE, O. O.**[2] and **AYANDIRAN T. A.**[2]

[1]Fisheries Technology Department, Oyo State College of Agriculture, P. M. B. 10 Igboora, Oyo State, Nigeria.
[2]Pure and Applied Biology Department, Ladoke Akintola University of Technology, P.M.B 4000 Ogbomoso, Oyo State, Nigeria.

The physiological growth of length and weight variables and the general welfare of *Hepsetus* odoe in Ogbomoso reservoir were investigated. The relationship regression coefficient indicated that the fish sampled showed algometric growth with regression factor of 0.88 in the natural population of fish. The mean K for 133 specimens was 1.5, while 1.5 and 1.6 were K based on sexes (male and female) respectively. Also, 1.789 ± 0.44 (F IV) and 1.880 ± 0.31(M IV) were the highest means K in female and male gonad development stages. The general well being of the sampled fish is affected by both gonads maturity and sexes (p>0.05) while habitat affinity, seasonal variations and size range do not (p<0.001).

Key words: Length-weight relationship, condition factors, resident water body, *Hepsetus odoe*.

INTRODUCTION

The geographical range of *Hepsetus odoe* covers the tropical region of West and Central Africa. It is found in most rivers in West Africa from the Senegal southwards to Botswana. The southern limit of its distribution is the Okavango Delta in Northern Botswana (Merron et al., 1990). Moriarty (1983) reported abundance of *H. odoe* in Kanji Lake, Nigeria. It is usually found near the banks of rivers in heavy vegetation, but also can be found in swampy environments, lagoons etc. In areas where one of its major predators is historically absent (tiger fish) *H. odoe* will venture into more open waters (Jackson, 1961).The colour of *H. odoe* varies and this is related to the wide range of distribution and developmental stages, that is, young to adults as observed by Jubb and Manning (1961). *H. odoe* is singled out of characins that venture into nest building and shows sexual dimorphism (Merron et al., 1990). Also, *H. odoe* is one of the economically important fishes of tropical African

freshwaters (Olaosebikan and Raji, 1998; Adedokun and Fawole, 2012).

The most striking feature of *H. odoe* is its dentition. Both upper and lower jaws are filled with sharp pointed teeth, but the lower jaw has two rows while the upper only has one. There are also two large canines in each jaw. Two pairs of dermal flaps can be found on the upper and lower jaws (Barnard, 1971; Adama and Abdul azeez, 2011). According to Merron et al. (1990), very little is known about the development of *H. odoe* and observed that the species undergoes direct development due to a very short larval period. Length- weight relationship is one of the parameters for growth ($W = al^b$), another growth parameter is condition factor (K). It expresses condition of fish in numerical forms, that is, the degree of well being or health or fatness of the fish) ($K = W \times 100/L^3$). The significance of these growth parameters will ensure a rational exploitation of the species in

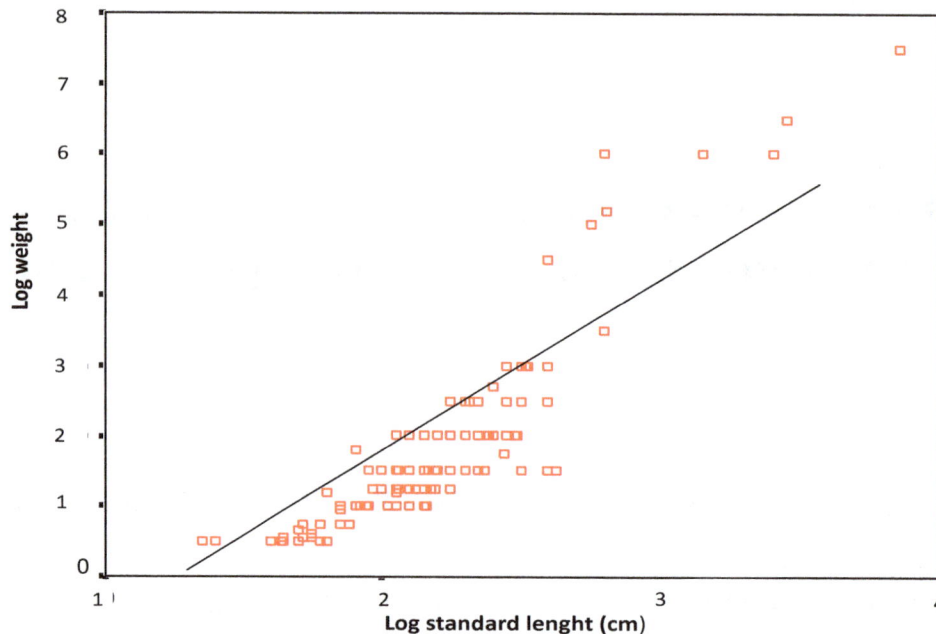

Figure 1. Log standard length - log weight relationship.

Ogbomoso reservoir. This work aims at providing information on: (i) the physiological growth (Length-weight variables) of African pike in the resident water body, and (ii) the effects of habitat, seasons, sex and size on the general welfare of African pike in the natural population of fish.

Study area

The area of study is Ogbomoso reservoir which was impounded in 1964 (Ojo, 2002). The major tributaries are Idekun, Eeguno, Kemolowo, Omoogun and Yakun streams. The reservoir has a catchment area of about 321 km^2, extending over longitude 8° 05^{1}N to 8° 10^{1}N and latitude 4° 10^1 E to 4° 15^1 E in Oyo State, Nigeria. The minimum depth is 0.83 m while the maximum depth is 16.36 m. The temperature of the reservoir is 28 and 32°C during dry and wet seasons respectively. The pH range of the reservoir was 5.8 to 7.1 and turbidity of 18 to 45 cm.

MATERIALS AND METHODS

During the period of study, 133 samples were collected and sampling for *H. odoe* commenced in April 2009 and extended till March 2010. The only fishing method employed was gill netting. The studied morphometric includes total length, standard length and weight of the fish in millimeters and grams by using measuring board and manual scale. Regression was used to analyse the log form of dependent variables:

W = a + bL (log .form)

The fish body weight expressed as a percentage of fish total length

cube was used as condition factor (K).

RESULTS

Length-weight relationship

The standard length of the specimen examined ranged from 13.5 to 39.0 cm while the weights ranged from 50 to 750 g. The mean standard length was 2.2 cm and mean weight was 183 kg. The equation describing the relationship of the two biological variables of fish is given as W = aL^b (Tesch, 1968); where W denotes weight (g), L = standard length in cm. A logarithm plot of weight against log length yields b as an exponent (with values between 2 and 4; often 3) and a (constant).

According to Tesch (1968), the value b=3 shows that the fish grows isometrically. Values obtained other than 3 indicate allometric growth (positive or negative). The equation above can be represented thus; log W = log a + blog L.

A regression line is drawn from the graph of logarithm weight against logarithm of length, where b is the regression co- efficient (slope of graph) and log a is the intercept of the regression line on the y- axis. The graph shows allometric growth for *H. odoe* in Ogbomoso reservoir and the value of b calculated from the graph is 4.11 with regression factor 0.88 (Figure 1).

Condition factor (K)

The condition factor (k) expresses the general well- being of fish in its habitat. The condition factor obtained in this

Table 1. Monthly mean condition factor for *H. odoe* in Ogbomoso reservoir.

Month	Number of fish	Mean K
Jan 2010	4	1.0250 ± 0.096
Feb 2010	5	1.6000 ± 0.837
Mar 2010	10	1.3000 ± 0.411
Apr 2009	11	1.3636 ± 0.425
May 2009	7	1.2571 ± 0.257
Jun 2009	9	1.2556 ± 0.188
Jul 2009	16	1.5125 ± 0.430
Aug 2009	12	1.4417 ± 0.309
Sep 2009	26	1.4846 ± 0.303
Oct 2009	14	1.6500 ± 0.293
Nov 2009	9	1.6000 ± 0.332
Dec 2009	10	1.6400 ± 0.263

Table 2. Condition factor in relation to gonad maturity in *H. odoe*.

Sex(gonad stages)	Number of fish	Mean K
F II	10	1.500 ± 0.123
F III	22	1.615 ± 0.298
F IV	28	1.789 ± 0.443
M II	3	1.480 ± 0.084
M III	11	1.437 ± 0.632
M IV	19	1.880 ± 0.312

study ranged from 0.8 to 2.7 with mean condition factor of 1.5. While male and female had condition factors 1.5 and 1.6 respectively.

Monthly condition factor

Table 1 shows monthly condition factor of *H. odoe* in Ogbomoso reservoir with mean condition factor greater than 1.

Condition factor in relation to the stage of gonad maturity

Table 2 shows average condition factor based on gonad development stages. As the gonad stages increased, there is development in size and weight of the gonads and the more significantly difference along the gonads development stages ($p > 0.05$).

Condition factor in relation to habitat and season

Table 3 shows the impacts of habitats (inshore and offshore) and seasons (dry and wet) on the general welfare of *H. odoe* in Ogbomoso reservoir. The condition factor values of sampled fish from different micro habitats

were 1.6, 1.7 and 1.5 for grassy, middle and open water habitats. Also for dry and wet seasons *H. odoe* had 1.5 and 1.5 condition factor respectively. Therefore, there is no significant difference in both microhabitats and between seasons ($p < 0.001$).

Condition factor in relation to sex and size

Table 4 shows that size has no negative effects but sex do on the physical health of the fish in the resident water body. The sizes of *H. odoe* caught during this period of study ranged from 13.5 to 18.0 cm S. L. as smaller fishes and 18.1 to 39.0 cm S. L. as larger fishes. As a result, fish sizes showed no significant difference ($p < 0.001$). While there is significant difference between male and female sexes ($p > 0.05$)

DISCUSSION

The length-weight relationship shows that *H. odoe* enjoyed positive allometric growth in Ogbomoso reservoir. This implies that, the fish was heavier than its length in the resident water body. This result is not in line with the findings of Fawole and Arawomo (1999) who reported negative allometric growth in *Tilapia galileaus* in Opa reservoir. Though, they are two different species with different foraging mode. The well being of *H. odoe* in the reservoir might be attributed to the suitable environment as indicated by the high condition factor (Table 1). The increase in weight of fish may be associated to the quality of the food which is the proximate factor determining growth rate which agrees with the works of Uwem et al. (2011).

The results indicate that, the condition factor of the fish is not affected by habitat and season variations (Table 3). This may not be unconnected with the availability of food and close proximity between prey and predator during falling water period. The significant difference between male and female and gonads maturation may be due to the differences in size and weight gain of male and female gonads; which is a reflection of early maturation of gonads in female than in its male counterpart. Meanwhile, the condition factor does not vary with the size of the fish (Table 4).

However, the productivity of the reservoir and high values of condition factors are reflection of the existing balance amongst the environmental elements, which is evidenced in the influx of nutrients from the adjoining upland and the symbiotic interplay of abiotic factors within the aquatic ecosystem.

Conclusion

The physical development and general welfare of *H.*

Table 3. Condition factor in relation to habitat and season.

Variables	Habitat		Season	
	Inshore	Offshore	Wet	Dry
Number of fish	74	59	81	52
Mean K	1.592±0.405	1.590±0.394	1.508±0.409	1.519±0.418

Table 4. Condition factor in relation to fish sex and size.

Variables	Sex		Size	
	Male	Female	Small (13.5-18.0 cm)	Large (18.1-39.0 cm)
Number of fish	46	63	24	109
Mean K	1.584±0.393	1.602±0.404	1.575±0.225	1.578±0.410

odoe in the resident water body are determined through studied growth parameters. The general well being of the sampled fish is not affected by habitats affinity, season variations and size range while gonad maturity and sexes indeed do.

RECOMMENDATION

Though, *H. odoe* specimens do not occur presently in large number in the commercial gillnets of 10.2 cm (4") mesh size; the ongoing commercial exploitation of this species probably with a lesser mesh size of 8.9 cm (3 ½") might enhance faster growth, improve socio-economic status of the local fishermen and help to sustain the much needed protein in man.

REFERENCES

Adedokun MA, Fawole OO (2012). Distributions and Habitations of African pike *Hesetus odoe* (Bloch, 1794), in Oba Reservoir, Ogbomoso, Nigeria (Actinopterygii: Hepsetidae). Munis Entomol. Zool. 7(2):708-713.

Adama SO, Abdul azeez A (2011). Composition of Fish Species and Fisheries activities of Tunga Kawo Dam, (Wushishi) Niger State. 26th Annual Conference and Fair of the Fisheries Society of Nigeria (FISON) at Federal University of Technology Minna, Niger State.

Barnard K (1971). A pictorial guide to South African fishes. Cape Town: Maskew miller Limited.

Fawole OO, Arawomo GA (1999). Fecundity of *Sarotherodon galilaeus* in the Opa reservoir Ile Ife, Nig. J. Sci. Res. 4(1):107-111.

Jackson PBN (1961). The impacts of predation especially by the tiger fish *(Hydrocynus sp)* on African freshwater fishes. Proc. Zool. Soc. London 136:603-622.

Jubb RA, Manning, S (1961). Freshwater fishes of the Zambezi River, lake Kariba, Pungive, Sabi Lundi and Limpopo. Cape Town Gothnic Printing Company 7:1-22.

Merron GS, Holden KK, Bruton MN (1990). The production biology and early development of the African pike, *Hepsetus odoe*, on the Okavango Delta, Botswana. Environ. Biol. Fish. 28:215-235.

Moriarty C (1983). The African pike *Hepsetus odoe*. Nig. Field. 47:212-222.

Ojo OO (2002). Artificial Lake fisheries Management in Oyo State; Pre-season Training of Officers handling the IFAD- Assisted Artisanal Fisheries. pp. 1-8.

Olaosebikan BD, Raji A (1998). The field guide to Nigerian freshwater fishes. FCFFT, P.M.B.1500, New- Bussa Niger State, Nigeria. Decency Enterprises and Stationeries Ltd.,

Tesch FW (1968). Methods for Assessment of fish production in freshwaters, (Ed. Ricker, E. W.) I. B. P. Handbook No. 3: IST Edition;. *Blackwell, Scientific Publications Oxford and Edinburgh.* pp. 93-123.

Uwem GU, Ekanem AP, George E (2011). Food and Feeding habits of *Ophiocephalus obscura* (African snake head) in the Cross River Estuary, Cross River State, Nigeria.

9

Study on the impacts of inter-basin water transfer: Northern Karun

Hossein Samadi-Boroujeni[1] and Mehri Saeedinia[2]

[1]Department of Water Engineering, Faculty of Agriculture, Shahrekord University, Shahrekord, Iran.
[2]Department of Irrigation and Drainage, Faculty of Water Sciences Engineering, Shahid Chamran University, Ahwaz, Iran.

This study aimed to investigate the effects of inter-basin water transfer projects from Northern Karun basin to Zayanderud basin in Iran, with emphasis to Beheshtabad water transfer project. For this purpose the situation of water resources of Northern Karun basin was modeled by using WEAP model. This model was calibrated by 10 years data (1995 to 2004) and then the model was run for a 30 years period. Results showed that Northern Karun basin water resources for transferring to Zayanderud basin should be limited up to 314 million m^3 per year (MCM/year), by assuming complication of the origin basin development and considering river environment water requirement in the future. This is 46% less than the value that has been allocated by allocation committee. Results obtained from the WEAP model also indicated that the useful storage required for regulating of water resources was obtained 600 MCM as maximum value. In order to minimize social and environmental impacts of Beheshtabad tunnel, a new alternative has been introduced.

Key words: Inter basin water transferring, Beheshtabad and Kuhrang basins, WEAP model.

INTRODUCTION

Water managers and policy makers are in need to have tools at their disposal that will support them in their decision-making. This is very important for inter-basin water transfer projects. There exist today a variety of generic simulation models incorporated within interactive graphics-based interfaces that are available for studying water related planning and management issues in river basins. Sechi and Sulis (2010) Compared application of five generic models for simulating water resource systems: AQUATOOL-SimWin (Andreu et al., 1996), MODSIM (Labadie et al., 2000), RIBASIM (Delft Hydraulics, 2006), WARGI-SIM (Sechi and Sulis, 2009) and WEAP (SEI, 2005) to a multireservoir and multiuse water system in Southern Italy. While each model has its own characteristics, the proposed application comparison does not identify all the features of each model, but rather gives general information on the identification and evaluation of operating policies with the aid of these simulation models. Because the water evaluation and planning system (WEAP), is an efficient and user friendly model and it requires no additional software or cost, in this study is used.

WEAP is an exemplary application linking supply and demand site requirements. Allowing scenario analysis, changes in supply and demand structures can be simulated in order to discover potential shortages and the effects of different management strategies (Yates et al., 2005). Evaluating scenarios requires validated model results. Therefore, a challenge of many studies in which WEAP was applied is the model validation at different spatial and temporal scales (Al-Omari et al., 2009; Yates et al., 2009). In recent years many researchers have applied WEAP model for water resources planning and management. Ospina-Noreña et al. (2011) used WEAP model for simulating of water resources of the Sinú-Caribe river basin in Colombia to create several baseline and adaptation strategy scenarios for water supply, use and demand, and to make projections for the future including the potential impacts of climate change. The results show that the supply requirement would increase and thus unmet demand would increase more quickly under climate change conditions.

In upper Guadiana basin in Spain's inland region of Castilla La Mancha, the research focuses on the analysis of water and agricultural policies aimed at conserving groundwater resources and maintaining rural livelihoods in a basin in Spain's central arid region by using WEAP model (Varela-Ortega et al., 2011). Results showed that the region's current quota-based water policies may contribute to reduce water consumption in the farms but will not be able to recover the aquifer and will inflict income losses to the rural communities. This situation would worsen in case of drought. In South Africa, water resources in the Olifants river basin, catchment management committees (CMCs), must therefore be able to get a rapid and simple understanding of the water balances at different levels in the basin and for this purpose WEAP model was used (Levite et al., 2003). Höllermann et al. (2010) by modeling the water balance of the Ouémé–Bonou catchment with WEAP, showed that the pressure on Benin's water resources will increase, leading to greater competition for surface water. The WEAP results offer a solid basis to assist planners in developing recommendations for future water resource management by revealing hot spots of action. Harma (2011) used WEAP model to consider future scenarios for water supply and demand in both unregulated and reservoir supported streams that supply the district of Peachland in British Columbia's Okanagan basin. Results demonstrate that anticipated future climate conditions will critically reduce stream flow relative to projected uses (societal demand and ecological flow requirements). The surficial storage systems currently in place were found unable to meet municipal and in stream flow needs during "normal" precipitation years by the 2050s. Yilmaz (2010) used WEAP model as a simulation and evaluation tool to assess the performance of possible management alternatives in Gediz River basin, which is measured by nine proposed indicators. The results of the study indicated that the Gediz River basin is quite sensitive to drought conditions, and the agricultural sector is significantly affected by irrigation deficits. Raskin et al. (1992) studied on the aral sea basin water resources management. The Aral Sea, a huge saline lake located in the arid south-central region of the former U.S.S.R., is vanishing because the inflows from its two feed rivers, the Amudar'ya and Syrdar'ya, have diminished. The loss of river flow is the result of massive increases   in   river

withdrawals in the basins. WEAP model has been applied for simulating current water balances and evaluating water management strategies in the aral sea region. The analysis provides a picture of an unfolding and deepening crisis (Raskin et al., 1992).

Purkey (2007) used WEAP model as a simulation tool to presents an overview of decision making processes ranked based on the application of a 3S: sensitivity, significance, and stakeholder support, standard, which demonstrates that while climate change is a crucial factor in virtually all water related decision making in California, it has not typically been considered, at least in any analytical sense. The authors will engage with stakeholders in these three processes, in hope of moving climate change research from the academic to the policy making arena (Purkey, 2007).

There are more studies for evaluating the effects of inter-basin water transfer projects of Northern Karun. In the past years, in order to transfer of Northern Karun basin water resources into Zayanderud basin by a total capacity of 550 MCM/year, two tunnels which named Kuhang-1 tunnel and Kuhang-2 tunnel, was constructed. These tunnels have been under operation since 1953 and 1985, respectively. In order to increase transferring the Northern Karun basin water resources, constructing of third Kuhrang tunnel with a length of 24 km had been started in 1994. Against to the first and second Kugrang tunnels, this tunnel caused many socio-economic and environmental impacts and problems because of more length and geological problems. It caused some springs such as Morvarid and Zarrin spring to be dried. Also, time and cost of the third tunnel project will be significantly increased. For example, third Kuhrang tunnel project has been under constructed since 1994 and the project cost has been increased 8 times of initial estimation. Unfortunately, despite some important springs near the tunnel have been dried due to digging of the tunnel, the tunnel has not finished yet.

In recent years forth Kuhrang tunnel which in this paper we call Beheshtabad project, has been defined to transfer of Northern Karun basin water resources to Zayanderud basin as an inter-basin water transfer project. For this purpose WEAP software has been used. Based on the obtained results from this study, some applicable recommendations are suggested in this paper.

MATERIALS AND METHODS

Study area

The study area is Kuhrang and Behesthabad basins with latitude of 3°, 50' to 32°, 35' and longitude of 50° to 51°, 25'. These basins wholly located in the area of Chaharmahal and Bakhtiyari province and they are as two sub-basin of Northern Karun basin. Beheshtabad basin with 3860 km² in area is the largest sub-basin in the Northern Karun basin comprising from 8 hydrologic units as shown in Figure 1 The widest plains Chaharmahal and Bakhtiyari province are located in Beheshtabad basin. The Kuhrang basin with 1230 km² in area is also another sub-basin of Northern Karun

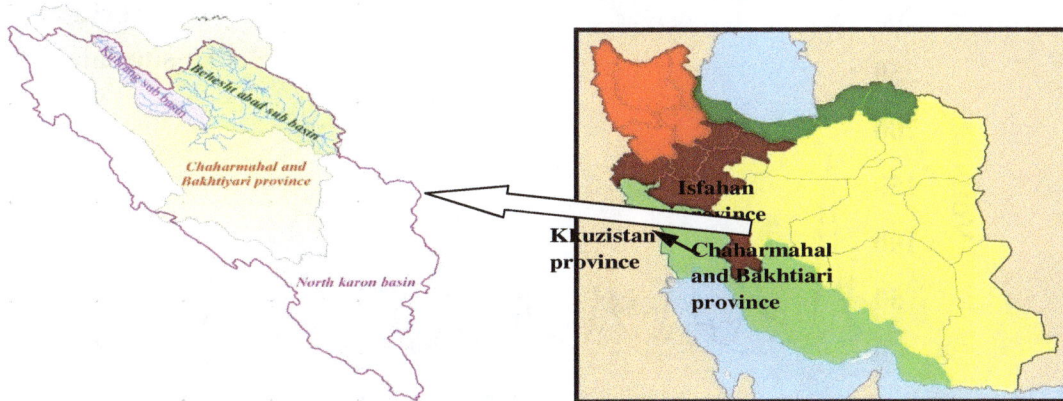

Figure 1. Study area (sub-basins of Behesht Abad and Kuhrang).

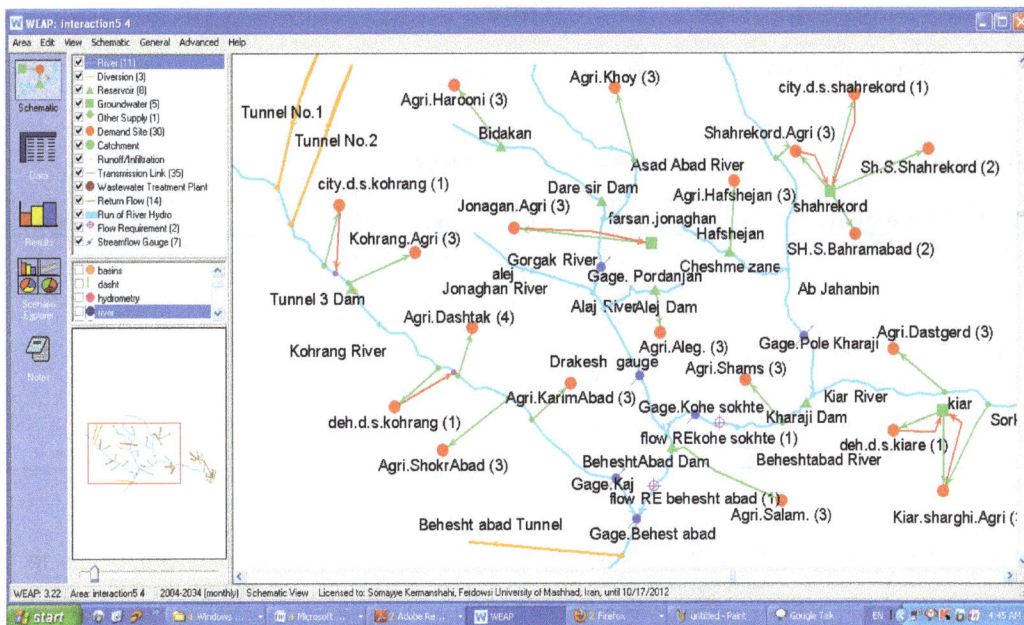

Figure 2. Configuration schematic of water resources system of Behesth Abad and Kuhrang basins.

basin. Due to high snowing, this sub-basin plays an important role in water resources of Karun and also Zayanderud basin because in the past years, there have been dug tunnels 1, 2 and 3 of Kuhrang for transferring the Kuhrang river flow to Zayandehrood basin (located in Isfahan province) with a capacity of 300, 250 and 270 MCM/year, respectively. Beheshtabad and Kuhrang rivers reach together near Ardal city in Chaharmahal and Bakhtiyari province and at this point the Karun River is formed.

In recent years the study of Beheshtabad water transfer project has been conducted for transferring an average of 600 MCM/year. This project is comprised a large dam at the point of connecting Beheshtabad and Kuhrang Rivers and a tunnel with 65 km in length which it will begin from the Beheshtabad dam site and end at Zayanderud River.

Configuration of water resources system

In WEAP model, there is used a nod-link network structure for

modeling of the basin water resources. In this study, for modeling the Beheshtabad and Kuhrang basins water resources, different sites for municipal, industrial and agricultural water demand, dams reservoirs, aquifers, environmental requirement were defined as nods and different paths such as water transfer network from water supply resources to the demand sites and water return water network were used as links. Figure 2 shows the components of model in the studied basins.

Input data and model application

WEAP applications generally include several steps: (i) create a geographic representation of the area, (ii) enter the data for the different supply and demand sites, (iii) compare results with observations and if required update data and calibrate model, (iv) define scenarios, and (v) compare and present the results of different scenarios.

In this study, there was chosen 2004 as reference year and there

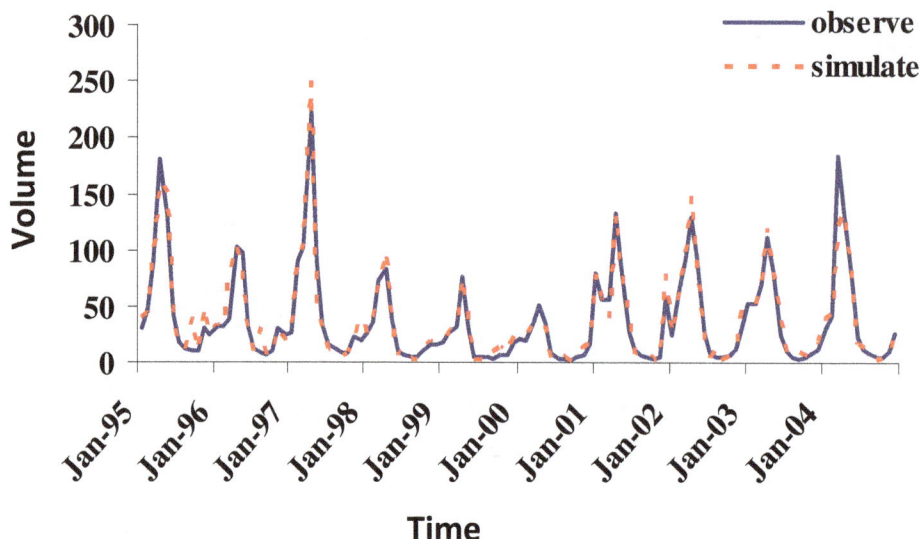

Figure 3. Observed and calculated monthly discharge.

Table 1. Error average of estimating the monthly discharge in hydrometery stations of study area.

Item	Name of hydrologic unit	River name	Hydrometery station	Error average (%)
1	Broujen-Fardanbeh	Kiar	Tang Dahano	1.25
2	Eastern Kiar	Est Kiar	Kharaji Bridge	5.52
3	Shalamzar	Shalamzar	Burnt Mountain	6.35
4	Soorashjan	Gorgak	Tang Pardanjan	7.43
5	Jonaghan	Jonghan	Darkesh Varkesh	3.08
6	Beheshtabad	Behesht Abad	Behesth Abad	2.59
7	Kuhrang	Kuhrang	Kaj	1.04

was selected 30 years statistical period (2004 to 2033) for simulations. Data for surface water resources, aquifers, dam reservoirs, water demand amount and users' priority were collected and entered in the model. In the simulation period it was assumed that inflow of rivers to be similar what happen in the past years, with due regards deterministic approach. It should be noted that the WEAP model uses a standard linear programming for resolving the water allocation problems.

It must be also mentioned that before forming the input file of the model, the existing hydro-climatology stations were homogenously tested and completed by regression method. The environmental requirement was considered with first priority in the water resources planning of river in comparing of municipal, industrial and agricultural sectors. The environmental requirement value was monthly determined as a 20% of average of river inflow (Tharme, 2003).

RESULTS AND DISCUSSION

Model calibration

For calibrating the model a 10 years period were used from 1995 to 2004. After forming the input file of model, it is necessary to calibrate the model based on the information of discharge of hydrometric stations and piezometric data of different plains in the studied area. The basis for calibration is that the percentage error of surface and groundwater water resources obtained from the model during 10 years of calibration must be less than an acceptable rate (less than 10%) in comparing of observed values.

For example, Figure 3 indicates observed and calculated monthly discharge of studied rivers. The error of estimating the monthly discharge has been calculated in hydrometery stations of study area by Equation 1 and related result indicated in Table 1.

$$E = \left| \frac{Q_{cal} - Q_{obs}}{Q_{obs}} \right| \times 100 \tag{1}$$

Where; Q_{obs} is the observed discharge, Q_{cal} is calculated discharge.

For the groundwater resources the volume of aquifer storage was also considered for calibrating of the model. The aquifer specific yield, value of groundwater recharge from precipitation, and hydraulic conductivity coefficient were considered as calibrating parameters. By changing these parameters, the error of calculating monthly groundwater storage of all aquifers within Beheshtabad

Table 2. Parameters of aquifer under study when calibration and average error of the model when estimating the volume of aquifer.

Item	Plain name	Initial volume of aquifer (M m³)	Plain infiltration coefficient	Plain special yield coefficient (%)	Mean monthly volume of aquifer (mm³)		Error average (%)
					Observation	Calculation	
1	Broojen	201	0.13	3	2009	204.3	4.94
2	Sefid Dasht	926	0.18	3	64.4	61.5	6.03
3	Farsan	88.1	0.27	2.2	85.8	86.6	2.66
4	Kiar	191.4	0.17	3	1788	181.3	4.5
5	Shahrekord	1332.1	0.17	4.7	1313.7	1297.8	5.2
6	Shalamzar	82	0.8	2.2	84.3	86.7	5.66

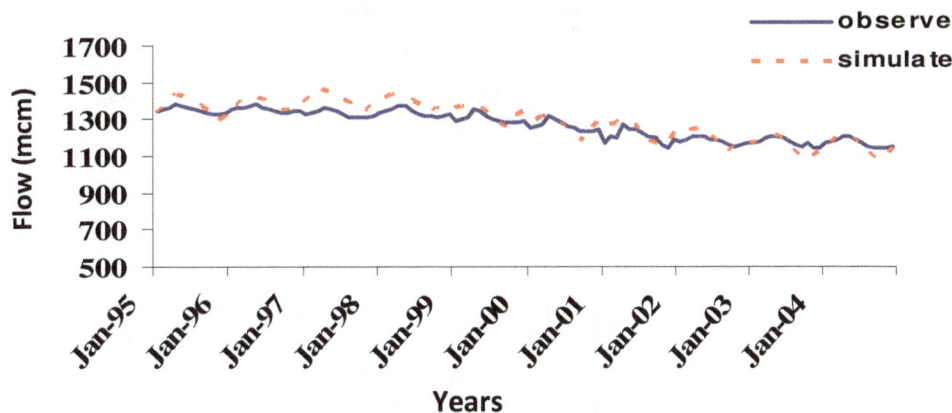

Figure 4. Observed and simulated reservoir volume in Shahr Kord plain.

basin in comparing with observed value was obtained less than an acceptable rate (less than 10%). Table 2 indicates the average model error for different plains and for example, Figure 4 shows monthly volume of the aquifer storage in Shahrkord plain obtained from the WEAP model and observed values.

Water resources evaluation

According to the calibration conducted, WEAP model was used for ultimate development scenario of the Beheshtabad and Kuhrang basins in a 30 years period (2004 to 2033) and its results were evaluated. In the ultimate development scenario, it was assumed that in the Beheshtabad basin, there were operated reservoir dams as well as artificial recharge projects as ultimate development scenario of the basin. Results obtained from the model, as illustrated in Figure 5, showed that the amount of transferable water in Beheshtabad project is averagely obtained about 470 MCM/year and it is limited to 314 MCM/year with a reliability of 90%. This is 46% less than the value that has been allocated by allocation committee. Also results showed that maximum 600 MCM storage facilities are needed for regulating of water resourced (Figure 6). It should be noted that in the

conducted project study, Beheshtabad dam reservoir volume is desired about 1800 MCM, which it is overestimate.

Another important matter in Beheshtabad project is socio-economic and environment impacts and problems due to digging of 65 km tunnel because of difficult geological condition and high over load. In order to reduce tunnel impacts, many replacement alternatives can be defined. One of these options is transferring of water resources by pipeline method. In this option a pump station with a capacity of 11 m³/s in flow rate and 26 Mw in power rate will be needed. At the end of pipeline, a hydroelectric plant with a capacity of 21 MW can be installed because of low elevation in Zayanderud river. Net consumed energy in this option was calculated about 117000 MWh/year. Full supply cost for this option was calculated equal to 15 US cent /m³ for a discount rate of 7%. This option has no significantly socio-economic and environment impacts. Also time of executive of the project will be shorted.

Conclusion

The results of this study was conducted by general modeling of water resources of Beheshtabad and

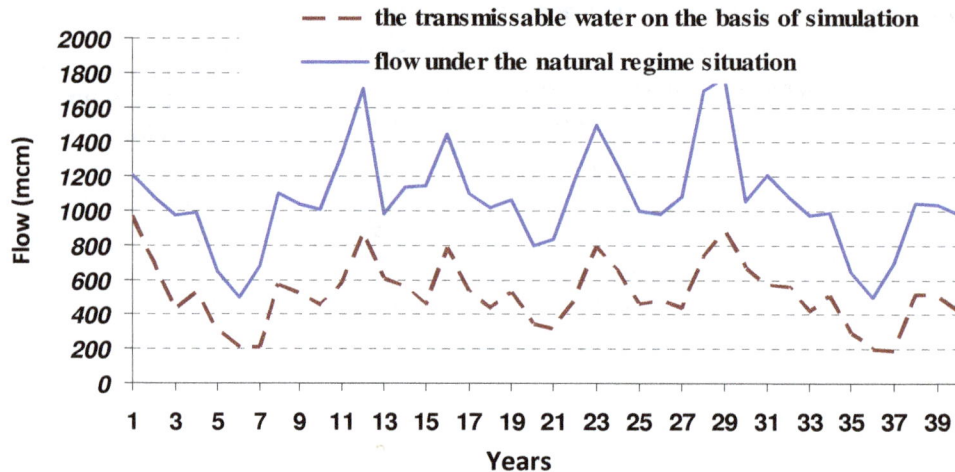

Figure 5. Changes of annual outflow of both rivers, Kuhrang and Behesht Abad in both conditions with and without ultimate development in the basin.

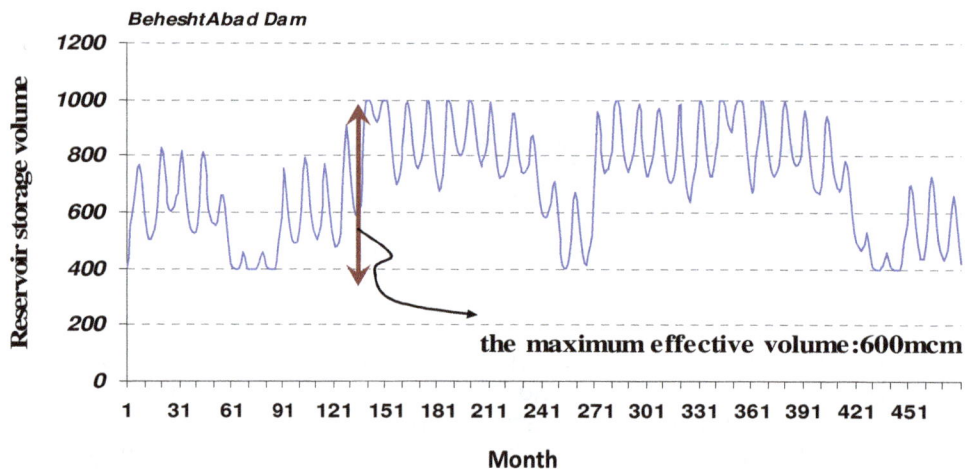

Figure 6. Changes in the water storage volume in the reservoir of Behesht Abad Dam in different months of simulation.

Koorang basins using WEAP model are as below:

1. The capacity of water transfer with a reliability of 90%. is obtained bout 314 MCM/year based on the results of model and considering the minimum Environmental requirement of downstream river and considering effect of all of the other water development projects.
2. Water transferring by tunnel will create many socio-economic and environment impacts. For example, it will drain the groundwater resources and dry the springs of region. There is considered the option of water transfer by pipeline.
3. Model results indicate that effective volume needed for storing and regulating the water is maximally equaled about 600 MCM and this figure is less than a third figure as considered by project consultant in the studies.
4. It is recommended that due to the more sensitivity of

inter-basin water transfer projects, in the trend of studies, the global experiences and standards may be applied for this project.

REFERENCES

Al-Omari A, Al-Quraan S, Al-Salihi A, Abdulla F (2009). A water management support system for Amman Zarqa Basin in Jordan. Water. Resour. Manage. 23:3165-3189.

Andreu J, Capilla J, Sanchìs E (1996). AQUATOOL, a generalised decision-support system for water-resources planning and operational management. J. Hydrol. 177(3-4):269-291.

Delft H, ydraulics (2006). River Basin Planning and Management Simulation Program. Proceedings of the iEMSs Third Biennial Meeting: "Summit on Environmental Modelling and Software", Voinov, A., A.J. Jakeman and A.E. Rizzoli (eds), International Environmental Modelling and Software Society, Burlington, Vermont, USA.

Harma KJ, Johnson MS, Cohen SJ (2011). Future Water Supply and

Demand in the Okanagan Basin, British Columbia: A Scenario-Based Analysis of Multiple, Interacting Stressors. Water. Resour. Manage. DOI 10.1007/s11269-011-9938-3.

Höllermann B, Giertz S, Diekkrüger B (2010). Benin 2025Balancing FutureWater Availability and Demand Using the WEAP 'Water Evaluation and Planning' System. Water. Resour. Manage. 24:3591-3613.

Levite H, Sally H, Cour J (2003). Testing water demand management scenarios in a water-tressed basin in South Africa: Application of the WEAP model. Phys. Chem. Earth 28:779-786.

Ospina-Noreña JE, Gay-García C, Conde AC (2011). Water availability as a limiting factor and optimization of hydropower generation as an adaptation strategy to climate change in the Sinú-Caribe River basin. Atmósfera 24(2):203-220.

Purkey DR, Huber-Lee A, Yates DN, Hanemann M, Herrod-Julius S (2007). Integrating a climate change assessment tool into stakeholder-driven water management decision-making processes in California. Water. Resour. Manage. 21:315-329.

Raskin P, Hansen E , Zhu J, Iwra M (1992). Simulation of water supply and demand in the Aral sea region. Water Int. 17:55-67.

Sechi GM, Sulis A (2009). Water system management through a mixed optimization simulation approach. J. Water. Resour. Plann. Manage. 135(3):160-170.

Sechi GM, Sulis A (2010). Intercomparison of Generic Simulation Models for Water Resource Systems. International Congress on Environmental Modelling and Software Modelling for Environment's Sake, Fifth Biennial Meeting, Ottawa, Canada.

SEI (Stockholm Environmental Institute) (2005). WEAP water evaluation and planning system, Tutorial, Stockholm Environmental Institute, Boston Center, Tellus Institute.

Tharme RE (2003). A Global Perspective on Environmental Flow Assessment: Emerging Trends in the Development And Application of Environmental Flow Methodologies For Rivers. River Res. Appl. pp. 397-441.

Varela-Ortega C, Blanco-Gutierrez I, Swartz CH, Downing TE (2011). Balancing groundwater conservation and rural livelihoods under water andclimate uncertainties: An integrated hydro-economic modeling framework. Glob. Environ. Change 21:604-619.

Yates D, Sieber J, Purkey D, Huber-Lee A (2005). WEAP 21-a demand-, priority-, and preference-driven water planning model. Part 1: model characteristics. Water. Int. 30:487–500

Yates D, Purkey D, Sieber J, Huber-Lee A, Galbraith H, West J, Herrod-Julius S, Young C, Joyce B, Rayej M (2009). Climate driven water resources model of the Sacramento Basin, California. J. Water Resour. Plan. Manage. 135(5):303–313.

Yilmaz B, Harmancioglu NB (2010). An Indicator Based Assessment for Water Resources Management in Gediz River Basin, Turkey. Water Resour. Manage. 24:4359–4379.

10

Physiological and yield parameters of noni (*Morinda citrifolia*) as influenced by organic manures and drip irrigation

S. Muthu Kumar and V. Ponnuswami

Horticultural College and Research Institute, Tamil Nadu Agricultural University, Periyakulam, Theni District-625604, Tamil Nadu, India.

An experiment was conducted in *Morinda citrifolia* to find out the effect of various organic manures and drip irrigation on physiological and yield parameters. The experiment was carried out in split plot design with irrigation regimes on main plot and organic manures on sub plot. Among the treatment combinations, M_2S_4 (100% crop water requirement through drip irrigation + 50% farmyard manure + 50% vermicompost) recorded the superior score for leaf area index (0.2597, 0.7755 and 2.0937), light interception (8.65, 11.97 and 17.68%), specific leaf weight (10.89, 12.98 and 12.23 mg cm^{-2}), chlorophyll 'a' (1.483, 1.732 and 1.946 mg g^{-1}), chlorophyll 'b' (0.705, 0.772 and 0.840 mg g^{-1}), total chlorophyll (2.223, 2.554 and 2.818 mg g^{-1}), soluble protein (13.68, 15.84 and 14.42 mg g^{-1}), total phenol (4.55, 6.79 and 5.63 mg g^{-1}) and nitrate reductase activity (38.92, 49.06 and 45.84 µg NO$_2$ g^{-1} h^{-1}) during vegetative, flowering and harvesting stages respectively. The same treatment combination exhibited the highest yield parameters viz, fruit weight (167.86 g), number of fruits per plant (198.12), yield per plant (33.26 kg).

Key words: *Morinda citrifolia*, farmyard manure, vermicompost, coir pith compost, drip irrigation, physiology, yield.

INTRODUCTION

The availability of irrigation water becomes dwindling day-by-day as such adoption of conventional methods of irrigation to crops leads to an acute scarcity of water and results in reduced production and productivity of crops. Therefore, it becomes imperative to go for alternate water saving methods and income for every drop of water through drip irrigation which provides continuous supply of required water in drops right at the root zone of the crop plant.

By adopting drip irrigation, it is possible to increase the yield potential of crops by three times with the same quantity of water, by saving about 45 to 50% of irrigation water and increasing the productivity by about 40% (Behera et al., 2012).

In the cultivation of modern crop varieties and appropriate management strategies, use of chemical fertilizers have contributed up to 50% raise in food grain output (Braun and Roy, 1983). Despite the key role played by fertilizers, a total dependence on them in achieving a contemplated productivity goal is not fully justified. Furthermore an unabated up rise in the use of chemical fertilizers can inflict irreparable damage to land and environment (Katyal, 1989). Skepticism of nature is being widely exposed in the regions where fertilizer use is already massive. At the same time, any saving in the consumption of chemical fertilizers without affecting the productivity is a social necessity. Measurable decrease in fertilizer consumption without compromising the yield and quality can also be made practically possible through organic inputs.

Large scale cultivation under organic conditions is gaining momentum to produce toxic free medicinal plant products (Padmanabhan, 2003). Organically grown herbal materials are more preferred in the herbal preparations since they are more effective and safe. Organic farming provides better and balanced environment, better food and living conditions to the human beings (Baby, 2012).

One upcoming botanical name, the fruit of *Morinda citrifolia* very popularly known as noni belongs to the family Rubiaceae. Indigenous to South East Asia, noni was domesticated and cultivated by Polynesians. The tree grows rigorously from India to Malaysia, Fiji and Polynesia (Mathivanan et al., 2005).

Noni is the biggest pharmaceutical unit in the universe because it has more than 160 nutraceuticals, vitamins, minerals, micro and macro nutrients that help the body in various ways from cellular level to organ level (Rethinam and Sivaraman, 2007). The fruit juice is in high demand in alternative medicine for different kinds of illnesses such as arthritis, diabetes, high blood pressure, muscle aches, menstrual difficulties, headaches, heart disease, AIDS, cancers, gastric ulcers, sprains, mental depression, senility, poor digestion, atherosclerosis, blood vessel problems and drug addiction (Wang et al., 2002).

The purpose of this medicinal herb will be fulfilled only if it is free from residual effects due to chemical farming. Hence, the study was undertaken to find out best organic nutrition schedule along with irrigation regimes for (Table 2) obtaining highest (Table 3) yield through improved physiological parameters.

MATERIALS AND METHODS

This study was conducted at Horticultural College and Research Institute, TNAU, Periyakulam, Tamil Nadu, India which is situated at 77°E longitude, 10°N latitude and at an altitude of 300 m above mean sea level.

Methodology

The plants were obtained from World Noni Research Foundation, Chennai, India. The experimental design is as shown in Table 1.

Treatment details

All organic manures were applied on equivalent weight of RDN (60 g/plant/year - on N equivalent basis). The treatments S1 to S6 are applied with Azospirillum (10 g/ plant) + phosphobacteria (10 g/ plant) + VAM (20 g/ plant).

Crop water requirement (WRc)

Crop water requirement was calculated by using the following formula before every irrigation.

WRc = $P_e \times K_p \times K_c \times A \times WP$ liter/plant/day

Where, P_e = Pan evaporation in mm; K_p = Pan Co-efficient (0.75); K_c = Crop factor (0.90 for vegetative stage, 0.95 for flowering and harvesting stage); A = Area occupied by the tree (3.6 m × 3.6 m), and WP = wetted percentage (40).

Observations

Physiological parameters

Leaf area index (LAI): The leaf area index was worked out by the method suggested by Williams (1946).

$$LAI = \frac{\text{Total leaf area per plant}}{\text{Ground area occupied}}$$

Light interception (LI): The percent light interception in the canopy was calculated by comparison with a Lux meter placed in the open sunny situation every day. The measurement was made between 12.00 and 14.00 hours as per the method suggested by Nelliat et al. (1974) and expressed in %.

$$LI (\%) = \frac{\text{Mean of light intensity at middle of the canopy and ground level}}{\text{Light intensity in the open}} \times 100$$

Specific leaf weight (SLW): The specific leaf weight was calculated using the formula postulated by Pearce et al. (1968) and expressed as mg cm^{-2}.

$$SLW = \frac{\text{Leaf dry weight}}{\text{Leaf area}}$$

Chlorophyll content: Fresh leaves from each treatment were collected and the chlorophyll 'a' and 'b' and total chlorophyll contents were determined by the method suggested by Yoshida et al. (1971) and expressed as mg g^{-1} of fresh weight.

Soluble protein: Soluble protein was extracted with phosphate buffer and estimated as per the method described by Lowry et al. (1951) and the same was expressed in mg g^{-1} fresh weight.

Total phenol content: The total phenol content of the fresh leaves was estimated by the method suggested by Bray and Thorpe (1954) using Folin Ciocalteu reagent and expressed as mg g^{-1} material.

Nitrate reductase activity: The nitrate reductase activity of leaf samples was estimated using the method described by Sinha and Nicholas (1981). The values were expressed as μg NO$_2$ g^{-1} h^{-1} on wet weight basis.

Yield parameters

Fruit weight: The whole weight of ten fruits was taken and their mean weight was expressed in grams (g).

Number of fruits per plant: The number of fruits harvested from each plant over several harvests were counted and expressed in number.

Fruit yield per plant: The yield was recorded after weighing fully matured fruits at each harvest, summed and expressed in kilograms per plant.

Table 1. Design of experiment.

Statistical design	Split plot design
Factors	2
Replications	2
Spacing	3.6 m × 3.6 m
Number of plants per replication	5

Table 2. Main plot (Irrigation).

M_1	75% WRc (Crop water requirement through drip irrigation)
M_2	100% WRc (Crop water requirement through drip irrigation)
M_3	125% WRc (Crop water requirement through drip irrigation)
M_4	Check basin method of irrigation (once in 5 days)

Table 3. Sub plot (Organic manures).

S_1	100% Farmyard manure (FYM)
S_2	100% vermicompost (VC)
S_3	100% coir pith compost (CPC)
S_4	50% FYM + 50% VC
S_5	50% FYM + 50% CPC
S_6	50% VC + 50% CPC
S_7	100% recommended dose of nitrogen (RDN) through inorganic fertilizers
S_8	Control (no manures and no fertilizers)

Statistical analysis

The statistical analysis of data was done by adopting the standard procedures of Panse and Sukhatme (1985). The AGRES software (version 3.01) was used for analysis of data.

RESULTS

Physiological parameters

Leaf area index (LAI)

There is a linear increase in LAI from vegetative to harvesting stage (Table 4). Among the main plot treatments application of 100% WRc through drip irrigation (M_2) was found to have significant influence on the LAI at various stages of crop growth. An increased LAI (0.1826, 0.5376 and 1.3427) in vegetative, flowering and harvesting stages respectively was observed in the treatment M_2 (100% WRc through drip irrigation) and this was followed by M_3 (125% WRc through drip irrigation) with LAI of 0.1778, 0.5273 and 1.3107. The LAI was found to be the lowest in the M_4 (check basin method of irrigation) during vegetative (0.0661), flowering (0.1780) and harvesting (0.5171) stages.

Between the sub plots, S_4 (50% FYM + 50% VC) recorded the superior performance for LAI (0.1777, 0.5260 and 1.3868) in vegetative, flowering and harvesting stages respectively and this was followed by S_2 (100% VC) with LAI of 0.1625, 0.4807 and 1.2258. The treatment S_8 (no manures and no fertilizers) registered the lowest score for LAI (0.0457, 0.1133 and 0.3211) in different stages of crop growth.

The experimental plots receiving 100% WRc through drip irrigation + 50% FYM + 50% VC (M_2S_4) produced the highest LAI of 0.2597, 0.7755 and 2.0937 in vegetative, flowering and harvesting stages respectively and this was followed by M_3S_4 (125% WRc through drip irrigation + 50% FYM + 50% VC) with LAI of 0.2408, 0.7239 and 1.8706. While the lowest LAI (0.0350, 0.0902 and 0.2559) was recorded from the treatment combination M_4S_8 (check basin method of irrigation + no manures and no fertilizers).

Light interception

There was a proportionate increase in light interception in all the treatments as the age of the crop advanced (Table 5). Among the main plots, the treatment M_2 (100% WRc through drip irrigation) recorded the highest light interception

Table 4. Effect of different water regimes and organic manures on leaf area index.

Treatment	Vegetative stage					Flowering stage					Harvesting stage				
	M_1	M_2	M_3	M_4	Mean	M_1	M_2	M_3	M_4	Mean	M_1	M_2	M_3	M_4	Mean
S_1	0.1055	0.1861	0.1799	0.066	0.1344	0.2808	0.5387	0.5396	0.1754	0.3836	0.8067	1.2851	1.3371	0.5218	0.9877
S_2	0.1231	0.2222	0.2306	0.0742	0.1625	0.3385	0.6776	0.6973	0.2094	0.4807	0.8959	1.6548	1.7490	0.6036	1.2258
S_3	0.0931	0.1549	0.1494	0.0600	0.1144	0.2496	0.4561	0.4309	0.1464	0.3208	0.7183	1.0681	1.0198	0.4399	0.8115
S_4	0.1285	0.2597	0.2408	0.0818	0.1777	0.3736	0.7755	0.7239	0.2310	0.5260	0.9280	2.0937	1.8706	0.6548	1.3868
S_5	0.0982	0.1706	0.1664	0.0641	0.1248	0.2691	0.4973	0.4834	0.1602	0.3525	0.7631	1.2011	1.1375	0.4758	0.8944
S_6	0.1168	0.2157	0.2090	0.0697	0.1528	0.3120	0.6401	0.6246	0.1878	0.4411	0.8426	1.5941	1.5453	0.5634	1.1364
S_7	0.1371	0.2003	0.1970	0.0778	0.1531	0.4032	0.5886	0.5968	0.2239	0.4531	0.9733	1.4718	1.4832	0.6216	1.1375
S_8	0.0465	0.0515	0.0496	0.0350	0.0457	0.1149	0.1267	0.1215	0.0902	0.1133	0.3125	0.3729	0.3432	0.2559	0.3211
Mean	0.1061	0.1826	0.1778	0.0661	0.1332	0.2927	0.5376	0.5273	0.1780	0.3839	0.7801	1.3427	1.3107	0.5171	0.9876
	M	S	M at S	S at M		M	S	M at S	S at M		M	S	M at S	S at M	
SE(d)	0.001	0.001	0.002	0.002		0.003	0.003	0.006	0.006		0.006	0.008	0.016	0.016	
CD at 5%	0.003	0.002	0.005	0.004		0.008	0.006	0.014	0.013		0.020	0.016	0.036	0.032	

Table 5. Effect of different water regimes and organic manures on light interception (%).

Treatment	Vegetative stage					Flowering stage					Harvesting stage				
	M_1	M_2	M_3	M_4	Mean	M_1	M_2	M_3	M_4	Mean	M_1	M_2	M_3	M_4	Mean
S_1	6.28	7.59	7.50	5.59	6.74	9.12	10.83	10.74	8.03	9.68	13.90	16.26	16.15	12.20	14.63
S_2	6.61	8.20	8.36	5.70	7.22	9.53	11.42	11.58	8.18	10.18	14.32	16.93	17.12	12.36	15.18
S_3	6.10	7.12	7.03	5.43	6.42	8.95	10.28	10.24	7.84	9.33	13.69	15.65	15.57	11.88	14.20
S_4	6.68	8.65	8.47	5.86	7.42	9.62	11.97	11.73	8.39	10.43	14.38	17.68	17.35	12.65	15.52
S_5	6.23	7.34	7.19	5.50	6.57	9.06	10.52	10.40	7.91	9.47	13.83	15.96	15.79	12.02	14.40
S_6	6.47	8.05	7.92	5.65	7.02	9.34	11.27	11.15	8.12	9.97	14.15	16.73	16.59	12.27	14.94
S_7	6.75	7.80	7.74	5.76	7.01	9.68	10.97	10.92	8.28	9.96	14.49	16.41	16.34	12.54	14.95
S_8	4.79	4.99	4.91	4.60	4.82	6.68	6.95	6.82	6.39	6.71	10.40	10.75	10.59	9.94	10.42
Mean	6.24	7.47	7.39	5.51	6.65	9.00	10.53	10.45	7.89	9.47	13.65	15.80	15.69	11.98	14.28
	M	S	M at S	S at M		M	S	M at S	S at M		M	S	M at S	S at M	
SE(d)	0.037	0.048	0.097	0.096		0.051	0.069	0.138	0.137		0.077	0.103	0.208	0.207	
CD at 5%	0.117	0.099	0.215	0.197		0.163	0.141	0.304	0.281		0.244	0.212	0.458	0.423	

(7.47, 10.53 and 15.80%) in vegetative, flowering and harvesting stages respectively. Among the main plots, M_4 (check basin method of irrigation) exhibited the lowest light interception (5.51, 7.89 and 11.98%) in all stages of crop growth.Among the sub plot treatments, S_4 (50% FYM + 50% VC) exhibited the superior performance with 7.42, 10.43 and 15.52% light interception during vegetative, flowering and harvesting stages respectively (Table 6). The lowest light interception of 4.82, 6.71 and 10.42% was recorded with no manures and no fertilizers treatment (S_8). Among the interactions, the treatment combination M_2S_4 (100% WRc through drip irrigation + 50% FYM + 50% VC) registered the highest light interception in vegetative (8.65%), flowering (11.97%) andharvesting (17.68%) stages which is on par with M_3S_4 (125% WRc through drip irrigation + 50% FYM + 50% VC) which showed light interception of 8.47, 11.73 and 17.35% in different stages of crop growth. Check basin method of irrigation + no manures and no fertilizers (M4S8) treatment combination exhibited the lowest light interception in vegetative (4.60%), flowering (6.39%) and harvesting (9.94%) stages.

Specific leaf weight (SLW)

Among the main plot, M2 (100% WRc through drip irrigation) exhibited the highest SLW (9.35, 11.12 and 10.49 mg cm^{-2}) during vegetative, flowering and harvesting stages. SLW was found to be the lowest in the M4 (check basin method of irrigation) during vegetative (7.20 mg cm^{-2}), flowering (8.14 mg cm^{-2}) and harvesting (7.57 mg cm^{-2}) stages.

Among the manure treatments, application of 50% FYM + 50% VC (S_4) had resulted in the highest SLW of 9.31, 11.00 and 10.32 mg cm^{-2}. While S_8 (no manures and no fertilizers) registered the least score for SLW (6.41, 7.06 and 6.71 mg cm^{-2}).

Between the interactions, the treatment combination, M_2S_4 (100% WRc through drip irrigation + 50% FYM + 50% VC) recorded the utmost SLW (10.89, 12.98 and 12.23 mg cm^{-2}) in vegetative, flowering and harvesting stages respectively and this was followed by M_3S_4 (125% WRc through drip irrigation + 50% FYM + 50% VC) with SLW of 10.21, 12.26 and 11.67 mg cm^{-2} in various crop growth stages. Whereas the lowest SLW in vegetative (6.02 mg cm^{-2}), flowering (6.43 mg cm^{-2}) and harvesting (6.28 mg cm^{-2}) stages was noticed from the treatment M_4S_8 (check basin method of irrigation + no manures and no fertilizers).

Chlorophyll 'a'

Among the main plot treatments provision of 100% WRc through drip irrigation (M_2) was found to have profound influence on the chlorophyll content in all the stages and this treatment recorded the highest chlorophyll'a' (Table 7)

content in vegetative (1.310 mg g^{-1}), flowering (1.516 mg g^{-1}) and harvesting (1.693 mg g^{-1}) stages. While the lowest chlorophyll content of 1.111, 1.244 and 1.365 mg g^{-1} was registered in the treatment M_4 (check basin method of irrigation).

Application of 50% FYM + 50% VC (S_4) recorded the highest chlorophyll content (1.329, 1.538 and 1.711 mg g^{-1}) during vegetative, flowering and harvesting stages respectively and this was followed by S_2 (100% VC) with chlorophyll content of 1.302, 1.496 and 1.662 mg g^{-1} in various stages of crop growth. The least score for chlorophyll content (0.917, 1.012 and 1.109 mg g^{-1}) was recorded with S_8 (no manures and no fertilizers).

Among the interaction effects, the treatment combination M_2S_4 (100% WRc through drip irrigation + 50% FYM + 50% VC) recorded the superior scores for chlorophyll 'a' content (1.483, 1.732 and 1.946 mg g^{-1}) in vegetative, flowering and harvesting stages respectively and this was followed by M_3S_4 (125% WRc through drip irrigation + 50% FYM + 50% VC) with 1.416, 1.644 and 1.826 mg g^{-1} in vegetative, flowering and harvesting stages. The treatment combination M_4S_8 (check basin method of irrigation + no manures and no fertilizers) exhibited poor performance for chlorophyll 'a' content (0.872, 0.961 and 1.057 mg g^{-1}) in various stages of crop growth.

Chlorophyll 'b'

Application of 100% WRc through drip irrigation (M_2) recorded the highest chlorophyll 'b' content in vegetative0.584 mg g^{-1}), flowering (0.624 mg g^{-1}) and harvesting (0.664 mg g^{-1}) stages. While the treatment M_4 (check basin method of irrigation) recorded the lowest content of chlorophyll 'b' with 0.409, 0.439 and 0.460 mg g^{-1} in vegetative, flowering and harvesting stages respectively.

Application of 50% FYM + 50% VC (S_4) recorded the highest content of chlorophyll 'b' with 0.591, 0.641 and 0.684 mg g^{-1} in vegetative, flowering and harvesting stages respectively. Whereas, the lowest chlorophyll 'b' content was recorded with S_8 (no manures and no fertilizers) with 0.303, 0.321 and 0.333 mg g^{-1} in various stages of crop growth.

Between the interaction, the treatment combination, M_2S_4 (100% WRc through drip irrigation + 50% FYM + 50% VC) recorded the utmost chlorophyll 'b' content (0.705, 0.772 and 0.840 mg g^{-1}) in vegetative, flowering and harvesting stages respectively and this was followed by M_3S_4 (125% WRc through drip irrigation + 50% FYM + 50% VC) with chlorophyll 'b' content of 0.677, 0.725 and 0.781 mg g^{-1} in various crop growth stages. Whereas the lowest chlorophyll 'b' in vegetative (0.246 mg g^{-1}), flowering (0.262 mg g^{-1}) and harvesting (0.275 mg g^{-1}) stages was noticed from the treatment M_4S_8 (check basin method of irrigation + no manures and no fertilizers).

Total chlorophyll

Among the irrigation regimes application of 100% WRc through drip irrigation (M_2) recorded the highest total chlorophyll content in vegetative (1.927 mg g^{-1}), flowering (2.172 mg g^{-1}) and harvesting (2.370 mg g^{-1}) stages while check basin method of irrigation (M_4) recorded the lowest score for total chlorophyll content (1.540, 1.690 and 1.832 mg g^{-1}) in the entire crop growth period (Table 9). Among the manure treatments, application of 50% FYM + 50% VC (S_4) recorded the highest total chlorophyll content (1.951, 2.206 and 2.412 mg g^{-1}) in vegetative, flowering and harvesting stages respectively and this was followed by application of 100% VC (S_2) with 1.908, 2.127 and 2.324 mg g^{-1} of total chlorophyll content in various crop growth stages. The lowest total chlorophyll content of 1.228, 1.340 and 1.449 mg g^{-1} was noticed from the treatment S_8 (no manures and no fertilizers) in vegetative, flowering and harvesting stages respectively. Among the interactions, the treatment combination M_2S_4 (100% WRc through drip irrigation + 50% FYM + 50% VC) registered the highest total chlorophyll content in vegetative (2.223 mg g^{-1}), flowering (2.554 mg g^{-1}) and harvesting (2.818 mg g^{-1}) stages and this was followed by M_3S_4 (125% WRc through drip irrigation + 50% FYM + 50% VC) with total chlorophyll content of 2.142, 2.407 and 2.624 mg g^{-1} in different stages of crop growth. Check basin method of irrigation + no manures and no fertilizers (M_4S_8) treatment combination exhibited the lowest total chlorophyll content in vegetative (1.129 mg g^{-1}), flowering (1.231 mg g^{-1}) and harvesting (1.339 mg g^{-1}) stages.

Soluble protein

In all the treatments soluble protein content increased from vegetative to flowering stage and decreased towards the harvesting stage (Table 10). Among the main plot, M_2 (100% WRc through drip irrigation) exhibited superior performance for soluble protein content (11.90, 13.80 and 12.55 mg g^{-1}) during vegetative, flowering and harvesting stages respectively. The soluble protein content was found to be the lowest in the treatment M_4 (check basin method of irrigation) during vegetative (9.23 mg g^{-1}), flowering (10.15 mg g^{-1}) and harvesting (9.45 mg g^{-1}) stages. In the sub plot, the treatment, S_4 (50% FYM + 50% VC) recorded the highest soluble protein content (12.02, 13.79 and 12.54 mg g^{-1}) in vegetative, flowering and harvesting stages respectively. While the treatment S_8 (no manures and no fertilizers) registered the lowest score for soluble protein content (7.48, 8.14 and 7.68 mg g^{-1}) in all stages of crop growth. In the interaction, the treatment combination M_2S_4 (100% WRc through drip irrigation + 50% FYM + 50% VC) recorded the highest soluble protein content in vegetative (13.68 mg g^{-1}), flowering (15.84 mg g^{-1}) and harvesting (14.42 mg g^{-1}) stages and this was followed by M_3S_4 (125% WRc through drip irrigation + 50% FYM + 50% VC) with soluble protein content of 13.21, 15.36 and 13.88 mg g^{-1} in various stages of crop growth. The lowest soluble protein content of 6.24, 6.71 and 6.39 mg g^{-1} was registered in the treatment combination M_4S_8 (check basin method of irrigation + no manures and no fertilizers).

Total phenol

Total phenol content increased from vegetative to flowering stage and decreased thereafter (Table 11). Among the main plot, M_2 (100% WRc through drip irrigation) registered the highest total phenol content in vegetative (3.92 mg g^{-1}), flowering (5.92 mg g^{-1}) and harvesting (4.76 mg g^{-1}) stages. The total phenol content was found to be the lowest (2.82, 4.07 and 2.76 mg g^{-1}) in the treatment comprising check basin method of irrigation (M_4).

Regarding the sub plot treatments, the highest total phenol content in vegetative (3.92 mg g^{-1}), flowering (5.80 mg g^{-1}) and harvesting (4.61 mg g^{-1}) stages was recorded from the treatment comprising 50% FYM + 50% VC (S_4). The lowest total phenol content of 2.33, 3.16 and 1.90 mg g^{-1} was recorded from the treatment S_8 (no manures and no fertilizers) during various crop growth stages.

Among the interaction effects, the treatment combination M_2S_4 (100% WRc through drip irrigation + 50% FYM + 50% VC) registered the highest total phenol content of leaves in vegetative (4.55 mg g^{-1}), flowering (6.79 mg g^{-1}) and harvesting (5.63 mg g^{-1}) stages and this was followed by M_3S_4 with 4.41, 6.62 and 5.49 mg g^{-1}. Whereas lowest total phenol content (2.13, 2.89 and 1.60 mg g^{-1}) was observed from M_4S_8 (check basin method of irrigation + no manures and no fertilizers).

Nitrate reductase

Among the main plot treatments, M_2 (100% WRc through drip irrigation) recorded the highest nitrate reductase activity (32.88, 39.61 and 35.00 µg NO_2 g^{-1} h^{-1}) in vegetative, flowering and harvesting stages respectively (Table 12). Whereas the lowest nitrate reductase activity (22.22, 23.89 and 18.61 µg NO_2 g^{-1} h^{-1}) was noticed from treatment comprising check basin method of irrigation (M_4).

Regarding the sub plots, application of 50% FYM + 50% VC (S_4) exhibited the superior scores for nitrate reductase activity (32.02, 38.28 and 33.35 µg NO_2 g^{-1} h^{-1}) followed by application of 100% VC (S_2) with 30.91, 36.28 and 31.31 µg NO_2 g^{-1} h^{-1}. While the treatment S_8 (no manures and no fertilizers) showed very poor performance for nitrate reductase activity with 16.79, 17.97 and 14.08 µg NO_2 g^{-1} h^{-1} in vegetative, flowering and harvesting stages respectively.

Among the interactions, the treatment combination

Table 6. Effect of different water regimes and organic manures on specific leaf weight (mg cm^{-2}).

Treatment	Vegetative stage					Flowering stage					Harvesting stage				
	M$_1$	M$_2$	M$_3$	M$_4$	Mean	M$_1$	M$_2$	M$_3$	M$_4$	Mean	M$_1$	M$_2$	M$_3$	M$_4$	Mean
S$_1$	8.26	9.55	9.43	7.30	8.64	9.52	11.29	11.23	8.22	10.07	9.01	10.68	10.60	7.68	9.49
S$_2$	8.54	10.02	10.09	7.44	9.02	10.03	12.10	12.15	8.48	10.69	9.36	11.49	11.55	7.79	10.05
S$_3$	8.07	8.94	9.02	7.13	8.29	9.31	10.89	10.92	8.10	9.81	8.79	10.16	10.23	7.54	9.18
S$_4$	8.60	10.89	10.21	7.53	9.31	10.12	12.98	12.26	8.62	11.00	9.48	12.23	11.67	7.88	10.32
S$_5$	8.12	9.22	9.29	7.22	8.46	9.26	11.04	11.16	8.14	9.90	8.83	10.30	10.44	7.60	9.29
S$_6$	8.32	9.89	9.80	7.38	8.85	9.69	11.79	11.65	8.40	10.38	9.07	11.19	11.12	7.73	9.78
S$_7$	8.39	9.64	9.74	7.59	8.84	9.80	11.44	11.52	8.73	10.37	9.22	10.91	10.96	8.06	9.79
S$_8$	6.46	6.63	6.54	6.02	6.41	7.11	7.40	7.29	6.43	7.06	6.71	6.96	6.87	6.28	6.71
Mean	8.10	9.35	9.27	7.20	8.48	9.36	11.12	11.02	8.14	9.91	8.81	10.49	10.43	7.57	9.32
	M	S	M at S	S at M		M	S	M at S	S at M		M	S	M at S	S at M	
SE(d)	0.046	0.061	0.123	0.122		0.054	0.072	0.145	0.144		0.051	0.068	0.136	0.135	
CD at 5%	0.146	0.125	0.272	0.251		0.171	0.147	0.319	0.294		0.163	0.138	0.301	0.277	

Table 7. Effect of different water regimes and organic manures on chlorophyll 'a' (mg g^{-1}) content of noni leaves.

Treatment	Vegetative stage					Flowering stage					Harvesting stage				
	M$_1$	M$_2$	M$_3$	M$_4$	Mean	M$_1$	M$_2$	M$_3$	M$_4$	Mean	M$_1$	M$_2$	M$_3$	M$_4$	Mean
S$_1$	1.227	1.329	1.342	1.135	1.258	1.409	1.542	1.554	1.279	1.446	1.573	1.729	1.734	1.410	1.612
S$_2$	1.253	1.392	1.405	1.159	1.302	1.432	1.621	1.633	1.296	1.496	1.603	1.807	1.817	1.422	1.662
S$_3$	1.207	1.289	1.280	1.103	1.220	1.395	1.503	1.492	1.243	1.408	1.558	1.688	1.662	1.356	1.566
S$_4$	1.244	1.483	1.416	1.172	1.329	1.446	1.732	1.644	1.329	1.538	1.614	1.946	1.826	1.458	1.711
S$_5$	1.215	1.318	1.309	1.119	1.240	1.386	1.520	1.538	1.256	1.425	1.532	1.706	1.716	1.376	1.583
S$_6$	1.239	1.384	1.372	1.146	1.285	1.424	1.610	1.598	1.270	1.476	1.586	1.794	1.780	1.398	1.640
S$_7$	1.264	1.353	1.361	1.179	1.289	1.453	1.569	1.574	1.314	1.478	1.629	1.746	1.752	1.443	1.643
S$_8$	0.918	0.932	0.947	0.872	0.917	1.012	1.031	1.042	0.961	1.012	1.109	1.129	1.142	1.057	1.109
Mean	1.196	1.310	1.304	1.111	1.230	1.370	1.516	1.509	1.244	1.410	1.526	1.693	1.679	1.365	1.566
	M	S	M at S	S at M		M	S	M at S	S at M		M	S	M at S	S at M	
SE(d)	0.007	0.009	0.018	0.018		0.007	0.010	0.021	0.020		0.008	0.011	0.023	0.023	
CD at 5%	0.021	0.018	0.039	0.036		0.024	0.021	0.045	0.042		0.026	0.023	0.050	0.046	

Table 8. Effect of different water regimes and organic manures on chlorophyll 'b' (mg g^{-1}) content of noni leaves.

Treatment	Vegetative stage					Flowering stage					Harvesting stage				
	M_1	M_2	M_3	M_4	Mean	M_1	M_2	M_3	M_4	Mean	M_1	M_2	M_3	M_4	Mean
S_1	0.503	0.593	0.604	0.418	0.530	0.541	0.637	0.642	0.468	0.572	0.566	0.676	0.679	0.492	0.603
S_2	0.527	0.648	0.659	0.449	0.571	0.554	0.693	0.709	0.476	0.608	0.579	0.740	0.758	0.501	0.645
S_3	0.485	0.563	0.552	0.395	0.499	0.532	0.607	0.594	0.419	0.538	0.560	0.638	0.627	0.438	0.566
S_4	0.520	0.705	0.677	0.460	0.591	0.570	0.772	0.725	0.498	0.641	0.596	0.840	0.781	0.519	0.684
S_5	0.497	0.579	0.570	0.403	0.512	0.522	0.611	0.619	0.431	0.546	0.549	0.645	0.654	0.452	0.575
S_6	0.512	0.640	0.629	0.431	0.553	0.548	0.680	0.671	0.465	0.591	0.571	0.725	0.716	0.486	0.625
S_7	0.538	0.617	0.621	0.471	0.562	0.579	0.653	0.659	0.494	0.596	0.607	0.693	0.702	0.513	0.629
S_8	0.310	0.326	0.331	0.246	0.303	0.329	0.342	0.350	0.262	0.321	0.341	0.353	0.362	0.275	0.333
Mean	0.487	0.584	0.580	0.409	0.515	0.522	0.624	0.621	0.439	0.552	0.546	0.664	0.660	0.460	0.582
	M	S	M at S	S at M		M	S	M at S	S at M		M	S	M at S	S at M	
SE(d)	0.003	0.004	0.008	0.008		0.003	0.004	0.001	0.008		0.003	0.004	0.009	0.009	
CD at 5%	0.009	0.008	0.017	0.016		0.010	0.008	0.018	0.017		0.010	0.009	0.019	0.018	

Table 9. Effect of different water regimes and organic manures on total chlorophyll (mg g^{-1}) content of noni leaves

Treatment	Vegetative stage					Flowering stage					Harvesting stage				
	M_1	M_2	M_3	M_4	Mean	M_1	M_2	M_3	M_4	Mean	M_1	M_2	M_3	M_4	Mean
S_1	1.754	1.963	1.989	1.579	1.821	1.961	2.214	2.230	1.756	2.040	2.148	2.409	2.414	1.911	2.221
S_2	1.813	2.083	2.107	1.627	1.908	1.996	2.349	2.382	1.779	2.127	2.196	2.568	2.601	1.929	2.324
S_3	1.715	1.884	1.869	1.512	1.745	1.937	2.133	2.102	1.670	1.961	2.124	2.331	2.294	1.801	2.138
S_4	1.797	2.223	2.142	1.643	1.951	2.028	2.554	2.407	1.834	2.206	2.219	2.818	2.624	1.986	2.412
S_5	1.741	1.924	1.912	1.548	1.781	1.919	2.162	2.186	1.694	1.990	2.089	2.364	2.380	1.836	2.167
S_6	1.778	2.061	2.037	1.602	1.870	1.982	2.320	2.306	1.740	2.087	2.166	2.537	2.512	1.890	2.276
S_7	1.836	2.011	2.022	1.678	1.887	2.057	2.267	2.278	1.812	2.104	2.245	2.443	2.469	1.963	2.280
S_8	1.236	1.264	1.283	1.129	1.228	1.349	1.380	1.398	1.231	1.340	1.457	1.489	1.510	1.339	1.449
Mean	1.709	1.927	1.920	1.540	1.774	1.904	2.172	2.161	1.690	1.982	2.081	2.370	2.351	1.832	2.158
	M	S	M at S	S at M		M	S	M at S	S at M		M	S	M at S	S at M	
SE(d)	0.009	0.013	0.026	0.026		0.011	0.014	0.029	0.029		0.012	0.016	0.032	0.031	
CD at 5%	0.030	0.026	0.057	0.053		0.034	0.030	0.064	0.059		0.037	0.032	0.069	0.064	

Table 10. Effect of different water regimes and organic manures on soluble protein (mg g^{-1}) content of noni leaves.

Treatment	Vegetative stage					Flowering stage					Harvesting stage				
	M$_1$	M$_2$	M$_3$	M$_4$	Mean	M$_1$	M$_2$	M$_3$	M$_4$	Mean	M$_1$	M$_2$	M$_3$	M$_4$	Mean
S$_1$	11.06	12.09	11.92	9.58	11.16	12.80	14.16	13.88	10.49	12.83	11.66	12.81	12.54	9.75	11.69
S$_2$	11.24	12.84	13.07	9.80	11.74	12.87	14.92	15.14	10.79	13.43	11.70	13.50	13.68	10.02	12.23
S$_3$	10.75	11.72	11.64	9.26	10.84	12.34	13.61	13.49	10.21	12.41	11.29	12.33	12.27	9.52	11.35
S$_4$	11.32	13.68	13.21	9.86	12.02	13.04	15.84	15.36	10.92	13.79	11.80	14.42	13.88	10.07	12.54
S$_5$	10.82	11.86	11.80	9.44	10.98	12.31	13.76	13.68	10.40	12.54	11.33	12.50	12.39	9.70	11.48
S$_6$	10.91	12.66	12.61	9.67	11.46	12.58	14.83	14.72	10.62	13.19	11.58	13.43	13.29	9.89	12.05
S$_7$	11.37	12.29	12.43	9.97	11.52	13.12	14.41	14.56	11.06	13.29	11.83	13.06	13.14	10.23	12.07
S$_8$	7.66	8.07	7.94	6.24	7.48	8.29	8.90	8.64	6.71	8.14	7.82	8.33	8.17	6.39	7.68
Mean	10.64	11.90	11.83	9.23	10.90	12.17	13.80	13.68	10.15	12.45	11.13	12.55	12.42	9.45	11.39
	M	S	M at S	S at M		M	S	M at S	S at M		M	S	M at S	S at M	
SE(d)	0.059	0.079	0.159	0.158		0.067	0.091	0.182	0.181		0.061	0.083	0.166	0.165	
CD at 5%	0.186	0.162	0.350	0.323		0.213	0.185	0.401	0.371		0.195	0.169	0.366	0.338	

Table 11. Effect of different water regimes and organic manures on total phenol content (mg g^{-1}) of noni leaves.

Treatment	Vegetative stage					Flowering stage					Harvesting stage				
	M$_1$	M$_2$	M$_3$	M$_4$	Mean	M$_1$	M$_2$	M$_3$	M$_4$	Mean	M$_1$	M$_2$	M$_3$	M$_4$	Mean
S$_1$	3.40	4.02	3.96	2.77	3.54	5.06	6.10	6.13	4.11	5.35	3.79	4.96	5.02	2.71	4.12
S$_2$	3.51	4.32	4.36	2.96	3.79	5.18	6.54	6.57	4.28	5.64	3.97	5.36	5.44	3.02	4.45
S$_3$	3.37	3.72	3.79	2.75	3.41	4.91	5.82	5.89	4.03	5.16	3.72	4.62	4.71	2.65	3.93
S$_4$	3.63	4.55	4.41	3.08	3.92	5.39	6.79	6.62	4.40	5.80	4.15	5.63	5.49	3.18	4.61
S$_5$	3.29	3.88	3.93	2.84	3.49	4.89	6.01	6.08	4.15	5.28	3.64	4.84	4.93	2.76	4.04
S$_6$	3.48	4.25	4.21	2.92	3.72	5.12	6.51	6.44	4.21	5.57	3.91	5.31	5.27	2.90	4.35
S$_7$	3.59	4.11	4.13	3.11	3.74	5.32	6.24	6.29	4.47	5.58	4.11	5.14	5.20	3.24	4.42
S$_8$	2.29	2.47	2.42	2.13	2.33	3.14	3.35	3.26	2.89	3.16	1.86	2.20	1.95	1.60	1.90
Mean	3.32	3.92	3.90	2.82	3.49	4.88	5.92	5.91	4.07	5.19	3.64	4.76	4.75	2.76	3.98
	M	S	M at S	S at M		M	S	M at S	S at M		M	S	M at S	S at M	
SE(d)	0.019	0.025	0.051	0.051		0.028	0.038	0.077	0.076		0.022	0.030	0.060	0.060	
CD at 5%	0.060	0.052	0.113	0.104		0.088	0.078	0.168	0.156		0.068	0.061	0.131	0.122	

Table 12. Effect of different water regimes and organic manures on leaf nitrate reductase activity ($\mu g\ NO_2\ g^{-1}\ h^{-1}$) at various stages.

Treatment	Vegetative stage					Flowering stage					Harvesting stage				
	M_1	M_2	M_3	M_4	Mean	M_1	M_2	M_3	M_4	Mean	M_1	M_2	M_3	M_4	Mean
S_1	28.06	34.62	34.53	23.40	30.15	32.10	42.17	42.06	24.88	35.30	25.55	36.57	36.43	19.17	29.43
S_2	28.76	35.62	35.78	23.49	30.91	32.40	43.85	43.79	25.07	36.28	26.08	39.68	39.16	20.32	31.31
S_3	27.48	32.68	32.63	22.56	28.84	30.86	37.09	37.02	24.45	32.36	24.07	32.74	32.81	18.43	27.01
S_4	29.35	38.92	36.04	23.77	32.02	33.18	49.06	45.72	25.16	38.28	27.20	45.84	40.17	20.18	33.35
S_5	28.43	34.18	34.10	22.09	29.70	32.17	40.38	40.31	24.21	34.27	25.64	34.65	34.19	18.20	28.17
S_6	28.60	34.81	34.87	22.75	30.26	32.59	42.53	42.66	24.63	35.60	26.53	36.91	36.98	19.03	29.86
S_7	29.27	35.18	35.11	23.86	30.86	33.06	43.60	43.47	25.92	36.51	26.82	39.05	38.90	20.56	31.33
S_8	16.90	17.05	17.38	15.82	16.79	18.03	18.16	18.89	16.79	17.97	14.13	14.56	14.68	12.96	14.08
Mean	27.11	32.88	32.56	22.22	28.69	30.55	39.61	39.24	23.89	33.32	24.50	35.00	34.17	18.61	28.07
	M	S	M at S	S at M		M	S	M at S	S at M		M	S	M at S	S at M	
SE(d)	0.151	0.211	0.423	0.422		0.182	0.248	0.497	0.495		0.159	0.211	0.425	0.422	
CD at 5%	0.481	0.432	0.928	0.865		0.579	0.507	1.094	1.014		0.505	0.432	0.938	0.864	

comprising 100% WRc through drip irrigation + 50% FYM + 50% VC (M_2S_4) registered the highest nitrate reductase activity of 38.92, 49.06 and 45.84 $\mu g\ NO_2\ g^{-1}\ h^{-1}$ in vegetative, flowering and harvesting stages respectively and this was followed by M_3S_4 (125% WRc through drip irrigation + 50% FYM + 50% VC) with 36.04, 45.72 and 40.17 $\mu g\ NO_2\ g^{-1}\ h^{-1}$.

The nitrate reductase activity was found to be the lowest (15.82, 16.79 and 12.96 $\mu g\ NO_2\ g^{-1}\ h^{-1}$) in the treatment combination M_4S_8 (check basin method of irrigation + no manures and no fertilizers) during vegetative, flowering and harvesting stages, respectively.

Yield parameters

Fruit weight

Between the main plot, higher fruit weight (138.73 g) was registered from M2 (100% WRc through drip irrigation) and this was on par with M_3 (125% WRc through drip irrigation) with fruit weight of 138.10 g (Table 13). The fruit weight was found to be the lowest in the treatment M_4 (check basin method of irrigation) with 97.75 g.

Among the sub plot treatments, S_4 (50% FYM + 50% VC) recorded the highest fruit weight of 143.11 g and this was followed by S_2 (100% VC) with 137.04 g. The treatment S_8 (no manures and no fertilizers) exhibited the lowest fruit weight of 64.27 g.

Pertaining to the interaction effects, the treatment combination M_2S_4 (100% WRc through drip irrigation + 50% FYM + 50% VC) produced the highest fruit weight (167.86 g) and this was followed by M_3S_4 (125% WRc through drip irrigation + 50% FYM+50% VC) with 163.44 g.

Among the interactions, the fruit weight was found to be the lowest (58.86 g) in the treatment combination comprising check basin method of irrigation + no manures and no fertilizers (M_4S_8).

Number of fruits per plant

The highest number of fruits were counted in the treatment M_2 (100% WRc through drip irrigation) with 176.04 (Table 13). The lowest number (142.43) was obtained from the treatment M_4 (check basin method of irrigation).

Between the sub plot treatments, an increase in number of fruits (177.63) were obtained from the treatment S_4 (50% FYM + 50% VC) followed by S_2 (100% VC) with 173.14. While the treatment S_8 (no manures and no fertilizers) registered the least number (113.78) of fruits per plant.

In the interaction, the treatment M_2S_4 (100% WRc through drip irrigation + 50% FYM + 50% VC) expressed greater number of fruits per plant (198.12) and this was followed by M_3S_4 (125% WRc through drip irrigation + 50% FYM + 50%

Table 13. Effect of different water regimes and organic manures on fruit weight (g), number of fruits per plant and yield per plant (kg).

Treatment	Fruit weight (g)					Number of fruits per plant					Yield per plant (kg)				
	M_1	M_2	M_3	M_4	Mean	M_1	M_2	M_3	M_4	Mean	M_1	M_2	M_3	M_4	Mean
S_1	120.24	141.77	143.69	100.88	126.65	161.55	180.94	182.05	145.72	167.57	19.41	25.64	26.14	14.70	21.47
S_2	125.96	156.78	159.74	105.68	137.04	164.84	188.32	190.54	148.86	173.14	20.75	29.51	30.42	15.74	24.11
S_3	115.28	134.26	133.27	95.63	119.61	155.69	171.63	170.52	142.07	159.98	17.96	23.02	22.72	13.57	19.32
S_4	129.25	167.86	163.44	111.88	143.11	165.32	198.12	193.21	153.88	177.63	21.35	33.26	31.59	17.22	25.86
S_5	117.77	140.44	138.72	97.52	123.61	157.75	178.42	175.36	142.88	163.60	18.56	25.08	24.34	13.94	20.48
S_6	123.27	152.43	150.66	104.26	132.66	162.32	187.75	186.83	146.25	170.79	20.02	28.60	28.16	15.24	23.01
S_7	131.69	147.53	148.89	107.25	133.84	167.18	185.34	186.12	152.63	172.82	22.04	27.33	27.70	16.35	23.36
S_8	63.12	68.74	66.35	58.86	64.27	113.88	117.82	116.28	107.12	113.78	7.18	8.10	7.73	6.31	7.33
Mean	115.82	138.73	138.10	97.75	122.60	156.07	176.04	175.11	142.43	162.41	18.41	25.07	24.85	14.13	20.62
	M	S	M at S	S at M		M	S	M at S	S at M		M	S	M at S	S at M	
SE(d)	0.326	0.452	0.906	0.904		0.429	0.588	1.180	1.176		0.057	0.078	0.157	0.157	
CD at 5%	1.036	0.926	1.989	1.852		1.364	1.204	2.594	2.408		0.182	0.160	0.346	0.321	

VC) with 193.21. While the lesser number was noted in the treatment M_4S_8 (check basin method of irrigation + no manures and no fertilizers) with 107.12.

Yield per plant

Concerning the main plot, M_2 (100% WRc through drip irrigation) registered the highest fruit yield per plant of 25.07 kg and this was followed by M_3 (125% WRc through drip irrigation) with 24.85 kg (Table 13. The fruit yield per plant was found to be the lowest in M_4 (check basin method of irrigation) with 14.13 kg.

Pertaining to the sub plot treatments, S_4 (50% FYM + 50% VC) produced the highest fruit yield per plant of 25.86 kg and this was followed by S_2 (100% VC) with 24.11 kg. While the treatment S_8 (no manures and no fertilizers) recorded the lowest fruit yield per plant of 7.33 kg.

Among the interactions, M_2S_4 (100% WRc through drip irrigation + 50% FYM + 50% VC) registered the superior score for fruit yield per plant of 33.26 kg and this was followed by M_3S_4 (125% WRc through drip irrigation + 50% FYM + 50% VC) with fruit yield per plant of 31.59 kg. Fruit yield per plant was found to be the lowest (6.31 kg) in the treatment combination M_4S_8 (check basin method of irrigation + no manures and no fertilizers).

DISCUSSION

Physiological parameters

Any crop management practice should aim in keeping the physiological process of the plants in an active stage so that the plants can produce biomass with the least destructive processes.

LAI is a measure of vegetative growth of plants and the assimilatory surface on which the production of dry matter takes place. LAI is a positive indication on plant growth with a direct influence. The LAI has a decisive role on physiological parameters like SLW and light interception. Combined application of 100% WRc through drip irrigation + 50% FYM + 50% VC (M_2S_4) has led to higher LAI. This may due to continuous and uninterrupted supply of water and nutrients. This finding was strengthened by previous work of Singh et al. (2004) and Umesha et al. (2011).

The plants maintain a turgid condition during the day time under drip irrigation as compared to check basin method of irrigation. There is a possibility of wide opening of stomata for longer period which might have resulted in high exchange of gases. Similarly, leaves might have remained turgid and produced more leaf surface. Thus, in turgor state helps in absorption of more sun light and solar radiation. It has resulted in higher rate of photosynthesis and increased the photosynthetic capacity, which ultimately might have resulted production of more LAI and in turn the light interception for photosynthesis. The crops experienced water stress period before each irrigation under check basin method irrigation due to the availability of moisture and nutrients were limited for the roots to absorb. As results, there is reduction in LAI and light interception.

The higher LAI and light interception as a result of maintenance of favorable soil moisture in the root zone and effective absorption by plants. The optimum P uptake and greater P mobility through frequent or continuous low volume irrigation can maintain three dimensional distribution patterns of water and nutrients and provide improved conditions for growth and water and nutrient uptake (Gal and Dudley, 2003).

Similarly the higher LAI and light interception may due to the optimum uptake of nutrients especially nitrogen, iron and magnesium from the soil resulted in more chlorophyll content which might have enhanced the photosynthetic rate and production of more leaf area inturn light interception for photosynthesis (Khandkar and Nigam, 1996).

The application of FYM in the treatment would have increased the friability, promoted aggregation of soil and increased the level of humus, thereby increasing the microbial activity. This in turn would have enhanced the increased production of more photosynthates due to enhanced photosynthesis leading to more accumulation of carbohydrates. This may be responsible for increased LAI, light interception and SLW.

Furthermore, inoculation of the biofertilizer, *Azospirillum* would have increased the activity of root exudates which in turn might have accelerated the activity of beneficial microbes by higher nitrogen fixation and secretion of growth promoting substances as reported by Okon and Kapulnik (1986) which owed to the luxuriant growth of vegetative parameters that was reflected on improved LAI.

In the present experiment 100% WRc through drip irrigation + 50% FYM + 50% VC (M_2S_4) exhibited the highest SLW. SLW was considered to be a good indicator of photosynthetic capacity of the leaf (Wallace et al., 1972). The highest SLW was mainly due to minimum water stress and better nutrient availability developed by low rate of evapotranspiration and minimum nutrient loss under drip irrigation system. These results are similar to the findings of Rao et al. (2000) in tomato and Shoba (2009) in black nightshade (*Solanum nigrum*).

A higher photosynthetic activity is a good indication of physiological efficiency in plants. This primarily depends upon the leaf chlorophyll content. The chlorophyll content in leaves indicates the efficiency of photosynthesis where the solar energy is converted into chemical energy. A slight fluctuation in chlorophyll is enough to trigger changes in physiological processes of the plants particularly photosynthesis.

The chlorophyll content was significantly higher in the treatment combination comprising 100% WRc through drip irrigation + 50% FYM + 50% VC (M_2S_4) as compared to check basin method of irrigation + no manures and no fertilizers (M_4S_8). The amount of chlorophyll present has a direct relationship with the rate of photosynthesis. Hence, an increase in biomass production was obtained by higher chlorophyll content in plants.

Such increase may be due to optimum water absorption and optimum uptake of nutrients which have close association with chlorophyll biosynthesis as reported by Raghavendra (2000). Similarly, the presence of humic substances in FYM was the additional source of polyphenols that might have acted as respiratory catalysts, which in turn enhanced the rate of respiration and metabolic activity of the plants and thereby increasing the chlorophyll content (Padmanabhan, 2003).

The reduced chlorophyll content in check basin method of irrigation + no manures and no fertilizers (M_4S_8) may due to high moisture stress and poor nutrient availability. This situation cause delay in chloroplast membrane synthesis (Hinningsen, 1970) which can be a reason for reduced total chlorophyll content. Also, water stress and nutrient deficient condition causes a reduction in synthesis of precursors of chlorophyll (Mukhmudov, 1983) which in turn reduce the chlorophyll content.

The soluble protein content in leaves indirectly indicates the photosynthetic efficiency of the crop. Since, it constitutes more than 70% of the RuBp carboxylase, the enzyme responsible for CO_2 fixation in photosynthesis (Nogle,1979). An increase in soluble protein content denotes the increasing ability of plants to fix CO_2 effectively. Hence, a level of soluble protein content is considered as an index for the assessment of photosynthetic efficiency. The treatment combination M_2S_4 (100% WRc through drip irrigation + 50% FYM + 50% VC) registered the highest soluble protein content. This might be due to high soil moisture status, thus maintaining normal cell integrity, cell elongation and functioning of biopolymers apart from optimum nutrient uptake.

Phenols are the physiologically active secondary compounds produced by all the higher plants and perhaps by each cell of the plant. They can be found in the cytoplasm, vacuoles and cell walls and the sites of their biosynthesis indicate the potential importance of these compounds in plant's life (Zaprometov, 1989). Hence deposition of more structural proteins and phenolics in the cell wall regions would directly influence the resistance mechanisms. The present study indicates that application 100% WRc through drip irrigation + 50% FYM + 50% VC (M_2S_4) resulted in increased phenol content. It could be attributed to the fact that phenols induce resistance to pathogens by the production of PR (plant resistance) proteins (Raskin, 1992).

Moreover, the humic substances present in FYM and VC are known to contain phenyl alanine; the precursor for several phenolic substances would also have contributed to the increase in the phenol content (Padmapriya, 2004).

Combined application of 100% WRc through drip irrigation + 50% FYM + 50% VC (M_2S_4) recorded the highest nitrate reductase activity.

Application of water at frequent time interval through drip irrigation maintains the soil moisture, prevents the plant from soil moisture stress and keeps the plant always in physiological active state which would have resulted in higher nitrate reductase activity. Similar trend of results have been documented by Sachdev et al. (1987) and Prakash (2010).

Mahadevan (1988) suggested that, high nitrate reductase activity indicates a higher level of protein synthesis and accumulation of soluble protein. This in turn indicates that nitrogenous compounds in the plants are well utilized for various metabolic activities. Major part of soluble protein consists of RUBISCO enzyme, which is carboxylation as well as oxygenation enzyme. This is very essential for the fixation of atmospheric CO_2 to produce carbohydrates. Therefore, if the soluble protein is high, photosynthetic efficiency will be more, which may result in better yield.

This may be also due to the fact that VC is a loosely packed, granular aggregate of enzymatically digested organic matter containing essential nutrients in easily available or mineralisable form. So, when VC was added to the soil, it enhanced the soil microbial activity and provided essential macro and micro nutrients for better crop growth. Besides, earthworm casts have been shown to stimulate nitrate reductase activity, which regulates the nitrate availability for the plants (Jat and Kumar, 2002).

Yield parameters

The ultimate goal of any management practice is to improve the yield level with minimal cost of production. The system has to be maintained as such without letting any sort of degradation in soil, water or environment besides obtaining quality produce. The experimental plots receiving 100% WRc through drip irrigation + 50% FYM + 50% VC (M_2S_4) recorded the highest fruit weight.

Drip irrigation at optimum level provides a consistent moisture regime in the soil due to which root remains active throughout the season resulting in optimum uptake of water and nutrients. This facilitates proper translocation of food materials which accelerates the fruit growth and development in noni. The highest fruit weight under drip irrigation might be ascribed to better water utilization, minimum losses of water through percolation and evaporation and excellent soil-water-air relationship with higher oxygen concentration in the root zone and optimum uptake of nutrients. These results are in agreement with the findings of Bafna et al. (1993) and Prakash (2010).

This could be also due to the slow release of nutrients in synchrony with improved physical properties of soil resulting in optimum uptake of nutrients, which might have facilitated improvement in fruit characters particularly fruit weight. In addition to this, FYM has favoured the supply of micronutrients through its own decomposition, besides acting as an additional source of ammoniacal nitrogen ultimately resulting in increased fruit weight. FYM improved soil physical structure and texture, decreased the bulk density and increased moisture retention. All these comprehensive changes paved the way for greater fruit weight.

Moreover addition of biofertilizers and soil beneficial rhizosphere microflora produces organic acids viz, malic and succinic acids which convert insoluble soil phosphates to more soluble compounds thereby increasing the availability of nutrients. This increased nutrient availability could have resulted in higher accumulation of carbohydrates in sink thereby exerting a remarkable increase in fruit dimensions.

Number of fruits is an important character that decides the yield of the crop. It was observed in the present study that the highest number of fruits per plant was obtained with the treatment combination comprising 100% WRc through drip irrigation + 50% FYM + 50% VC (M_2S_4). Better availability and utilization of nutrients and water may be the possible reason for the promotary effects.

Roots can easily translocate absorbed water from the soil where available soil moisture content was optimum at 100% WRc through drip irrigation. Required energy for water absorption was less under this treatment and ultimately led to easy energy translocation to the reproductive parts. Application of FYM and VC had increased the soil organic matter and improved the soil structure and biological activity of the soil. This would have reduced the loss of nitrogen by increased cation and anion exchange capacities in soil thereby enhancing the fruit development and yield (Ayisha, 1997).

The pronounced yield improvement in organic treatment might be due to sustained availability of nitrogen in the soil throughout the growing phase and also due to enhanced carbohydrate synthesis and effective translocation of these photosynthates to the sink, that is, fruit, while at lower fertility levels plants remained stunted resulting in decreased yield.

Drip irrigation maintains the soil moisture around the field capacity between two irrigation intervals. On the other hand, check basin method of has high fluctuation of moisture between field capacity and permanent wilting point. This might have resulted in lower fruit yield under check basin method of irrigation. These results collaborate with the findings of Aladakatti et al. (2012) and Behera et al. (2012).

The reduced moisture availability in M_4S_8 (check basin method of irrigation + no manures and no fertilizers) creates water stress condition. As water stress increases, CO_2 assimilation per unit leaf area decreases. When soil moisture stress intensifies, photosynthesis gets restricted to few hours and peak rate reduces. As a result biomass accumulations become slower (Suanez et al., 1989).

Water stress generate active oxygen species which are extremely reactive and cytotoxic and it can affect the respiratory activity in mitochondria, can cause pigment break down and thereby less of carbon fixing capacity of chloroplasts (Scandalios, 1993). As result there is reduction in fruit number and fruit yield.

From the present study, it could be concluded that application 100% WRc through drip irrigation + 50% FYM + 50% VC (M_2S_4) produced the highest and sustainable crop yield through improved physiological efficiency of noni plants.

REFERENCES

Aladakatti YR, Palled YB, Chetti MB, Halikatti SI, Alagundagi SC, Patil PL, Patil VC, Janawade AC (2012). Effect of irrigation schedule and planting geometry on growth and yield of stevia (Stevia rebaudiana Bertoni.). Karnataka J. Agric. Sci. 25(1):30-35.

Ayisha TP (1997). Yield and quality of Piper longum L. under different spacing and manurial regimes in coconut gardens. MSc Thesis, Kerala Agricultural University, Thrissur, Kerala, India.

Baby PT (2012). Influence of ultra high density planting and organics in enhancing leaf production and protein content in annual moringa (Moringa oleifera Lam.) cv.PKM-1. MSc Thesis, Tamil Nadu Agricultural University, Coimbatore, Tamil Nadu, India.

Bafna AM, Daftardar SY, Khade KK, Patel PV, Dhotre RS (1993). Utilization of nitrogen and water by tomato under drip irrigation system. J. Water Manage. 1:1-5.

Behera MS, Mahapatra PK, Verma OP, Singandhupe RB, Kumar A (2012). Drip fertigation impact on yield and alkaloid content of Withania somnifera (L.) Dunal. Medicinal Plants. 4(3):133-137.

Braun H, Roy RN (1983). Maximizing the efficiency of mineral fertilizers. Proceedings of Symposium on efficient use of fertilizers in agriculture development in plant and soil Sciences held at Hague, Netherlands. pp. 251-273.

Bray HG, Thorpe MU (1954). Analysis of phenolic compounds of interest in metabolism. Meth. Biochem. Anal. 9:27-52.

Gal AB, Dudley LM (2003). Phosphorus availability under continuous point source irrigation. Soil Sci. Soc. Am. J. 67:1449-1456.

Hinningsen KW (1970). Macro molecular physiology of plastids. J. Cell. Sci. 7:587-621.

Jat RS, Kumar M (2002). Vermicompost – A way towards organic revolution. Farmer's Forum. 2(4):12-14.

Katyal JC (1989). Fertilizer use and impact on environment. Proceedings of national seminar on fertilizer, agriculture and national economy held at New Delhi. pp. 1-8.

Khandkar UR, Nigam KB (1996). Effect of farmyard manure and fertility level on growth and yield of ginger (Zingiber officinale). Indian J. Agric. Sci. 66(9):549-550.

Lowry OH, Brought LA, Farr RJ, Randall (1951). Protein measurement with folin, phenol measurement with folin phenol reagent. J. Biol. Chem. 193:265-275.

Mahadevan VC (1988). Effect of foliar nutrition of NPK on banana cv. Nendran (AAB). MSc Thesis, Tamil Nadu Agricultural University, Coimbatore, Tamil Nadu, India.

Mathivanan N, Surendiran G, Srinivasan K, Sagadevan E, Malarvizhi K (2005). Review on the current scenario of Noni research: Taxonomy, distribution, chemistry, medicinal and therapeutic values of Morinda citrifolia. Intl. J. Noni Res. 1(1):1-16.

Mukhmudov HA (1983). A study on chlorophyll formation in wheat leaves under moisture stress. Fld. Crop. Abst. 39(3):1753.

Nelliat EV, Bavappa KV, Nair KR (1974). Multistoryed cropping - a new direction in multiple cropping for coconut. World Crops. 26(6):223-226.

Nogle F (1979). Introductory plant physiology. Prentice Hall of India Pvt. Ltd. New Delhi.

Okon Y, Kapulnik Y (1986). Development and function of Azospirillum inoculated roots. Plant Soil. 90:3-16.

Padmanabhan K (2003). Effect of organic manures on growth, root yield and alkaloid content of ashwagandha (Withania somnifera (L.) Dunal) 'Jawahar'. MSc Theis, Tamil Nadu Agricultural University, Coimbatore, Tamil Nadu, India.

Padmapriya S (2004). Studies on effect of shade, inorganic, organic and bio fertilizers on growth, yield and quality of turmeric (Curcuma longa L.) genotype CL 147. PhD Thesis, Tamil Nadu Agricultural University, Coimbatore, Tamil Nadu, India.

Panse VG, Sukhatme PV (1985). Statistical methods for agricultural workers, Indian Council of Agricultural Research, New Delhi.

Pearce RB, Brown RH, Blaster RE (1968). Photosynthesis of alfalfa leaves as influenced by age and environment. Crop Sci. 8:677-680.

Prakash K (2010). Studies on influence of drip irrigation regimes and fertigation levels on mango var. Alphonso under ultra high density planting. MSc Theis, Tamil Nadu Agricultural University, Coimbatore, Tamil Nadu, India.

Raghavendra AS (2000). Photosynthesis - A comprehensive treatise, Cambridge University Press, London.

Rao NS, Bhatt RM, Sadashiva AT (2000). Tolerance of water stress in tomato cultivars. Pho. Synth. 38(3): 465-467.

Raskin (1992). Salicylate - a new plant hormone. Plant Physiol. 99:799-803.

Rethinam P, Sivaraman K (2007). Noni (Morinda citrifolia L) the miracle fruit - A holistic review. Intl. J. Noni Res. 2(1-2):4-37.

Sachdev P, Debe DL, Rastogi DK (1987). Effect of varying levels of zinc and molybdenum on plant constituents and enzyme activity at different growth stages of wheat. J. Nuclear Agric. Bio. 16(4):187-196.

Scandalios JG (1993). Oxygen stress and super oxide dismutase. Plant Physiol. 1010:7-12.

Shoba K (2009). Biomass production, partitioning and root architecture in black nightshade (Solanum nigrum L.) as influenced by nutrigation. PhD Thesis, Tamil Nadu Agricultural University, Coimbatore, Tamil Nadu, India.

Singh HN, Narda K, Chawla JK (2004). Efficacy of phosphorus through trickle fertigation of potato (Solanum tuberosum). Indian J. of Agri. Sci. 74(8):476-478.

Sinha SK, Nicholas DJ (1981). Physiology and biochemistry of drought resistance. Academic Press, Sydney, Australia. pp. 145-168.

Suanez IP, Hsiao TC, Fereres E, Henderson DW (1989). Water stress effects on carbon exchange rates of three upland cotton (Gossipium hirsutum) cultivars in the field. Field Crop Res. 21:85-89.

Umesha K, Soumya SP, Smitha GR, Sreeramu BS (2011). Influence of organic manures on growth, yield and quality of makoi (Solanum nigrum L.). Indian J. Hort. 68(2):235-239.

Wallace DH, Ozbun JL, Mungar HM (1972). Physiological genetics of crop yield. Adv. Agron. 41:171-190.

Wang MY, West B, Jensen CJ, Nowicki D, Su C, Palu AK, Anderson G (2002). Morinda citrifolia (Noni): a literature review and recent advances in noni research. Acta Pharmacologica Sinica. 23:1127-1141.

Williams RF (1946). The physiology of plant growth with special reference to the concept of net assimilation rate. Ann. Bot. 10:41-72.

Yoshida S, Forno DA, Cock JH (1971). Laboratory manual for physiological studies of rice. IRRI, Philippines. pp. 36-37.

Zaprometov MN (1989). The formation of the phenolic compounds in plant cell and tissue cultures and the possibility of its regulation. Adv. Cell Cult. 7:201-260.

11

Deficit irrigation scheduling for potato production in North Gondar, Ethiopia

Nigus Demelash

Gondar Agricultural Research Center, P. O. Box 1337, Ethiopia.

Scarcity of water is the most severe constraint for development of agriculture in arid and semi-arid areas. Under this condition, irrigation management has to be improved while still achieving high yield. This study was conducted in North Gondar zone in 2010/2011 to investigate deficit irrigation scheduling on potato yield and yield components. Nine treatment combination: Irrigation frequency based on Cropwat model output, two modified irrigation frequency and two deficit irrigations of 25 and 50% were laid out with three replications in randomized complete block design (RCBD) in factorial combinations. It was found that irrigation scheduling significantly affected yield and some yield components. The highest marketable tuber yields was obtained from 0% deficit irrigation and frequency of F1-2 days (T9) which was 26.33 t ha^{-1}, whereas 25% deficit irrigation with F1-2 days frequency (T8) gave 25.68 t ha^{-1}. The lowest marketable yield was 3.4 t ha^{-1} from T4. The highest water use efficiency (WUE) were obtained from T7 and T8 which were 6.61 and 5.59 kg mm^{-1}, respectively. Therefore, applying 75% of full irrigation depth throughout the whole growing season with frequency based on Cropwat model scheduling -2 days resulted better potato yield and saved significant depth of water which improved WUE.

Key words: Potato, deficit irrigation scheduling, water use efficiency, water saving.

INTRODUCTION

Irrigation is an age-old art-perhaps as old as civilization. Nevertheless, the increasing need for crop production due to the growing population in the world is necessitating a rapid expansion of irrigated agriculture throughout the world. As population increases and development calls for, the allocations of ground and surface water for the domestic, agriculture and industrial sectors increased; as a result the pressure on water resources intensifies. The increasing stress on freshwater resources brought about by ever rising demand of water is of serious concern (FAO, 2008). Despite the increase in water use by sectors other than agriculture, irrigation continues to be the main water user on a global scale. Irrigated agriculture consumes more than 70% of the water drawn from the rivers of the world and for the developing world; the proportion can reach 80% (FAO, 2002). In addition, the demand on the limited finite water supplies is increasing from time to time. Different economic sectors other than agriculture such as, hydroelectric production, industries, fishery, recreation or tourism, river, or lake navigation etc., also depend on water. Further, the trend for maintaining the natural river flows aimed at maintaining the natural environment has been increasing implying that the water that could be be abstracted for irrigation has to be regulated. Improving the efficiency of water use will contribute to saving water for irrigation development, thus, improving the productivity of irrigation, and to the expansion of other water dependant economic sectors.

The situation is no more different in Ethiopia. It has been clearly and loudly stated that if Ethiopia is to feed its ever increasing population, lessen risk of catastrophes caused by drought, and increase population density in the arid and sparsely populated areas, continuous and extensive effort need to be made towards developing irrigated agriculture and intensifying agricultural production.

For a country like Ethiopia that follows Agricultural Development Lead Industrialization (ADLI), there is no readily identifiable yield increasing technology other than improved seed-irrigation, fertilizer approach. Irrigation will, therefore, play an increasingly important role now and in the future both to increase the yield from already cultivated land and to permit the cultivation of what is today called marginal or unusable land due to moisture deficiency (FAO, 2002). In addition, production intensification without irrigation in the face of vagaries of weather cannot be imagined.

Moreover, under the traditional rain dependent agricultural system, improving crop management under rainfed condition could not be expected to provide a reliable output satisfying the ever increasing demands for food. Therefore, use of irrigated agriculture together with other productivity enhancing technologies such as deficit irrigation is the way out to ensure productivity which can meet the growing demand for agricultural produces.

The Federal and Regional Governments in Ethiopia have given due emphasis to irrigated agriculture to ensure food self-sufficiency. For instance, the Amhara Regional State has many irrigation development projects undergoing within the region. The Koga, Rib and Megech projects are among those in the region. In most irrigable lands, horticultural crops play an important role in contributing to the household food security. The horticultural crops such as garlic, onion, carrot being cash crop with nutritional value generate income for the poor households. Higher profits can be achieved by increasing productivity and production throughout the year when efficient irrigation system is used.

With increasing municipal and industrial demands for water, its allocation for agriculture is decreasing steadily. The major agricultural use of water is for irrigation, which, thus, is affected by decreased supply. Therefore, innovations are needed to increase the efficiency of use of the water that is available. There are several possible approaches. Irrigation technologies and irrigation scheduling may be adapted for more-effective and rational uses of limited supplies of water. It is necessary to develop new irrigation scheduling approaches, not necessarily based on full crop water requirement, but once designed to ensure the optimal use of allocated water with deficit irrigation scheduling. Deficit (or regulated deficit) irrigation

is one way of maximizing water use efficiency (WUE) for higher yields per unit of irrigation water applied: the crop is exposed to a certain level of water stress either during a particular period or throughout the whole growing season. The expectation is that any yield reduction will be insignificant compared with the benefits gained through diverting the saved water to irrigate additional land. The grower must have prior knowledge of crop yield responses to deficit irrigation (Shock et al., 1998).

Many researchers have evaluated the feasibility of deficit irrigation and whether significant savings in irrigation water are possible without significant yield penalties (Shock et al., 1998; Zhang et al., 2006; Shahnazari et al., 2007; Serhat and Abdurrahim, 2009). Similar works on potato (Trebejo and Midmore, 1990; Minhas and Bansal, 1991) and on many other crops have demonstrated the possibility of achieving optimum crop yields under deficit irrigation practices by allowing a certain level of yield loss from a given crop with higher returns gained from the diversion of water for irrigation of other crops. This new concept of irrigation scheduling has different names, such as regulated deficit irrigation, pre-planned deficit evapotranspiration, and deficit irrigation (English et al., 1990).

In North Gondar zone, Megech irrigation area, irrigation is typically applied on a routine basis without scheduling and inadequate management of irrigation water has been an important limiting factor to potato production. The farmers generally lack knowledge on aspects of soil-water-plant relationship and they apply water to the crop regardless of the plant needs. They seem to relate irrigation occurrence to days after planting with fixed intervals and water amounts rather than to crop stage progress. Therefore, irrigation scheduling based on deficit irrigation requires careful evaluation to ensure enhanced efficient use of increasingly scarce irrigation water in this area. The knowledge of proper irrigation scheduling, when to irrigate and how much water to apply, is essential to optimize crop production per unit water and for sustaining irrigated agriculture on permanent footing (Anac et al., 1999). In many deve-loping countries like Ethiopia, irrigation interval turns are mutually agreed and fixed among growers according to a pre-fixed schedule, this situation is no more different in North Gondar zone. However, this method does not give due consideration to crop water requirement, soil and water relations, yield responses, scarcity of water and climatic conditions. It is therefore, with this rationale that this study was conducted with objectives to determine optimum deficit irrigation depth and frequency relations, and to estimate WUE of potato for the study area.

MATERIALS AND METHODS

General description of the study area

The study area is located in the northwest part of Amhara National Regional State; North Gondar zone (12.50°N latitude and 37.24°E

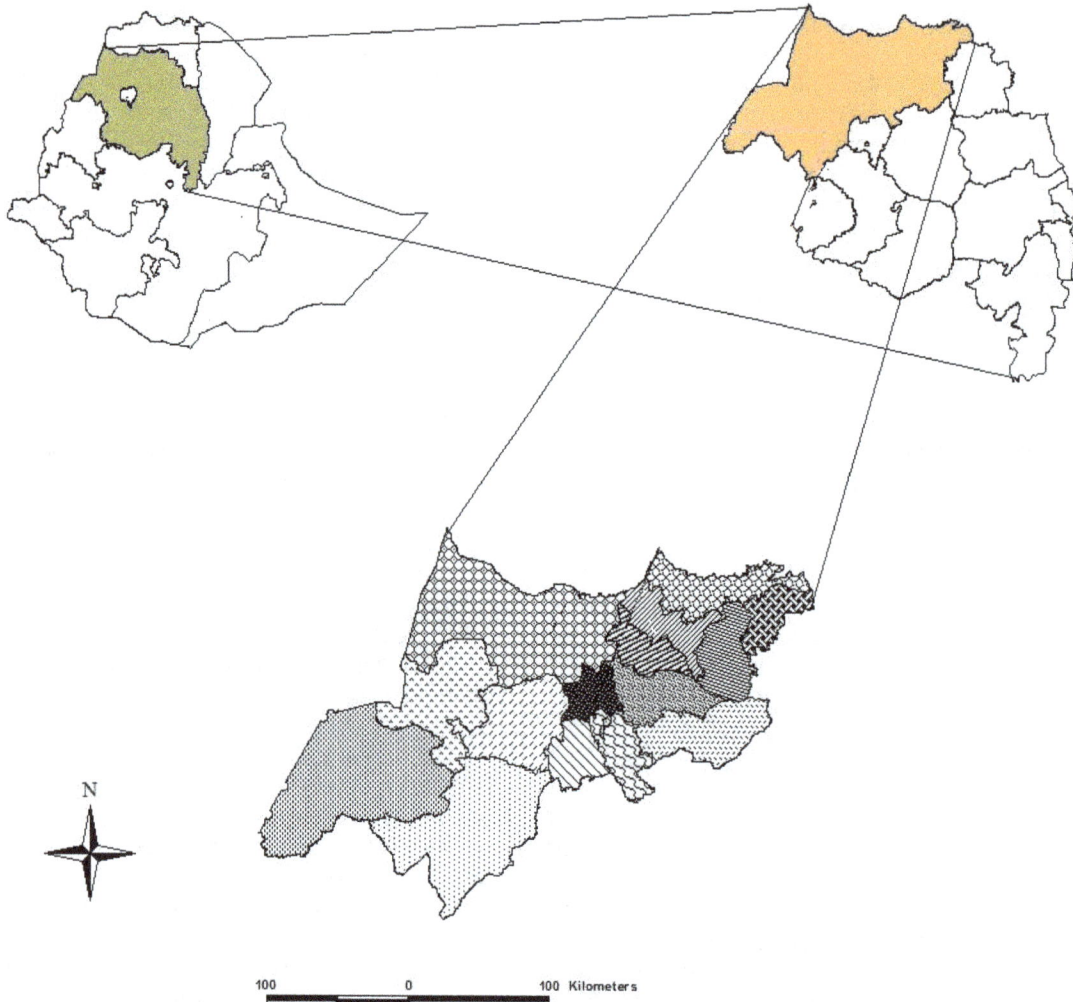

Figure 1. Location map of the study area.

longitude). The altitude of the experimental site is at about 2380 m above mean sea level (Figure 1). Rainfall is seasonal, varying in depth, space and time. The mean annual rainfall (1979 to 2009) in the area is about 1101 mm (Table 1) and it is erratic and uneven in distribution. The mean maximum temperature of the area is about 30.6°C, while the mean minimum temperature is about 12°C. The mean annual rainfall (1979 to 2009) in the area is about 1101 mm (Table 1) and it is erratic and uneven in distribution (NMSA, 2009).

Experimental design and treatments

The study was conducted from 07/12/2010 to 05/04/2011. The experiment involved a factorial combination of two deficit and one control irrigation depth and three irrigation frequencies laid out in randomized complete block design (RCBD) with three replications. The treatments combinations were:

T1 = F1L1 - (Irrigation frequency based on Cropwat model output and application of 50% deficit irrigation)
T2 = F1L2 - (Irrigation frequency based on Cropwat model output and application of 25% deficit irrigation)
T3 = F1L3 - (Irrigation frequency based on Cropwat model output and application of 0% deficit irrigation)

T4 = F2L1 - (Irrigation frequency based on Cropwat model output + 2 days and application of 50% deficit irrigation)
T5 = F2L2 - (Irrigation frequency based on Cropwat model output scheduling + 2 days and application of 25% deficit irrigation)
T6 = F2L3 - (Irrigation frequency based on Cropwat model output + 2 days and application of 0% deficit irrigation)
T7 = F3L1 - (Irrigation frequency based on Cropwat model -2 days and application of 50% deficit irrigation)
T8 = F3L2 - (Irrigation frequency based on Cropwat model -2 days and application of 25% deficit irrigation)
T9 = F3L3 - (Irrigation frequency based on Cropwat model -2 days and application of 0% deficit irrigation)

Well-sprouted potato tubers were planted on prepared ridges in rows with 70 cm spacing between rows and 30 cm between plants. Each plot consisted of four ridges and five furrows. Each furrow bed had 40 cm width at the bottom. The furrow had a parabolic shape with an average depth of 8 cm and width of 20 cm at the top. Fertilizer was applied on the prepared ridges in a band form and incorporated into the soil. The rate of fertilizer applied was 111 kg N ha^{-1} and 90 kg P$_2$O$_5$ ha^{-1}. Half of the rate of nitrogen (N) fertilizer and full rate of phosphorus (P) fertilizer were applied at sowing. The second half of the N fertilizer as urea was applied at 45 days after planting.

Table 1. Climate data and reference evapotranspiration at Gondar (1979 to 2009).

Month	Rainfall (mm)	Minimum temperature (°C)	Maximum temperature (°C)	Relative humidity (%)	Wind speed (m s^{-1})	Sunshine hour (h)	ETo (mm day^{-1})
January	5.0	19.4	27.6	36.2	1.4	9.5	4.64
February	4.5	19.9	30.6	33.5	1.4	9.0	5.17
March	18.2	21.7	29.7	34.7	1.6	8.1	5.61
April	35.9	17.7	29.6	37.9	1.5	7.7	5.48
May	87.3	15.7	28.3	48.6	1.7	7.0	4.84
June	151.2	14.1	25.3	67.1	1.7	4.5	3.72
July	297.3	13.3	22.7	79.2	1.1	4.4	3.14
August	278.5	13.2	22.8	79.4	1.1	5.0	3.30
September	119.1	12.8	24.9	70.1	1.2	7.0	3.90
October	73.3	12.7	26.2	55.4	1.1	7.6	3.93
November	22.4	12.2	27.0	46.9	1.1	8.8	3.83
December	8.7	18.6	27.2	42.1	1.0	9.0	4.12
Total	1101.4						
Average	91.78	15.9	26.8	52.5	1.3	7.9	4.31

Table 2. Crop input data of potato.

Parameter	Growth stage				Total growing period
	Initial	Development	Mid	Late	
Length of growing season (days)	25	30	35	30	120
Crop coefficient (Kc)	0.5	0.83	1.15	0.75	
Rooting Depths (m)	0.30	0.45	0.60	0.60	
Depletion level (p)	0.25	0.2	0.15	0.50	
Yield response factor (Ky)	0.45	0.80	0.80	0.30	

Source: FAO Cropwat model (Smith et al., 2002).

Determination of crop and irrigation water requirement of potato

Crop water requirement of potato was determined using the Cropwat model based on the climatic data of the area, the crop to be grown (potato). Input data for the model were obtained from the National Meteorological Services Agency, Soil laboratory results and FAO publications. Thirty (30) years (1979 to 2009) meteorological data was used to estimate crop water requirement and the data was obtained from Bahir dar National Meteorological Station. Calculations of water and irrigation requirements utilize inputs of climate, crop and soil data, as well as method of irrigation and rainfall data. Reference evapotranspiration was calculated from temperature, humidity, sunshine and wind speed data, according to the FAO Penman-Monteith method (Allen et al., 1998).

As per the output of the model, the optimum seasonal irrigation requirement was found to be 517.46 mm. As per the Cropwat program, the anticipated effective rainfall during the growing season was about 23.17 mm. However, there was no rainfall during the experiment period. Analysis of monthly reference evapotranspiration (ETo) calculated from meteorological data of 30 years data (1979 to 2009) meteorological station shows that the minimum reference ETo occurred in July (3.14 mm day^{-1}) and the maximum in March (5.61 mm day^{-1}) (Table 2).

The soil data include information on total available soil water content and maximum infiltration rate. These were determined by gravimetric method and double ring infiltrometer, respectively. The initial soil water content at the start of the season was also needed and was determined to be 156 mm m^{-1} of soil depth. Through the soil moisture content and evapotranspiration rates, the soil water balance was calculated on weekly bases by the Cropwat model.

Soil analysis

In the laboratory, soil samples were analyzed for bulk density, soil moisture, field capacity, permanent wilting point, soil texture, and soil pH at the Adet Agricultural Research Centre.

Soil texture of the field was determined in the laboratory using Hydrometer method. Soil bulk density was determined from undisturbed soil sample taken using a metal cylinder (core sampler) of known volume (100 cm^3) that was driven into the soil of desired depth and calculated as the ratio of oven dry weight of soil to a known cylinder core sampler volume. Since bulk density varies considerably spatially, the measurements were taken at three different soil depths of the soil profile and three samples across the experimental site. The gravimetric method was used to determine the soil moisture content and calculated as a dry weighed fraction (Michael, 2008).

The water content at field capacity was determined in the laboratory by using a pressure (porous) plate apparatus by applying

Table 3. Irrigation depth and number of irrigation under each treatment.

Treatment	Number of irrigation (days)	Net irrigation (mm)
T1	17	238.0
T2	17	356.5
T3	17	475.0 (optimum)
T4	14	183.4
T5	14	274.5
T6	14	365.8
T7	25	346.6
T8	25	507.3
T9	25	664.3

-1/3 bar to a saturated soil sample. When water is no longer leaving the soil sample, the soil moisture was taken as field capacity. Permanent wilting point was also determined by using pressure membrane apparatus by applying -15 bar to a saturated soil sample. When water is no longer leaving the soil sample, the soil moisture was taken as permanent wilting point. Soil pH was determined from saturation pest extract using pH meter.

Plant and yield parameters measured

Data were recorded on plot basis and extrapolated in hectare basis. All parameters were determined and calculated from the middle two rows. Plant height, number of stems, days to physiological maturity, average number of tubers per plant, marketable tuber, unmarketable tuber yield and total tuber yield were recorded. Irrigation IWUE was computed as the ratio of the yield (kg ha^{-1}) to the depth of irrigation applied (m^3) (Michael, 2008).

Depth of irrigation under the different treatments

All treatments were set according to the initially planned framework and received the required irrigation depth and frequency (Table 4). Table 3 shows the net depths of water applied to each treatment and the number of irrigation in the experiment period. It is to be stressed that there was no rainfall during the experiment period.

Data analysis

The data collected from the field study were subjected for analysis of variance (ANOVA) using SAS software. Whenever treatment effects were significant, mean comparison were made using least significant difference (LSD).

RESULTS AND DISCUSSION

Characterization of the experimental soil

The result of the textural analysis of the soil from the experiment site (Table 5) showed that the composition of clay, silt and sand percentage were 33.76, 38.62 and 27.62, respectively. Thus, as per the USDA soil textural classification, the soil was classified as clay loam soil.

The bulk density of soil of the area showed variation with depth (Table 6). It varied between 1.19 and 1.33 g

cm^{-3} and generally the surface soil had slightly lower bulk density than the subsurface soil. The soil had an average bulk density of 1.24 g cm^{-3}. The soil pH of the experimental field also varied with depth. The pH of the experimental site ranged from 5.57 to 6.35 in the 0 to 60 cm depth. The average pH of the soil was 6.07 which showed that the soil of the site was suitable for potato crop production with regard to soil pH.

The water content at field capacity and permanent wilting point of the soil were determined to be 32.21 and 19.52%, respectively. The moisture content at field capacity varied with depth between 35.63 and 29.10% on mass basis. The top 0 to 20 cm had a larger average water content of field capacity value of 35.63%, while the subsurface 40 to 60 cm had a lower value of field capacity that was 29.10%. The moisture content at permanent wilting point also showed variation with depth with values of 21.56% at the top (0 to 20 cm) and at the subsurface (40 to 60 cm) the value of permanent wilting point was 17.89%.

Total available water (TAW) which is the depth of water that a crop can extract from its root zone is directly related to variation in field capacity and permanent wilting point. The total average available soil moisture was 156.67 mm m^{-1} of soil depth and the maximum infiltration rate of the soil was 40 mm h^{-1}. As a result, high value of TAW is found in top soil; whereas lower values are found in the subsurface soil (Table 6).

Agronomic response of potato

Plant height

Plant height was significantly (P < 0.05) influenced by variation in the depth and frequency of water application. Irrigation depth × irrigation frequency interaction was found to be significant with respect to plant height (Table 7). T9 which received the maximum depth irrigation with frequency of F1-2 days of irrigation water and T8 which received the 75% of full irrigation depth with frequency of F1-2 days of irrigation water had the highest plant height. Whereas the most stressed crops that is, T4 had the

Table 4. Irrigation scheduling of potato throughout the growing season.

Treatment 1		Treatment 2		Treatment 3		Treatment 4		Treatment 5		Treatment 6		Treatment 7		Treatment 8		Treatment 9	
F1	L 1	F 1	L2	F 1	L3	F2	L 1	F 2	L 2	F 2	L 3	F3	L1	F3	L 2	F3	L 3
7	7.2	7	10.7	7	14.3	9	7.2	9	10.7	9	14.3	5	7.2	5	10.7	5	14.3
8	8.8	8	13.1	8	17.5	10	8.8	10	13.1	10	17.5	5	7.2	5	10.7	5	14.3
8	9.5	8	14.2	8	18.9	10	9.5	10	14.2	10	18.9	6	8.8	6	13.1	6	17.5
7	11.7	7	17.5	7	23.3	9	11.7	9	17.5	9	23.3	6	8.8	6	13.1	6	17.5
6	12.1	6	18.2	6	24.2	8	12.1	8	18.2	8	24.2	6	9.5	6	14.2	6	17.5
6	12.8	6	19.1	6	25.5	8	12.8	8	19.1	8	25.5	7	11.7	7	17.5	7	23.3
5	15.1	5	22.6	5	30.1	7	15.1	7	22.6	7	30.1	7	11.7	7	17.5	7	23.3
5	14.2	5	21.3	5	28.4	7	14.2	7	21.3	7	28.4	5	12.1	5	18.2	5	24.2
5	14.7	5	22.1	5	29.4	7	14.7	7	22.1	7	29.4	5	12.1	5	18.2	5	24.2
5	15.0	5	22.5	5	30.0	7	15.0	7	22.5	7	30.0	4	12.8	4	19.1	4	25.5
5	15.3	5	22.9	5	30.5	7	15.3	7	22.9	7	30.5	4	15.1	4	19.1	4	25.5
5	15.5	5	23.2	5	30.9	7	15.5	7	23.2	7	30.9	4	16.1	4	22.6	4	25.5
5	14.7	5	22.0	5	29.3	7	14.7	7	22.0	7	29.3	3	14.2	3	21.3	3	28.4
5	16.8	5	25.1	5	33.5	7	16.8	7	25.1	7	33.5	3	14.7	3	22.1	3	28.4
7	15.9	7	23.9	7	31.8	-	-	-	-	-	-	3	14.7	3	22.5	3	28.4
8	17.5	8	26.3	8	35.0	-	-	-	-	-	-	3	15.0	3	22.9	3	28.4
10	21.2	10	31.8	10	42.4	-	-	-	-	-	-	3	15.3	3	23.2	3	28.4
-	-	-	-	-	-	-	-	-	-	-	-	3	16.5	3	22	3	28.4
-	-	-	-	-	-	-	-	-	-	-	-	3	14.7	3	22	3	28.4
-	-	-	-	-	-	-	-	-	-	-	-	4	16.8	4	25.1	4	33.5
-	-	-	-	-	-	-	-	-	-	-	-	4	18.6	4	23.9	4	33.5
-	-	-	-	-	-	-	-	-	-	-	-	4	15.9	4	23.9	4	33.5
-	-	-	-	-	-	-	-	-	-	-	-	5	17.5	5	26.3	5	35.0
-	-	-	-	-	-	-	-	-	-	-	-	5	18.5	5	26.3	5	35.0
-	-	-	-	-	-	-	-	-	-	-	-	8	21.2	8	31.8	8	42.4

L, Irrigation depth in mm; F, irrigation frequency in days.

Table 5. Particle size distribution.

Soil depth (cm)	Particle size distribution (%)		
	Clay	Silt	Sand
0 - 20	36.07	35.09	28.84
20 - 40	33.23	38.62	28.15
40 - 60	32.12	42.16	25.72
Average	33.76	38.62	27.62

Table 6. Results of laboratory analysis for samples from the experimental site.

Soil depth (cm)	Bulk density (g cm^{-3})	pH	FC (%)	PWP (%)	TAW (mm m^{-1})
0 - 20	1.19	5.57	35.63	21.56	167.43
20 - 40	1.20	6.29	31.89	19.10	153.48
40 - 60	1.33	6.35	29.10	17.89	149.09
Total average	1.24	6.07	32.21	19.52	156.67

shortest plant height (Table 7). Stressing the crop by 25% deficit throughout the growing season with frequency of F1-2 days had a comparable plant height to 0% deficit with the same frequency. As shown in Table 7, plant

Table 7. The influence of full and deficit irrigation on plant height (cm) of potato.

Irrigation frequency (F)	Irrigation levels (L)			Mean (F)
	L1	L2	L3	
F1	41.0^{ef}	43.0^{de}	47.3^{b}	43.8^{b}
F2	40.3^{f}	42.3^{def}	44.0^{dc}	42.2^{c}
F3	46.3^{bc}	57.0^{a}	58.6^{a}	55.3^{a}
Mean (L)	42.5^{c}	47.4^{b}	51.3^{a}	
LSD (0.05) for mean F			1.26	
SE±			1.04	
LSD (0.05) for mean L			1.26	
SE±			1.04	
LSD (0.05) for F × L		2.33		
SE±		1.1		

Means followed by the same letter within a column or row are not statistically significantly different at 5% level of significance.

Table 8. The influence of full and deficit irrigation on stem number, days to maturity, average tuber number per hill and unmarketable tuber yield of potato.

Treatment	Number of stems per hill	Days to maturity	Average tuber number per hill	Unmarketable yield (t ha^{-1})
T1	4.01	107	4.8^{f}	3.74^{e}
T2	2.89	109	5.6^{cde}	4.56^{de}
T3	4.41	110	6.5^{cd}	5.87^{b}
T4	3.57	107	4.0^{f}	3.25^{f}
T5	3.89	107	5.4^{ef}	2.86^{g}
T6	4.07	109	5.8^{cde}	4.46^{c}
T7	3.7	111	6.7^{c}	4.17^{cd}
T8	4.3	110	8.2^{b}	2.68^{g}
T9	4.5	116	10.5^{a}	8.12^{a}
LSD (0.05)	NS	NS	1.59	0.42
CV (%)	8.3	1.9	9.5	5.7
SE±	0.50	3.3	0.56	0.2

Means followed by the same letter within a column or row are not statistically significantly different at 5% level of significance.

height increased consistently from 40.33 cm for an irrigation depth of 50% of full irrigation depth and F1+2 frequency of application of water requirement to 62.6 cm to that of full irrigation and F1-2 days irrigation frequency of application.

The main effect of irrigation depth and frequency on plant height was also showed significant (p < 0.05) difference (Table 7). Thus, it is possible to conclude from the present observation that irrigation frequency is helpful in amplifying the effect of irrigation depth on plant height.

Stem number

Both irrigation depth and frequency showed non-significant difference in stem number (Table 8). Despite the fact that stem number is one of the most important yield components in potato, the results of the present study showed that the influences of irrigation depth and frequency on stem number were statistically non-significant. This could be due to the fact that stem number is determined very early in the ontogeny of the plant. It could also be due to the case that this trait was not influenced much by irrigation depth and frequency unless the depth of irrigation is too low.

Days to physiological maturity

Data on days to physiological maturity of potato was influenced by variation in the moisture level of the soil. However, there was no statistically significant difference in different levels of irrigation water on days to maturity between treatments T1 to T8 (Table 8). However, Treatment 9 with the maximum depth and frequency of irrigation application experienced a prolonged time to

Table 9. The influence of full and deficit irrigation on marketable yield (t ha^{-1}) of potato.

Irrigation frequency (F)	Irrigation levels (L)			Mean (F)
	L1	L2	L3	
F1	7.96^e	15.12^d	20.81^b	14.63^b
F2	3.40^f	3.98^f	8.16^e	5.18^c
F3	18.74^c	25.68^a	26.33^a	22.58^a
Mean (L)	9.03^c	14.93^b	18.43^a	
LSD (0.05) for mean F			0.61	
SE±			0.50	
LSD (0.05) for mean L			0.61	
SE±			0.50	
LSD (0.05) for F × L		1.13		
SE±		0.54		

Means followed by the same letter within a column or row are not statistically significantly different at 5% level of significance.

maturity. Similar result was also observed by Marouelli and Silva (2007).

Average tuber number per hill

The average tuber number results obtained are presented in Table 8. There was statistically significant (P < 0.05) difference in average number of tuber among the treatment. The highest numbers of tubers were recorded from treatment T9. However, there was a considerable size variation in this treatment. Extra large and too small tuber sizes were observed. The lowest average number of tubers was observed for T4 indicating that there was considerable variation in the number of tubers among the treatments.

There were significant (P < 0.05) responses to application irrigation depth and frequency in actual tuber numbers (Table 8). The highest average number of tuber (10.46) was found in treatment T9. The lowest number (4.47) was found in the treatment T4. Application of optimum irrigation at T8 gave relatively better tuber size with relatively higher tuber number as compared with the treatments. T3 and T7 also showed better performance in tuber size and number.

Marketable tuber yield

The difference observed in marketable tuber yield between treatments was statistically significant (p < 0.05). However, there was no statistically significant difference in marketable tuber yield between T9 and T8 which gave 26.33 and 25.68 t ha^{-1}, respectively. The lowest marketable yield was observed in T4, which was 3.4 t ha^{-1} (Table 9).

Decreasing irrigation depth from 100% application to 75% application with frequency of F1-2 days throughout the growing season decreased marketable tuber yield from 26.33 to 25.68 t ha^{-1} (Table 9). The difference in marketable yield was only 0.65 t ha^{-1}, which was statistically non significant. However, increasing irrigation depth from 50% application with irrigation frequency of F1+2 days to 100% application with F1-2 days application frequency and 75% application with F1-2 days application frequency increased marketable tuber yield from 3.40 to 26.33 t ha^{-1} and 25.68 t ha^{-1}, respectively (Table 9).

The yield increment due to irrigation depth and irrigation frequency was statistically significant (p < 0.05) between T4 and T9, and between T4 and T8. This result is similar to other study (Serhat and Abdurrahim, 2009). Potato needs frequent-irrigation for a good growth and yield. Yield is considerably affected by storage quality; disease resistance, the time, depth and frequency of irrigation. Water excess in the soil decreases the oxygen diffusion rate in the root zone (Wan and Kang, 2006) affecting crop yield increment.

As shown from Table 9, the interaction effect of irrigation depth (L) with irrigation application frequency (F) on marketable tuber yield was found to be significant (p < 0.05). The main effect of irrigation depth and frequency on marketable tuber yield was also statistically significant (p < 0.05). Improper irrigation depth and frequency can substantially reduce yields by increasing the proportion of rough, misshapen tubers and smaller sized tubers.

Widely fluctuating soil water contents and/or too frequent application of irrigation with maximum irrigation depth creates the greatest opportunity for developing these tubers defects results in increasing unmarketable tubers. Growth cracks are also associated with wide fluctuations in soil water availability and corresponding changes in tuber turgidity and volume of internal tissues (King and Stark, 1997).

Table 10. The influence full and deficit irrigation on total tuber yield (t/ha) of potato.

Irrigation frequency (F)	Irrigation levels (L)			Mean (F)
	L1	L2	L3	
F1	11.70[f]	19.68[e]	26.67[c]	19.35[b]
F2	6.64[g]	6.84[g]	12.62[f]	8.70[c]
F3	22.91[d]	28.36[b]	34.45[a]	27.57[a]
Mean (L)	12.75[c]	18.29[b]	24.58[a]	
LSD (0.05) for mean F			0.72	
SE±			0.59	
LSD (0.05) for mean L			0.72	
SE±			0.59	
LSD (0.05) for F × L		1.33		
SE±		0.63		

Means followed by the same letter within a column or row are not statistically significantly different at 5% level of significance.

Unmarketable tuber yield

As shown in Table 8, significant ($p < 0.05$) difference in unmarketable tuber yield was observed among the treatments. Similarly, there was statistically significant ($p < 0.05$) difference in unmarketable tuber yield between T9 and T8 which was 8.12 and 2.67 t ha[-1], respectively. This result seemed to suggest that unmarketable tuber yield can be controlled more effectively by irrigation depth and frequency management. The unmarketable tuber yield was significantly reduced from 8.12 to 2.67 t ha[-1]. Decreasing irrigation depth from 100% application to 75% application with the same irrigation frequency of F1-2 days throughout the growing season reduced unmarketable tuber yield from 8.12 to 2.67 t ha[-1], which was 5.45 t ha[-1]. The result showed in Table 8 that, improper irrigation depth and frequency substantially reduce yields by increasing the proportion of rough, misshapen tubers. Widely fluctuating soil water contents create the greatest opportunity for developing these tubers defects (Serhat and Abdurrahim, 2009; King and Stark, 1997).

Total tuber yield

Total potato tuber yield remarkably increased with the application of water. Application of optimum irrigation scheduling significantly increased total tuber yield (Table 10). Applying the right depth of irrigation and frequency of application increased the total tuber yield of potato production. This result is comparable to those obtained in other studies (Faberio et al., 2001; Ferreira and Carr, 2002). The highest total tuber yield (34.45 t ha[-1]) was recorded in the treatment T9. The lowest total tuber yield (11.70 t ha[-1]) on the other hand was recorded in the treatment T4. Total potato tuber production in the experiment was proportional to the availability of water but as stress intensity increased total tuber yield decreased (Table 10). Total tuber yield was affected severely when stress was imposed throughout the growing season with low irrigation frequency.

As depicted in Table 10, the yield response of potato to variations in soil moisture was statistically highly significant ($P < 0.05$). This result is also in agreement with the finding of Ferreira and Carr (2002). Treatments 9 and 8 with maximum and optimum depth of irrigation water and frequency of application resulted in highest total yield. Whereas the stressed plant with minimum irrigation depth and frequency resulted in lowest yield.

As shown in Table 10, the L and F interaction effect was highly significant in increasing total tuber yield. The result was evident that as the level of irrigation water depth and time of application increases towards the optimum value, tuber yield also increased. Similarly, increasing the level of frequency highly significantly increased total tuber yield from 13.8 to 24.9 t ha[-1].

Above ground biomass

The ANOVA showed that irrigation scheduling significantly affected the above ground biomass ($p < 0.05$). As shown in Table 11, the irrigation depth and frequency interaction effect was significant ($p < 0.05$) in increasing above ground biomass from 14.54 to 46.19 t ha[-1] as the depth of irrigation increased from 50% deficit irrigation to 0% deficit irrigation. The positive response of above ground biomass with increasing frequency of irrigation and the level of irrigation depth could be attributed to vegetative growth. The results also revealed that interaction of irrigation depth and irrigation frequency was statistically significant ($p < 0.05$) on above ground biomass (Table 11). The highest biomass was obtained when 0% deficit irrigation in combination with F1-2 days irrigation frequency (T9). However, the difference between these treatments was not statistically significant. It is also interesting to note that the least above ground

Table 11. The influence of full and deficit irrigation on biomass yield of potato

Irrigation frequency (F)	Irrigation levels (L)			Mean (F)
	L1	L2	L3	
F1	14.54[e]	24.56[d]	32.17[b]	23.76[b]
F2	8.75[g]	9.45[f]	15.63[e]	11.28[c]
F3	29.68[c]	34.10[b]	46.19[a]	36.66[a]
Mean (L)	17.66[c]	22.70[b]	31.33[a]	
LSD (0.05) for mean F			1.27	
SE±			1.04	
LSD (0.05) for mean L			1.27	
SE±			1.04	
LSD (0.05) for F × L		2.32		
SE±		1.1		

Means followed by the same letter within a column or row are not statistically significantly different at 5% level of significance.

Table 12. WUE of potato.

Treatment	Total tuber yield (kg ha^{-1})	Net irrigation (mm)	WUE (kg mm^{-1})	Water Saved[1] (mm)	Water saved[2] (mm)
T1	1170	238.0	4.92	426.8	237.5
T2	1968	356.5	5.52	308.1	118.7
T3	2667	475.0	5.61	189.3	0
T4	664	183.4	3.62	481.0	291.6
T5	684	274.5	2.49	389.8	200.5
T6	1262	365.8	3.45	298.5	109.2
T7	2291	346.7	6.61	317.7	128.4
T8	2836	507.3	5.59	157.0	-32.3
T9	3445	664.3	5.19	0	-189.3

Water Saved[1] = Saved water in relative to the maximum water depth (T9), Water saved[2] = Saved water in relative to the Cropwat estimate (T3).

biomass was recorded with the lowest irrigation amount and frequency.

Water use efficiencies

The result of this study showed that WUE was variable depending on the treatments applications (Table 12). The highest WUE was obtained from T7 and T8 which was 6.61 and 5.59 kg mm^{-1}, respectively. However, T8 (25.68 t ha^{-1}) gave highest marketable yield than T7, and 18.74 t ha^{-1}, respectively. The lowest value was obtained from T5 and T4. These results are similar to that reported by Onder et al. (2005).

Applying irrigation water 75% of full irrigation depth throughout the whole growing season with frequency of F1-2 days improved WUE. The proper application of deficit irrigation practices can generate significant savings in irrigation water allocation without significant yield reduction. The difference in marketable yield between T8 and T9 was only 0.65 t ha^{-1}, which was statistically insignificant in terms of yield change. However, significant depth of water was saved (Table 12).

The results in Table 12 showed that significant depth of water was saved without significant yield reduction. 157.0 mm water was saved in T8 relative to T9. Hence, diverting the saved water to increase the area irrigated may compensate for decreases in crop yields.Experimental results from field trials confirmed that with deficit irrigation strategies it is possible to increase WUE and save water for irrigation at T6 and T7. This could be especially important for areas facing with drought and limited water resources for agricultural production.Compared with full irrigation, the deficit irrigation treatments saved significant depth of water to irrigation, leading to increase of WUE. Similar data were obtained by other authors (Liu et al., 2005; Shahnazari et al., 2007). Water productivity increases under deficit irrigation, relative to its value under full irrigation, as shown experimentally for many crops (Fan et al., 2005; Zwart and Bastiaanssen, 2004).

Conclusions

One of the irrigation management practices which could

result in water saving is deficit irrigation. This experiment was conducted to study the effects of different irrigation levels and irrigation frequencies on yield and water productivity of potato (*Solanum tuberosum* L.) in North Gondar zone, Ethiopia. This study results confirmed that with deficit irrigation strategies it is possible to increase yield, water productivity and save significant depth of water for irrigation. The highest marketable yield was obtained from T9 which was 26.33 t ha^{-1} and T8 gave 25.68 t ha^{-1}. The lowest marketable yield was in T4 (3.4 t ha^{-1}). Since, the yield difference was between T8 and T9 was insignificant, 25% deficit irrigation with frequency based on Cropwat model scheduling -2 days irrigation interval was suitable to recommend. High WUE values were obtained from in T7 (6.61 kg mm^{-1}) and T8 (5.59 kg mm^{-1}). However, T8 gave higher marketable yield than T7 which were 25.68 and 18.74 t ha^{-1}, respectively. The lowest value was obtained from T5 and T4. Therefore, T8 (25% deficit irrigation and F1-2 days interval of irrigation application) was better in performance than T7 and the water productivity of potato in the study area was estimated as 4.54 to 5.06 kg m^{-3}.

Irrigation systems best suited to this task were capable of light, uniform, and frequent water applications which was 75% full irrigation and F1-2 days interval of irrigation application. Timing and depth of water application was determined as 75% of full irrigation in combination with F1-2 days irrigation frequency which was applied to minimize soil water fluctuations throughout the growing season. Potato needs frequent-irrigation for a good growth and yield and for high water productivity. It can, therefore, recommended that applying irrigation water 75% of full irrigation depth throughout the whole growing season of potato with frequency based on Cropwat model scheduling -2 days resulted better yield, saved significant depth of water and improved WUE which can be taken as optimum irrigation depth and frequency.

ACKNOWLEDGEMENTS

The author greatly appreciates the Amhara Region Agricultural Research Institute (ARARI), Gondar Agricultural Research Center (GARC) for granting the financial support for the research. All staff members of the GARC who offered unforgettable support are appreciated.

REFERENCES

Allen R, Pereira L, Raes D, Smith M (1998). Crop evapotranspiration: guidelines for computing crop water requirements. FAO Irrigation and Drainage. Rome, Italy: FAO. p. 56.

Anac MS, Ali UI, Tuzel IH, Anac D, Okur B, Hakerlerler H (1999). Optimum irrigation scheduling for cotton under deficit irrigation conditions. In: Kirda, C., Moutonnet, P., Hera, C. and Nielsen, D.R. (eds). Crop yield response to deficit irrigation, Dordrecht, The Netherlands, Kluwer Academic Publishers.

English MJ, Musick JT, Murty VV (1990). Deficit irrigation. In: Hoffman GJ, Towell TA, Solomon KH. (eds). Management of farm irrigation systems, St. Joseph, Michigan, United States of America, ASAE.

Faberio C, Olalla FMS, Juan JA (2001). Yield size of deficit irrigated potatoes. Agric. Water Manage. 48:255-266.

Fan T, Stewart BA, Payne WA, Wang Y, Song S, Luo J, Robinson CA (2005). Supplemental irrigation and water: yield relationships for plasticulture crops in the loess plateau of China. Agron. J. 97:177–188.

FAO (Food and Agriculture Organization) (2002). Crops and Drops. Land and Water Development Division. Rome, Italy.

FAO (Food and Agriculture Organization) (2008). Crops and Drops. Land and Water Development Division. Rome, Italy.

Ferreira TC, Carr MKV (2002). Response of potatoes (*Solanum tuberasum L.*) to irrigation and nitrogen in a hot, dry climate: I. water use. Field Crops Res. 78:51-64.

King BA, Stark JC (1997). Potato irrigation management for on-farm potato research. Potato research and extension proposals for cooperative action. University of Idaho, College of Agriculture. pp. 88-95.

Liu F, Jensen CR, Shahnazari A, Andersen MN, Jacobsen SE (2005). ABA regulated stomatal control and photosynthetic water use efficiency of potato (*Solanum tuberosum* L.) during progressive soil drying. Plant Sci. 168:831–836.

Marouelli W, Silva W (2007). Water tension thresholds for processing tomatoes under drip irrigation in Central Brazil. Irrigation Sci. 25:411–418.

Michael AM (2008). Irrigation Theory and Practice. Indian Agriculture Research Institute. New Delhi. India. pp. 427-429.

Minhas JS, Bansal KC (1991). Tuber yield in relation to water stress at stages of growth in potato (*Solanum tuberosum* L.). J. Indian Potato Assoc. 18:1-8.

NMSA (National Meteorological Service Agency) (2009). Climate and Agro-climate resource of Ethiopia. National Meteorological Service Agency of Ethiopia, Bahir dar.

Onder S, Caliskan ME Onder, Caliskan S (2005). Different irrigation methods and water stress effects on potato yield and yield components. Agric. Water Manage. 73:73–86.

Serhat A, Abdurrahim K (2009). Water-Yield Relationships in Deficit Irrigated Potato. J. Agric. Fac. Uludag University, Bursa.

Shahnazari A, Liu F, Andersen MN, Jacobsen SE, Jensen CR (2007). Effects of partial root-zone drying on yield, tuber size and water-use efficiency in potato under field conditions. Field Crop Res. 100:117-124.

Shock CC, Feibert EG, Saunders LD (1998). Potato yield and quality response to deficit irrigation. Horticultural Sci. 33:655-659.B

Smith M, Kivumbi D and Heng L K (2002). Use of the FAO CROPWAT model in deficit irrigation studies. In: Deficit irrigation practice. Water report . Rome, Italy. p. 22

Trebejo I, Midmore DJ (1990). Effects of water stress on potato growth, yield and water use in a hot and cool tropical climate. J. Agric. Sci. 114:321-334.

Wan S, Kang Y (2006). Effect of drip irrigation frecuency on radish (*Raphanus sativus* L.) growth and water use. Irrigation. Sci. 24:161-174.

Zhang B, Li FM, Huang G, Cheng ZY, Zhang Y (2006). Yield performance of spring wheat improved by regulated deficit irrigation in an arid area. Agric. Water Manage. 79:28–42.

Zwart SJ, Bastiaanssen WGM (2004). Review of measured crop water productivity values for irrigated wheat, rice, cotton and maize. Agric. Water Manage. 69:115–13

12

Irrigation withholding time management in four rice varieties at Guilan paddy fields (North Iran)

Naser Mohammadian Roshan, Maral Moradi, Ebrahim Azarpour and Hamid Reza Bozorgi

Department of Agriculture, Lahijan Branch, Islamic Azad University, Lahijan, Iran.

In order to determine the best irrigation withholding time in four varieties of rice, an experiment in factorial statistical format based on complete randomized block design was conducted. This experiment took place in National Rice Research Institute situated in Rasht township (Guilan province, North Iran) in 2008. First factor included four rice varieties (v_1 = Khazar, v_2 = Sepeed Roud, v_3 = Hassani, and v_4 = Binaam). The second factor included three drought periods (d_1 = 1 week after flowering, d_2 = 2 weeks after flowering, and d_3 = 3 weeks after flowering). The statistical analysis results almost in most measured traits showed significant difference at 1% probability level. The highest grain yield of 5635.8 kg/ha was obtained of Sepeed Roud variety. Also, the interaction effect of V_2d_3 with 6000 kg/ha was recorded the maximum grain yield. Since one of the goals for conducting this study was to determine a variety with short-water-necessity period in water deficiency years of cultivation, it is recommended that V_2d_1 level with grain yield of 5325 kg/ha be used even though it has a lower yield.

Key words: Rice varieties, irrigation withholding time, yield, yields components, Iran.

INTRODUCTION

Rice (*Oryza sativa* L.) is the most important cereal crop in the world and it is the primary source of food and calories for about half of mankind (Khush, 2005). More than 75% of the annual rice supply comes from 79 million ha of irrigated paddy land. Thus, the present and future food security of Asia depends largely on the irrigated rice production system. However, rice is a profligate user of water. It takes 3,000 to 5,000 L to produce 1 kg of rice, which is about 2 to 3 times more than the amount needed to produce 1 kg of other cereals such as wheat or maize (Bouman et al., 2002). Irrigation water is an important production factor in rice systems but is no longer available in unlimited rice-growing areas (Bindraban, 2001). In recent years, due to unprecedented growth of demand for water consumption both in domestic and industrial sectors and because lower water content in underground reservoirs due to human consumption, the volume of water for irrigation of paddy fields has significantly declined. According to climate conditions, Iran lies among semi-dry to dry belts of the world. The Guilan province has a high annual rainfall but experiencing water shortage problems in recent years. Rice, a major farming product of Guilan is a lowland crop. Therefore, water shortage means a rapid decline in its growth and yield as an agricultural product. This has several benefits. First, it prohibits addition of excess water to paddy fields after rice physiological water satisfaction. Second, it reduces production expenses and determines those varieties that are drought resistant with high qualitative and quantitative yield for cultivation in years to come. The future of rice production will depend heavily on developing and adopting strategies and practices that will use water efficiently in irrigation system. Numerous studies conducted on the

Table 1. Soil analysis results of the experimental sites.

Depth (cm)	0-30	SP%	69.7
Organic matter (%)	7.1	PH	7.1
Clay (%)	17.81	E.c.d. s/m	6.82
Sand (%)	22	N%	0.18
Silt (%)	60.19	P (ppm)	37.4
Texture	Silty loam	K (ppm)	29.1

manipulation of depth and interval of irrigation to save on water use without any yield loss have demonstrated that continuous submergence is not essential for obtaining high rice yield (Guerra et al., 1998). One method for reducing water consumption in rice cultivation is irrigation withholding at optimum time and blockage of flooding field for all duration of irrigation withholding. Drought stress due to irrigation withholding causes yield decrease and stress in flowering stage cause increase in unfilled grain percentage. According to several researches, there is a significant reduction in tillers and panicles numbers as well as plant height and grain yield due to water stress imposed at tillering stage. On the other hand, moisture stress at late vegetative and reproductive stage results to reduction in number of panicles per plant, percentage of filled grains and 1000-grain weight. Also, the reduction in grain yield was noted when plants were exposed to water stress at panicle initiation stage, while the moisture stress at the milk ripe or dough ripe had significant effects on grain yield (Bahattacharjee et al., 1973; De Datta et al., 1973; Krupp et al., 1971). Nour et al. (1994) reported that exposing rice plant to water stress for 36 days without flush irrigation during both tillering and panicle initiation significantly reduce plant height, number of tillers per plant, total dry matter, crop growth rate and grain yield. Boonjung and Fukai (1996) reported that drought stress during grains filling period results in acceleration of ripening time, causing reduction in growth period duration and filling grains. Abou-Khalifa (2010) in their study with three levels of irrigation withholding time: (w1) irrigation withholding at complete heading, (w2) irrigation withholding after 10 days from complete heading, (w3) irrigation withholding after 20 days from complete heading on two varieties of rice viz: H1 hybrid rice and Giza 177 inbred rice, found that the highest amounts of grain yield (10.59 t/ha), panicle length (18.80 cm) and number of grains per panicle (144) obtained of w3 treatment. This study has been conducted to find the best irrigation withholding time for rice cultivars in Guilan province, Iran.

MATERIALS AND METHODS

In order to study the determination of best irrigation withholding time in four varieties of rice in paddy fields of Guilan province (North Iran), an experiment in factorial statistical format based on complete randomized block design on a land parcel of 1000 m^2 area in National Rice Research Institute situated in Rasht township (Guilan province) with 37° 12'5'' N latitude and 49° 38'30'' E longitude in 2008 was conducted. The soil texture was silty loam, PH: 7.1, N: 0.18%, P: 37.4 ppm, K: 29.1 ppm. First factor included four rice varieties namely Khazar (v_1), Sepeed Roud (v_2), Hassani (v_3) and Binaam (v_4). The second factor included three drought periods namely; irrigation withholding 1 week after flowering (d_1), irrigation withholding 2 weeks after flowering (d_2), and irrigation withholding 3 weeks after flowering (d_3). Each replicate was designed from 48 experimental units and each unit with a distance of 30 cm away from one another. These units included 12 treatments and 7 lines with each line having 6 m long and 20 cm apart. Based on soil analysis (Table 1), phosphorous and potash fertilizer were applied for all treatments. Sowing in nursery was done 15, April and transplanted to field 22, May. All options consist of irrigation, weeding, fighting with pests and diseases up to harvest stage have been done. Characteristics to be evaluated are: grain yield, straw yield, harvest index, panicle length, number of grains per panicle, unfilled grain percentage, number of productive tillers, number of non-bearer tillers, 1000 grain weight and plant height. The yield and yield components were analyzed by using SAS software. The Duncan's multiple range tests was used to compare the means at 1% of significant.

RESULTS AND DISCUSSION

Results of variation analysis showed that, in more studied traits, effect of variety and irrigation withholding time has a significant difference at 1% probability level (Table 2). The effect of variety on traits such as grain yield, straw yield, harvest index, panicle length, number of grains per panicle, number of bearer tillers, percentage of unfilled grain per panicle, 1000 grain weight and plant height showed a significant different at 1% probability level. The highest amounts of grain yield, straw yield, harvest index and number of bearer tillers per square meter, respectively were 5635.8 kg/ha, 7870.4 kg/ha, 41.7% and 323/m^2 was obtained of V_2 (Sepeed Roud). The lowest grain yield, straw yield and harvest index was recorded by V_4 (Binaam), respectively were 2609.2 kg/ha, 4506.8 kg/ha and 36.6%. The V_1 (Khazar) with 164 tillers/m^2 was recorded the lowest number of tillers/m^2 (Table 3). Yield is a factor that its importance respectively depend on number of tillers per square meter, panicle length, number of grains per panicle and 1000 grain weight, the Seeped Roud variety due to maximum number of bearer tillers per square meter obtained the highest grain yield, while the Khazar variety due to lowest amounts of mentioned factors showed minimum grain yield. Number of tillers per square meters and plant height are two factors affecting

Table 2. The variance analysis of studied traits in rice varieties under irrigation withholding times.

S.O.V	df	Grain yield (kg/ha)	Straw yield (kg/ha)	Harvest Index (%)	Panicle length (cm)	The number of grains per panicle	The number of bearer tillers	percentage of unfilled grains	1000 grain weight (g)	Plant height (cm)
Block	3	39547.2223	59266.25	3.4858	0.027	198.92	1742.25	24.06	0.48	147.74**
Effect of variety	3	18989497.22**	25524936.25**	65.6612**	12.59**	1132.19**	53378.73**	402.48**	147.67**	5364.53**
Effect of irrigation withholding time	2	295581.25	1320700.725**	0.40305	0.195	74.94	516.51	24.37*	8.83**	0.34
Interaction effect	6	1350570.138**	162841.258**	6.7308	1.92	140.88	865.87	17.25*	0.31	21.09
Error	33	125879.0404	83075.136	4.1092	0.89	73.96	1234.25	5.31	0.39	32.84
CV(%)		8.96	4.69	5.23	3.74	8.83	1234.25	12.25	2.37	4.48

** and * respectively significant in 1 and 5% area.

Table 3. Yield and yield components of four rice varieties under three irrigation withholding time in north of Iran.

Treatment	Grain yield (kg/ha)	Straw yield (kg/ha)	Harvest Index (%)	Panicle length (cm)	The number of grain per panicle	The number of bearer tillers	percentage of unfilled grains	1000 grain weight (g)	Plant height (cm)
Variety									
V_1	3977.5b	6688.3ab	37.1b	23.7b	111a	164c	27.3a	23.8ab	123ab
V_2	5635.8a	7870.4a	41.7a	25.2ab	95a	323a	18.1ab	24.1ab	99b
V_3	3622.5b	5492.9bc	39.6a	25.9a	89b	210ab	14.4b	31.4a	143a
V_4	2609.2c	4506.8c	36.6ab	25.7a	92a	222ab	15.4b	26.1a	146a
Irrigation withholding time									
d_1	4043.8a	6310a	38.9a	25.3a	95a	235a	19.9a	25.6b	127.8a
d_2	4035.6a	5808ab	38.8a	25.1a	99a	213a	19.2a	26.3b	127.8a
d_3	3804.4a	6301a	38.6a	25.2a	98a	240a	17.5a	27.1a	127.5a

Within each column, treatments that carry the same superscript letter are not significantly different at P < 0.05.

straw yield, highest straw yield in v_2 treatment was due to the maximum number of tillers per square meter and high size of plant height. Because of higher grain yield and better transfer of photosynthetic materials to grains in v_2 treatment, the highest harvest index was showed in this level. One of the breeding rice varieties characteristics for maximum yield is more tillers production to compare with local rice varieties. This characteristic was observed in Sepeed Roud variety. Only the Khazar was the breeding rice variety that has lowest number of tillers than even the local varieties (Mojtahedi, 1989). With attention to Table 3, the V_3 (Hassani) resulted to highest panicle length with 25.9 cm and 1000 grain weight with 31.4 g. The lowest panicle length and 1000 grain weight was recorded from V_1 (Kazar) with 23.7 cm and 23.8 g, respectively. The grains distance on panicle in Hassani variety genetically is high. As a result, the longest panicle was shown in this treatment. Due to less grain in panicle of Hassani variety, the photosynthetic materials per each grain become more and as a result 1000 grain weight increased. The maximum number of grains per panicle with 111 was recorded from v_1 (Khazar) and the minimum with 89 was recorded from v_3 (Hassani). The highest percentage of unfilled grains per panicle of 27.3% and the lowest percentage of unfilled grains per panicle of 14.4%, respectively were obtained in

v_1 and v_3. It seems that high percentage of unfilled grains per panicle in Khazar variety is due to, first; for genetic problems in this variety and second, further number of grains in this variety. As a result, the proportion of unfilled grains to filled grains in this level is high. In the Hassani variety due to less number of grains, materials transition well division to all grains. In the other side, the proportion of unfilled grains to filled grains was low in this treatment because of less grain. The tallest plant height was recorded from v_4 (Binaam) with 146 cm and the shortest height of 99 cm was recorded from v_2 (Sepeed Roud). One of the clear characteristics of local rice varieties are that the taller plant height trait was shown in Binaam variety. In the other hand, plant height usually is shorter in the breeding varieties.

Rice grain filling and ripening are affected by many environmental factors, including water, temperature, radiation and soil nutritional conditions (Yoshida, 1981). With attention to Table 2, the effect of irrigation withholding time on straw yield and 1000 grain weight significant at 1% probability level and on percentage of unfilled grains per panicle showed a significant difference at 5% probability level. Other traits were not significantly different. Comparison of mean between irrigation withholding times show that (Table 3), the straw yield, harvest index, panicle length, percentage of unfilled grains per panicle and plant height was recorded from d_1 (irrigation withholding 1 week after flowering), respectively were 6310 kg/ha, 38.9%, 25.3 cm, 19.9% and 127.8 cm. d_3 treatment (irrigation withholding 3 weeks after flowering), having the lowest amounts of harvest index, percentage of unfilled grains per panicle and plant height, respectively with 38.6%, 17.5% and 127.5 cm was recorded. The minimum amounts of straw yield and panicle length respectively was 5808 kg/ha and 25.1 cm which was recorded from d_2 (irrigation withholding 2 weeks after flowering). In the first 10 days after flowering, cell division and expansion in the endosperm of most grains ends and starch deposition begins (Hoshikawa, 1967; Egli, 1998). With soon irrigation withholding in d_1 level due to decrease of photosynthesis process and photosynthetic materials production, transition of this materials to grains were reduced causing increase in percentage of unfilled grains in this level, whereas, in d_3 level this case was not shown. Vegetative growth in rice with flowering start become end, therefore not added anymore to plant height and tillering process come to maximum stage. Thus, with irrigation withholding 1 week after flowering (d_1) production of generative organs decreased, the materials resulting from photosynthesis, more transition to vegetative organs and in addition to tillers preservation, leaf area index increased, and as a result the maximum straw yield was obtained. But in d_2, treatment (irrigation withholding 2 weeks after flowering) resulted to more generative organs compared with d_1 level produced; further photosynthetic matters give to grains and in harvest time less straw yield was shown. The

highest 1000 grain weight of 27.1 g was obtained in d_3. In the other hand, the lowest 1000 grain weight was found from d_2 with 26.3 g and d_1 with 25.6. The 1000 grain weight is one of the factors in plant that is influenced more by genetic factors with little influence by environment factors. Water is one of the important soil and environmental factors that have a major influence on 1000 grain weight. If water exists sufficiently in all stages of plant growth, nutrients transition to plant and grains filling is well enhanced resulting to increased 1000 grain weight as it was the case from d_3 level. The maximum number of grains per panicle of 99 was found from d_2 and the minimum amount of this trait of 95 grains per panicle was found from d_1. Results of variation analysis showed that, the interaction effect of variety and irrigation withholding time on traits grain yield and straw yield in 1% probability level and on percentage of unfilled grains per panicle in 5% probability level have significant different (Table 2), The highest grain yield of 6000 kg/ha was recorded from v_2d_3 and the lowest one with 2307.5 was in v_4d_2 (Table 4). Since the Sepeed Roud variety genetically had high yield factors in the other hand, irrigation withholding 3 weeks after flowering gave sufficient time to plant for suitable photosynthesis and high photosynthetic materials production causing all flowers inoculation and transition of materials to grains was higher than grains materials receive to maximum amounts that causing high yield in v_2d_3 level compare with v_4d_2. In v_4d_2 treatment, due to lower yield factors and also with irrigation withholding 2 weeks after flowering resulted to insufficient time for transition of photosynthetic materials to all grains resulting to low yield. The maximum yield of straw of 8315 kg/ha was found from v_2d_3 and the minimum of this trait was found from v_4d_2 with 4144 kg/ha (Table 4). Water existence under rice bush is one of the factors that casing to better vegetative growth in this plant. As a result, in Sepeed Roud variety because of more tiller and partly taller plant height and also leaf area duration was increased and the highest straw yield obtained in v_2d_3 level. But in v_4d_2 in addition to low tiller in Binnan variety, irrigation withholding 2 weeks after flowering in this variety caused an increase in generative organs. As a result, more photosynthetic materials were send to generative organs and straw yield in harvest time at this level decreased. V_1d_1 treatment produced the highest percentage of unfilled grains per panicle with 30.1% and the lowest one with 12.2% was found from v_3d_1 (Table 4). Irrigation withholding 1 week after flowering in Khazar variety (v_1d_1) due to high number of grains per panicle in this variety, prevent from filling all grains. As a result, percentage of unfilled grains in this level increased. But in v_3d_1 level (Hassani variety, along with irrigation withholding 1 week after flowering), water withholding had low influence on percentage of unfilled grains because of sufficient photosynthetic materials production for filling the grains per panicle. Although the v_2d_3 treatment with 6000 kg/ha was obtained as the highest grain yield, but since

Table 4. The interaction effects of varieties and irrigation withholding times yield and yield components

Treatment	Grain yield (kg/ha)	Straw yield (kg/ha)	Percentage of unfilled grains
V_1d_1	4492.5^b	7293^b	30.1^a
V_1d_2	3102.5^{cde}	5523^{cd}	25.4^a
V_1d_3	4337.5^b	7250^b	26.2^a
V_2d_1	5325^a	7763^{ab}	15.9^{bcd}
V_2d_2	5582.5^a	7534^b	19.6^b
V_2d_3	6000^a	8315^a	18.3^{bc}
V_3d_1	3412.5^c	5275^{de}	15.8^{bcd}
V_3d_2	4225^b	6030^c	16.4^{bdc}
V_3d_3	3230^{cd}	5174^{de}	12.2^d
V_4d_1	2945^{cde}	4910^{ef}	17.2^{bc}
V_4d_2	2307.5^e	4144^g	15.3^{bcd}
V_4d_3	2575^{de}	4467^{fg}	13.5^{cd}

Within each column, treatments that carry the same superscript letter are not significantly different at $P<0$.

one of the objectives for conducting this study was to determine a variety with short- water-necessity period in water deficiency years for cultivation, it is recommended that v_2d_1 level with grain yield of 5325 kg/ha be used even though for cultivation it has a lower yield.

REFERENCES

Abou-Khalifa AA (2010). Response of some rice varieties to irrigation withholding under different sowing dates. Agriculture and Biology J. North Am. 1(1):56-64.

Bahattacharjee DP, Krishnayya GR, Ghosh AK (1973). Analyses of yield components and productive efficiency of rice varieties under soil moisture deficit. IN: Indian J. Agron. 16(3):314 -343.

Bindraban PS (2001). Water for food: converting inundated rice into dry rice. In: Hengsdijk H, Bindraban PS, editors. Water-saving rice production systems. Proceedings of an international workshop on water-saving rice production systems at Nanjing University, China, 2-4 April 2001. Plant Research International (PRI), Report 33. Wageningen, Netherlands. pp. 5-14.

Boonjung H, Fukai S (1996). Effects of soil water deficit at different growth stages on rice growth and yield under upland conditions. Field Crops Res. 48:47-55.

Bouman BAM, Hengsdijk H, Hardy B, Bindraban PS, Tuong TP, Ladha JK (2002). Water-wise rice production. Proceedings of the International Workshop on Water-wise Rice Production, 8-11 April 2002, Los Baños, Philippines. Los Baños (Philippines): International Rice Research Institute. p. 356.

De Datta SK, Abilay WP, Kalwar (1973). 1.Water stress effect on flooded tropical rice. 2.Water management in Philippines irrigation system research and operation. pp. 16-36.

Egli DB (1998). Seed biology and the yield of grain crops.Wallingford, UK: CAB Intl.

Guerra LC, Bhuiyan SI, Tuong TP, Barker R (1998). Producing more rice with less water from irrigated system. SWIM paper 5. IWMI/IRRI, Colombo, Sri Lanka.

Hoshikawa K (1967). Studies on the development of endosperm in rice: I. Processes of endosperm tissue formation. Proc. Crop Sci. Soc. Jpn. 36:151-161

Khush GS (2005). What it will take to Feed 5.0 Billion Rice consumers in 2030. Plant. Mole. Biol, 59:1 – 6.

Krupp HK, De Datta SK, Balaoing SN (1971). The effect of water stress at different growth stages on growth and yield of lowland rice. In: Proceeding of the Second Annual Scientific Meeting. Crop Sci. Soc. Philippines. pp. 398-411.

Mojtahedi A (1989). Rice. Research, education and propagation of agriculture. Iran.

Nour MA, Abd El-Wahab AE, Mahrous FN (1994). Effect of water stress at different growth stage on rice yield and contribu - ting variables. Rice Research and Training Center. 1996. Annual Agronomy Report.

Yoshida S (1981). Fundamentals of rice crop science. Los Banos, Philippines: IRRI. 6/18/2009.

13

Validating remote sensing derived evapotranspiration with the soil and water assessment tool (SWAT) model: A case study in Zhelin Basin, China

Chongliang Sun, Dong Jiang, Juanle Wang and Yunqiang Zhu

State Key Laboratory of Resources and Environmental Information System, Institute of Geographical Sciences and Natural Resources Research, Chinese Academy of Sciences, China.

Evapotranspiration (ET) is a major component in the water and heat balance of terrestrial ecosystems as well as in the water, energy and carbon cycles on the Earth's surface. A growing number of studies have focused on the retrieval of ET from remote sensing (RS) data. However, the RS-derived ET results could not be validated by station-observed data directly for the difference of the scale. The objective of this study is to present an operational approach to validation of RS-derived ET under the support of a distributed hydrological model: soil and water assessment tool (SWAT). Five years (2000-2004) evapotranspiration data of Zhelin Basin, the study area, were prepared. RS-derived ET and other data (DEM, land-use data, soil data, etc) were processed together in SWAT to simulate the hydrological cycle. The output monthly runoff is compared with observed runoff data. The RS-derived ET was then validated based on the results of those comparison (R^2=0.8516, RMSE=26.0860, MBE=-8.6578). It indicated that the method presented in the paper was an operational and feasible way for validation of ET data from remote sensing retrieval.

Key words: Distributed hydrological model, evapotranspiration, remote sensing, soil and water assessment tool (SWAT).

INTRODUCTION

Evapotranspiration (ET), including the evaporation from soil surface and the vegetation transpiration, is a major component in the water and heat balance of terrestrial ecosystems as well as in the water, energy and carbon cycles on the Earth's surface (Drexler et al., 2004; Gao, 2008; Hussey and Odum, 1992; Parasuraman et al., 2007; Zhou and Zhou, 2009).

Various ET studies have been conducted especially in arid regions on the basis of meteorological data with several main methods, such as Penman-Monteith method, Priestley-Taylor method, and the Hargreaves method, etc. (Amatya et al., 1995; Garcia et al., 2004;

Gavilán et al., 2006; Lopez-Urrea et al., 2006; Mohan, 1991; Zhang et al., 2008). For daily ET calculation, the Penman-Monteith method requires the daily meteorological data, including the maximum and the minimum air temperature, the relative humidity, the solar radiation and the wind speed, as the input. The Priestley-Taylor method also requires multiple climate parameters to estimate ET, while the Hargreaves equation requires air temperature data to estimate ET. Since 1980s, with the emergence and rapid development of distributed hydrological models, remote sensing (RS) based approaches have been regarded as the preferred

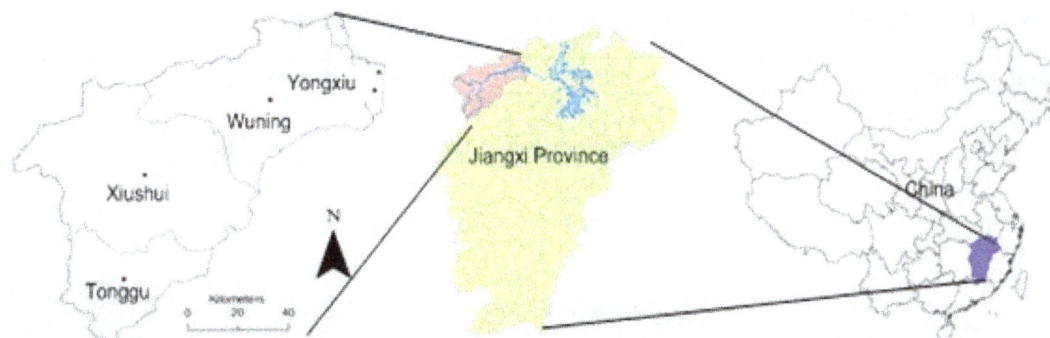

Figure 1. Location of the study area: Zhelin basin, Jiang Xi Province, China.

methods for estimating ET in large area with relative high spatial resolution (Overgaard et al., 2006). Numerous physical and empirical RS-based models have been developed for ET estimation in many different fields (Allen et al., 2005; Bastiaanssen, 2000; Bastiaanssen et al., 1998a, b; Granger, 1996, 2000; Jacob et al., 2002; Wang and Jiang, 2005). Most of RS methods for estimating ET are based wholly or partially on the energy balance principle, with net radiation adopted as the principal driving parameter (Jabloun and Sahli, 2008), which has led to a breakthrough in the high-resolution ET acquisition.

Unfortunately, problem exists in the validation of RS-derived ET data independently (Kite and Droogers, 2000). Since ET could not be observed directly, the 'ground truth' ET data were usually derived from Penman-Monteith method (etc.) using observed ET data, including the large aperture scintillomiter data (Jiang and Wang, 2003; Wang and Jiang, 2005) and the eddy covariance data (Boegh et al., 2009; Heilman et al., 2009; Kite and Droogers, 2000; Kustas and Norman, 1999; Sun and Song, 2008; Wu et al., 2006; Zhou and Zhou, 2009). The limitations are obvious for the in situ stations are rather limited in amount even in developed countries (Gavilán et al., 2006; Kite and Droogers, 2000), and the RS-derived ET could not be compared with station-observed data directly because the difference of the scale (Jabloun and Sahli, 2008; Liu et al., 2010). Additionally, the assumption that field methods are probably the most reliable is hard to justify, because different field methods differ considerably (Kite and Droogers, 2000).

Accordingly, proper method is urgently needed for validating the ET data obtained from RS retrieval. The objective of this study is to present an operational approach to validation of RS-derived ET under the support of a distributed hydrological model. The RS-derived ET, together with other auxillary data, were tansfered into runoff data by distributed hydrological model. The output runoff data could be compared with in situ observed runoff data, and then the ET could be validated accordingly. The soil and water assessment

tool (SWAT) was adopted in this study. Among the advanced distributed hydrological models, SWAT has been widely used around the world, and some related researches on the calibration and sensitivity (Immerzeel and Droogers, 2008; Kannan et al., 2007a) as well as the climate change sensitivity have already existed (Ficklin et al., 2009). The method presented in this paper were tested and evaluated in the study area.

Study area

Zhelin Basin, the study area, is located in the upper and middle reaches of the Xiuhe River Basin in the Northeast Jiangxi province, China. It is one of the branches of the Yangtze River, with the Zhelin Reservoir in its lower reaches, which is located in east longitude 115.5 and north latitude 29.2°. The Zhelin Basin is in a strip shape, that is about 176 km from west to east and more than 84 km from south to north on average, with the altitudes within the range of 10 to 1200 m. The area of basin is about 9340 km^2, and the main river is about 353 km long with bending coefficient of 1.69. The basin is surrounded by mountains on three sides, that is, the Mufu Mountains in the north, the Dawei Mountains in the west and the Jiuling Mountains in the south. In this way, a closed watershed is formed. The land type composition of the basin is: 60% of mountains, 30% of hills, 7% of hillocks and the rest 3% of valley plains (Figure 1). The basin contains abundant ground vegetation, with dense forests full of firs and pines. Only in the middle reaches, scanty bald hills can be found, while sparse grassland is distributed in a few regions in the lower reaches.

The rainfall from April to June accounts for 50% of annual amount in Zhelin basin. In this area, the mean annual temperature is 16 to 17 DEG C, with the maximum temperature of about 29 DEG C emerging in July. The minimum mean monthly temperature of about 6 to 7 DEG C centers in January. The mean annual humidity is 80%, dispersing evenly in the whole area. The mean annual wind speed is at Grade 2.1; in terms of the spatial distribution, the value is lower in the upper

Table 1. The spatial resolution and purpose of the three spectra of GMS image.

Band's name	Visible band	Thermal infrared band	Vapour band
Spatial resolution(km)	1.25	5	5
Purpose	Evapotranspiration retrieval	Evapotranspiration retrieval	Validation

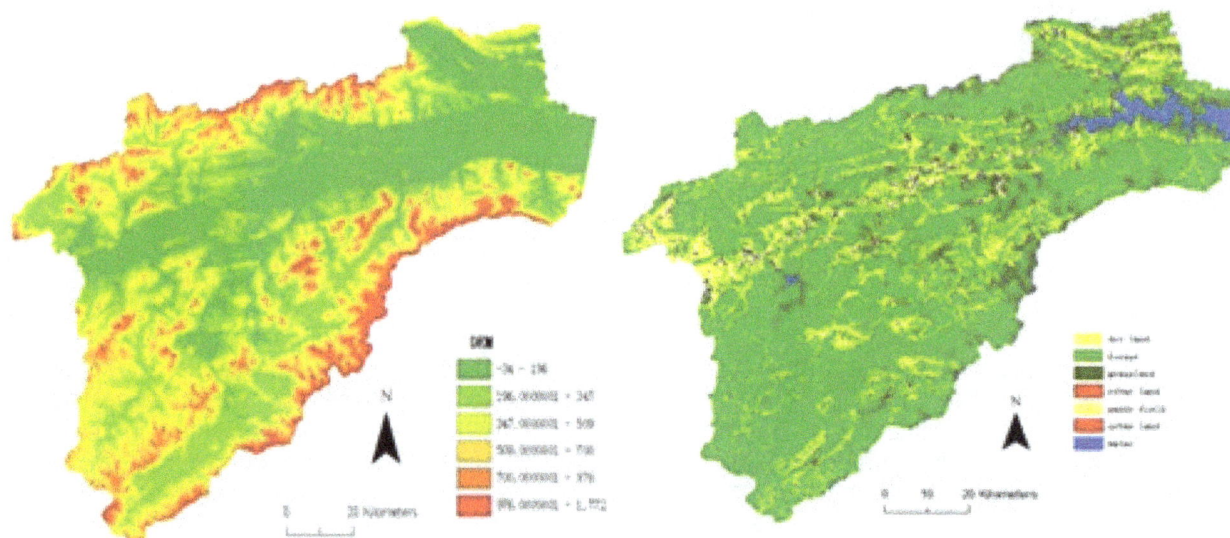

Figure 2. The DEM and land use of Zhelin basin.

reaches and higher in the lower reaches.

Data acquisition

RS data

Geostationary meteorological satellite data (GMS-5 data) were adopted as the data source for the ET retrieval. GMS-5 data can be easily acquired with relatively high temporal resolution (1 h). The GMS-5 data consists of three types of bands: (1) the visible band (VIS), with the spatial resolution of 1.25 km, and the spectrum range from 0.55 µm to 1.05 µm; (2) the thermal infrared band (TIR), with the spatial resolution of 5 km, and the spectrum range from 10.5 to 12.5 µm; (3) the water vapor band (WV), with the spatial resolution of 5 km, and the spectrum range from 6.2 to 7.6 µm. The visible and thermal infrared bands were employed for the ET retrieval, and the water vapor band was used for calibration and validation, as shown in Table 1.

Meteorological data

The meteorological data including daily precipitation, ET (obtained from retrieval), daily maximum temperature,

daily minimum temperature, daily relative humidity, daily solar radiation and daily wind speed, were derived from National Resources and Environmental Database presented by Resources and Environmental Scientific Data Center (RESDC), Chinese Academy of Sciences.

DEM data

DEM data were used in the process of SWAT-based hydrological simulation. It was supplied by State Bureau of Surveying and Cartography (SBSC). The resolution of the DEM adopted in this study was 90 m, as shown in Figure 2.

Land use data

Land use data for 2005 were achieved from Landsat TM data through human-computer interactive interpretation, presented by Resources and Environmental Scientific Data Center (RESDC). Six land use types were identified including (1) cultivate land; (2) woodland; (3) grass land (4) water; (5) urban and rural settlements; (6) barren land. The scale of the land use map was 1:100,000. For the SWAT model, the attribute codes of the land use data were converted to the U.S. version, as shown in Figure 2.

Validating remote sensing derived evapotranspiration with the soil and water assessment tool...

101

Table 2. Datasets and their sources adopted in this study.

Data type	DEM	Land use data	Property data	Soil data
Sources	http://www.geodata.cn	http://www.geodata.cn and RESDC,CAS	thttp://www.geodata.cn and RESDC,CAS	http://www.geodata.cn
Usages	Input for SWAT	Input for SWAT; ET retrieval	Input for SWAT	Input for SWAT
Data type	Meteorological data	Runoff data	ET	GMS-5 RS data
Sources	The China meteorological administration; the information center of the Institute of Water Resources and Hydropower Research and the information center of Zhelin HydroPower Corp	the information center of the Institute of Water Resources and Hydropower Research and the information center of Zhelin HydroPower Corp	Retrieved in this study. And the ET should be processed into the data type and format as SWAT needed	http://satellite.cma.gov.cn/
Usages	Input for SWAT	Analysis for the simulated runoff	Input for SWAT	ET retrieval

All of the datasets are shown in Table 2 in details.

METHOD

ET retrieval model

A method based on the energy-balance theory developed by Wang and Jiang (2005) was adopted for ET retrieval. The RS-based latent heat flux was treated as the residual of the surface energy balance equation through model calculation (Boegh et al., 2002; Kustas et al., 1994; Moran et al., 1994). Accordingly, the energy balance equation can be expressed as:

$$LE = I_n - H - G - B \tag{1}$$

Where I_n is the net solar radiation flux, with W/m^2 as the unit; H is the sensible heat flux, with W/m^2 as the unit; LE is the latent heat flux, with W/m^2 as the unit; G is the soil heat flux, with W/m^2 as the unit; B is the energy absorbed by vegetation, with W/m^2 as the unit.

According to the above energy-balance principle, the following steps are required to retrieve ET from RS data: first, the surface albedo (a), the vegetation index (NDVI) and the surface temperature (T_0) should be acquired through a specific RS channel; then, I_n, H, G and B as well as the instantaneous evpotranspiration are calculated, and the daily ET is calculated based on LE$_x$.

$$I_n = (1-\alpha)I_g - L_\uparrow + L_\downarrow \tag{2}$$

Where Ig is the total solar radiation, a is the albedo, L$_\uparrow$ is the long-wave radiation of the earth surface, and L$_\downarrow$ is the long-wave radiation of the atmosphere.

$$I_g = tS\cos(i_s) \tag{3}$$

Where S is the solar constant (W/m^2), t is the transmission coefficient of the atmospheric radiation, and i_s is the solar altitude angle.

$$L_\uparrow = \varepsilon_0 \sigma T_0{}^4 \tag{4}$$

$$L_\downarrow = \varepsilon_0 \varepsilon_a \sigma T_a{}^4 \tag{5}$$

Where T_0 is the surface temperature, Ta is the air temperature, ε_0 is the land surface emissivity, ε_a is the air emissivity, and σ is the Stefan-Boltzman constant.

According to the energy-balance principle, Brown and Rosenberg developed an impedance model for the sensible heat flux; Penmam-Monteith (Monteith, 1973) deduced a formula for H calculation:

$$H = -\rho C_p (T_a - T_0)/r_a \tag{6}$$

Where, T_0 is the ET surface temperature; T_a is the air temperature (at the attitude of 2.0 m); r$_a$ is the surface roughness.

The heat flux G_0 (Z=0) of the surface soil can be calculated according to the following formula:

$$G_0 = R_n \cdot [\Gamma_c + (1-f_c) \cdot (\Gamma_s - \Gamma_c)] \tag{7}$$

Where, Γ_c is the canopy proportion coefficient; Γ_s is the bare-soil proportion coefficient; f_c is the vegetation coverage.

SWAT model

SWAT is a distributed hydrological model providing the spatial coverage of the integral hydrological cycle, including the atmosphere, plants, unsaturated zone, groundwater and surface water (Arnold et al., 1993; Neitsch et al., 2001). The model is comprehensively described in literatures (Arnold et al., 1998; Neitsch et al., 2002) and widely used around the world. According to the water-balance principle, the principle of SWAT can be described as:

$$SW_t = SW_0 + \sum_{i=1}^{t}(R_{day} - Q_{surf} - E_a - w_{seep} - Q_{gw}) \tag{8}$$

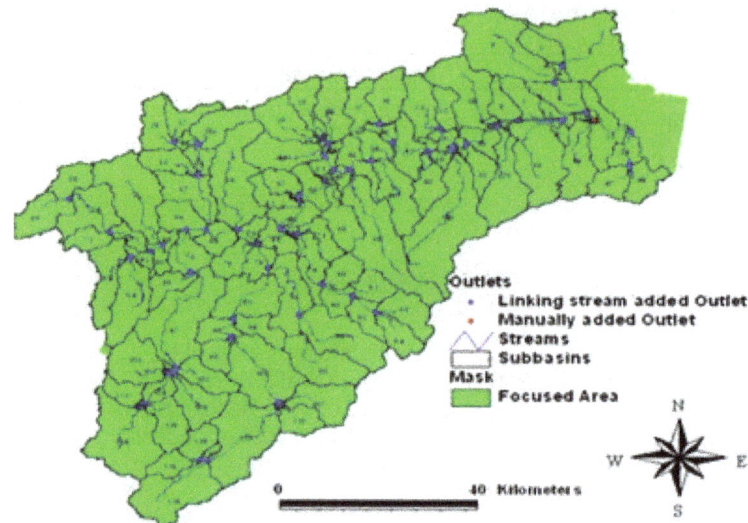

Figure 3. The map of the sub basin produced by SWAT.

Where, SW_t and SW_0 are the initial and the terminal water contents on day_i ; t is the time with a day as the unit; R_{day} is the rainfall on day_i ; Q_{surf} is the surface runoff on day_i ; E_a is the ET on day_i; w_{seep} is the infiltration amount; Q_{gw} is the runoff contribution from the groundwater .

Under SWAT, conceptually, the catchment is subdivided into sub basins and a river network based on DEM. SWAT integrates the simulation of weather, crop growth, ET, surface runoff, percolation, return flow, erosion, nutrient transport, groundwater flow, pond and reservoir storage, channel routing, field drainage, the water consumption of plants and other supporting processes . The tile drainage estimation is a function of the drain depth, the time needed for tile drains to bring the soil layer to field capacity and a drainage lag parameter. In SWAT, sub-catchments are divided into Hydrological Response Units (HRUs) as the unique combination of soil and land covers. The flow is not routed between HRUs, instead, the routing is used for flow in the channel network (Kannan et al., 2007b).

The input parameters of SWAT concern the ET studied in this paper, DEM, land use data, soil data, property data, the observed data for the outlet of the basin, meteorological data such as daily precipitation, daily maximum and minimum temperature, wind speed and relative humidity, as well as the runoff data on controlled sites and geographical materials, etc. The data required in SWAT for simulating the runoff and the data sources are shown Table 2.

The DEM of the catchment was prepared using the SRTM data with the spatial resolution of 90 m in the study area. Detailed land use information, which was acquired from RESDC and CAS, was used to draw the land use map and the soil map of the catchment. The Arc View-SWAT interface (AVSWAT-2000) was employed to delineate the catchment boundaries, and the burning-in option was used to acquire the drainage network. A visual inspection of the derived drainage network and the network delineation on the paper map showed good agreement. The multiple HRU options available in the AV-SWAT interface were applied with the objective of representing each field as a separate HRU. As a result, the study area was divided into 119 HRUs, as shown in Figure 3.

Validation of RS- derived ET

The study presented a new method of validating the

remote-sensing retrieval of evapotranspiration under the support of the SWAT model. The RS-derived ET could not be evaluated by station-observed data directly because the difference of the scale. We suggested that the RS-derived ET could be evaluated by comparison of RS-computed Runoff (with SWAT) and the observed Runoff:

1. RS-derived ET is used as one of the input factors for SWAT,
2. RS-derived ET and other data (Digital Elevation Model (DEM), land use data , soil data, etc) are processed together in SWAT to simulate the hydrological cycle,
3. The Runoff is output from the SWAT,
4. Output Runoff is compared with observed runoff data,
5. RS-derived ET is evaluated based on the results of 4). Three indications, including the Root Mean Square Error (RMSE), the mean deviation error (MBE) and R^2, were employed for data analysis in this study.

The analysis flowchart are shown in Figure 4.

RESULTS AND ANALYSIS

Spatial-temporal variation of RS-derived ET

The monthly ET results from 2000 to 2004 in the Zhelin Basin obtained with RS retrieval method were analyzed. The whole Zhelin Basin involved 238 RS pixels. The daily and monthly ET results from 2000 to 2004 were shown in Figures 5 and 6.

The spatial distribution of the monthly ET was analyzed. In March, when the rainfall was relatively less throughout the whole year, the maximum ET emerged in the area with plenty of water, such as reservoirs and paddy fields, while the minimum ET appeared in the upper reaches of the basin and the area with high altitudes, as shown in Figure 5, which indicated that the primary factor affecting ET is the water capacity in this season. In June and September when the rainfall is

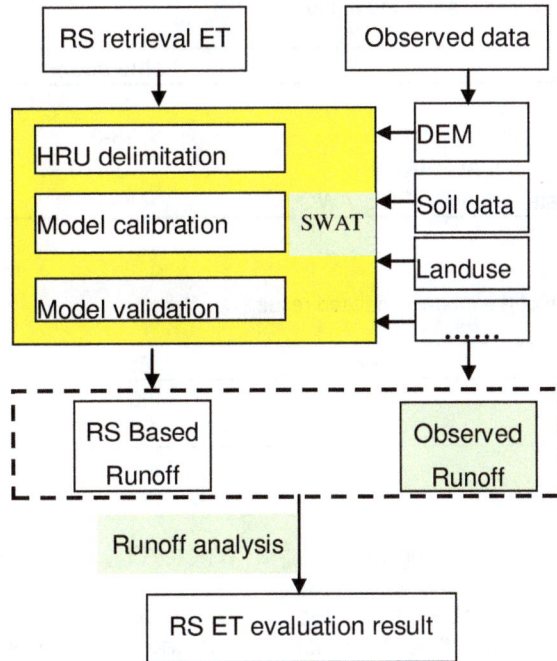

Figure 4. The flowchart of ET validation.

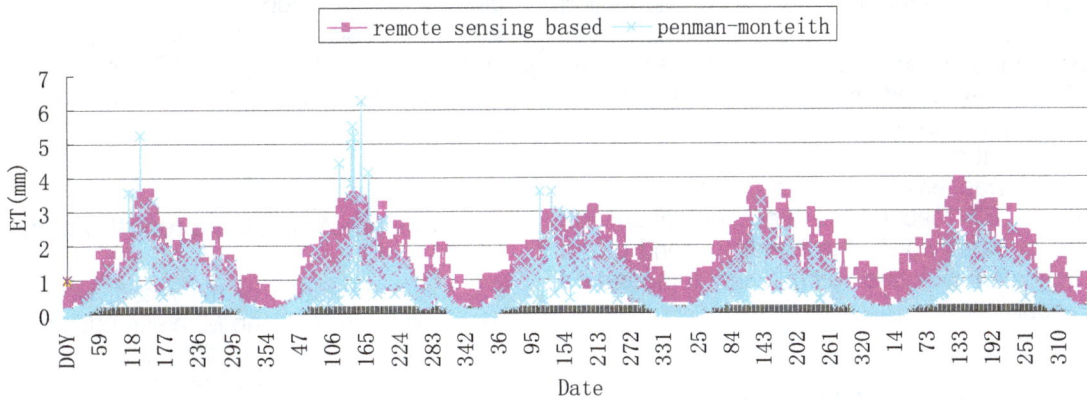

Figure 5. The daily ET result from the remote sensing based method and penman-monteithmethod from the year of 2000 to 2004.

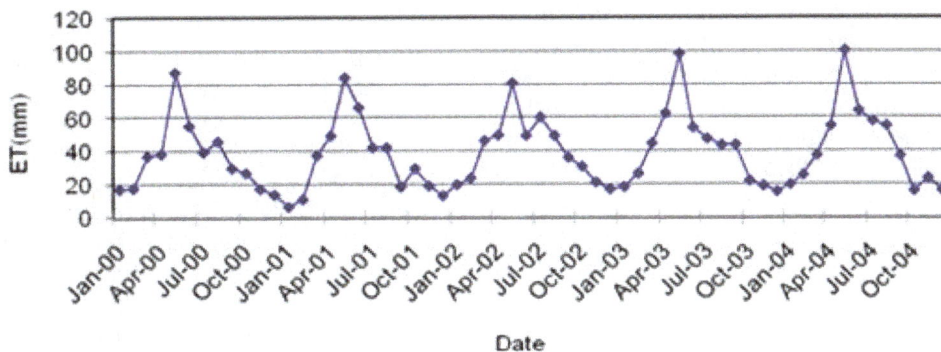

Figure 6. The monthly ET result from the remote sensing based method from the year of 2000 to 2004.

Table 3. Calibrated parameters of the model.

Parameter	Value scope	Value
ESCO	0-1	0.1
SCS runoff curve: CN_2	-8-8	-7
Baseflow a coefficient: ALPHA_BF	0-1	0.041
Soil available water capacity: SOL_AWC	0-1	0.04

Table 4. Estimation of the model simulated result.

Period of time	Re	Ens	R^2
Calibration	0.119	0.875	0.946
Validation	0.101	0.808	0.925

abundant and the temperature is high throughout the year, the ET distribution was uniform in the basin, as shown in Figures 5b and 5c. In December when the temperature was the lowest and the rainfall is the least among the year, as shown in Figure 5, the maximum ET emerged in the upper reaches of the basin and the area with high altitudes, and the total ET was the least among the year due to the low temperature; in this period, the altitude was the major factor affecting the ET distribution in the whole basin.

The seasonal variation of ET was analyzed, as shown in Figure 6. The ET exhibited an obvious rule of seasonal variation, that is, the maximum ET emerged in summer when the temperature is highest and the rainfall is abundant, the ET was less in spring and autumn than in summer, and the ET reached the least throughout the year in winter. In other words, the variation trend of the ET was coordinated with the rainfall and temperature variation in a year.

Runoff simulation with the SWAT model

Due to the regional adaptability, the SWAT model should be calibrated and validated before using. In this study, the data from 2001 to 2004 were adopted for calibration, and the data of 2000 were adopted for validation. According to Nash and Sutcliffe (1970), the model would be evaluated with the following parameters: model efficiency coefficient Ens, mean error Re and correlation coefficient R^2 which was calculated using excel, while Re and Ens were calculated according to the following formula:

$$R_e = \frac{Q_{sim,i} - Q_{obs,i}}{Q_{obs,i}} \times 100\% \quad Ens = 1 - \frac{\sum_{i=1}^{n}(Q_{obs,i} - Q_{sim,i})^2}{\sum_{i=1}^{n}(Q_{obs,i} - \overline{Q}_{obs})^2}$$

(9)

Where, $Q_{obs,i}$ is the observed runoff, $Q_{sim,i}$ is the simulated runoff, and $\overline{Q}_{obs}$ is the average observed runoff.

In the SWAT model, the whole basin was divided into 119 HRUs, as shown in Figure 6. The accepted range after calibration was shown in Table 3.

After the calibration and validation, the evaluation indicators of SWAT were shown in Table 4, with good general performance.

On the basis aforementioned, the data of the monthly simulated runoff using SWAT and the observed data in the Zhelin Basin from 2000 to 2004 were shown in Figure 7.

Comparison of simulated runoff with observed data

For the comparison of the result, we also calculate the runoff with the Penman-Monteith option which embedded in SWAT, as well as the runoff calculation with the remote-sensing retrieval result as the input. The monthly runoff data and the simulated runoff data with SWAT using the two ET methods, especially the remote-sensing retrieval result, in the Zhelin Basin from 2000 to 2004 were shown in comparison in Figure 8.

From the figure and table above, the correlation (R^2=0.8516) between the simulated results based on ET retrieval and the observed data was higher than that (R^2=0.8411) between the results simulated with PM-based ET and the observed data after data fusion. The RMSE (RMSE=26.0860) between the simulated runoff based on ET retrieval and the observed data was obviously smaller than that (RMSE=35.71904) between the runoff simulated with PM-based ET and the observed data. The MBE (MBE=-8.6578) between the simulated runoff based on ET retrieval and the observed data was obviously superior to that (MBE=-22.7313) between the simulated runoff based PM-based ET and the observed data.

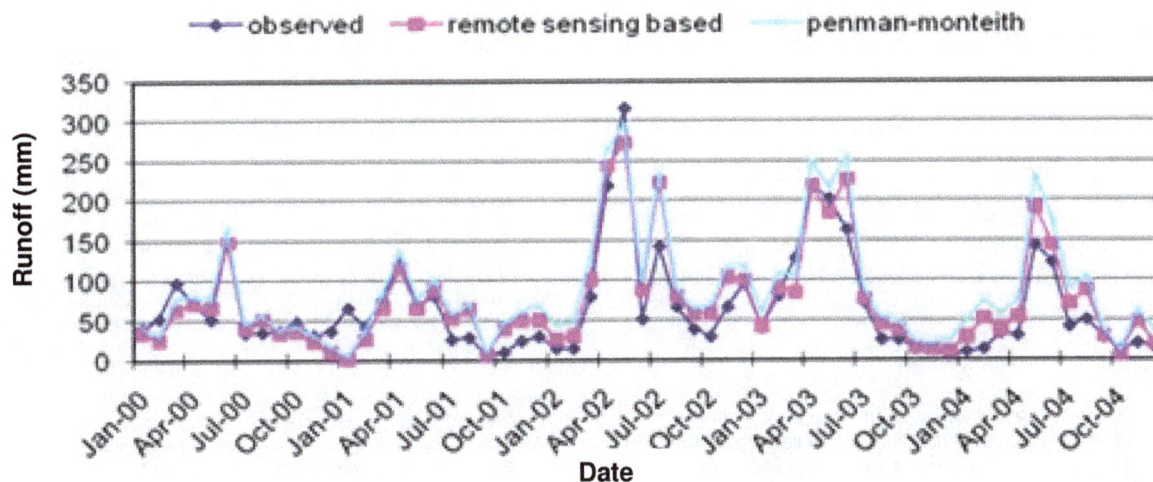

Figure 7. The monthly runoff simulated by SWAT from RS-derived ET, the SWAT embedded penman-monteith ET versus the observed runoff from the year of 2000 to 2004.

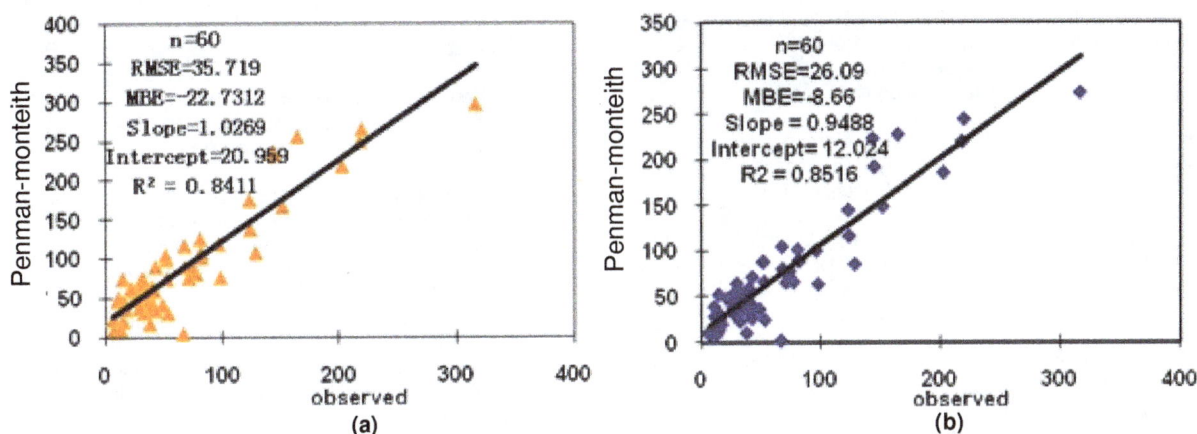

Figure 8. The monthly runoff simulated from RS-derived ET and Penman-monteith ET with SWAT versus the observed runoff data. (a) The simulated runoff with Penman-monteith model versus the observed runoff; (b) The simulated runoff with remote sensing based model versus the observed runoff.

Currently, it is difficult to validate the retrieval results pixel by pixel directly. However, the underlying conditions of the basin can be taken into account in the distributed hydrological model with high sensitivity to the accuracy of the input data, so that the retrieval results can be validated indirectly through this method and with the support of the distributed hydrological model.

Conclusions

This study presented a new method to validate the ET results from RS retrieval with the support of a distributed hydrological model- the SWAT model. Five years (2000-2004) evapotranspiration data of Zhelin Basin, the study area, were prepared. RS-derived ET and other data (DEM, land-use data, soil data, etc) were processed together in SWAT to simulate the hydrological cycle. The runoff data were then output from the SWAT model. When monthly total runoffs were compared with observed data, the model-estimated data had a RSME of 26.0860 and a R^2 of 0.8516. And the runoff data obtained from RS retrieval of ET was better than those from PM-based ET which was embedded in SWAT in terms of the parameters: RMSE, MBE and R^2. The RS-derived ET was then validated based on the results of comparison.

It indicated that the method presented in the paper was an operational and feasible way for validation of ET data derived from remote sensing data. The subsequent research work of this study should be focused on improving the temporal/spatial resolution of the RS data and enhance the improvement of ET retrieval according to the results of validation.

ACKNOWLEDGEMENTS

This paper was funded by the Earth System Science Data Sharing Program of China (2005DKA32300, http://www.geodata.cn);the National Natural Science Foundation Program of China (40801180、40771146).

REFERENCES

Allen R, Tasumi M, Morse A, Trezza R (2005). A Landsat-based energy balance and evapotranspiration model in Western US water rights regulation and planning. Irrigat. Drain Syst. 19(3):251-268.

Amatya D, Skaggs R, Gregory J (1995). Comparison of methods for estimating REF-ET. J. Irrigat. Drain. Eng. 121(6):427-435.

Arnold J, Srinivasan R, Muttiah R, J Williams (1998). Large area hydrologic modeling and assessment partl: model development1. JAWRA 34(1):73-89.

Arnold. J, Allen P, Bernhardt G (1993). A comprehensive surface-groundwater flow model. J. Hydrol. 142(1-4):47-69.

Bastiaanssen W (2000). SEBAL-based sensible and latent heat fluxes in the irrigated Gediz Basin, Turkey, J. Hydrol. 229(1-2):87-100.

Bastiaanssen W, Menenti M, Feddes R, Holtslag A (1998a). A remote sensing surface energy balance algorithm for land (SEBAL). 1. Formulation. J. Hydrol. 212:198-212.

Bastiaanssen W, Pelgrum H, Wang J, Ma Y, Moreno J, Roerink G, Van der Wal T (1998b). A remote sensing surface energy balance algorithm for land (SEBAL). Part 2: Validation. J. Hydrol. 212:213-229.

Boegh E, Poulsen R, Butts M, Abrahamsen P, Dellwik E, Hansen S, Hasager C, Ibrom A, Loerup J, Pilegaard K (2009). Remote sensing based evapotranspiration and runoff modeling of agricultural, forest and urban flux sites in Denmark: From field to macro-scale. J. Hydrol. 377(3-4):300-316.

Boegh E, Soegaard H, Thomsen A (2002). Evaluating evapotranspiration rates and surface conditions using Landsat TM to estimate atmospheric resistance and surface resistance. Remote. Sens. Environ. 79(2-3):329-343.

Drexler J, Snyder R, Spano D (2004). A review of models and micrometeorological methods used to estimate wetland evapotranspiration. Hydrol. Process. 18(11):2071-2101.

Ficklin D, Luo Y, Luedeling E, Zhang M (2009). Climate change sensitivity assessment of a highly agricultural watershed using SWAT. J. Hydrol. 374(1-2):16-29.

Gao Y (2008). Intercomparison of remote sensing-based models for estimation of evapotranspiration and accuracy assessment based on SWAT. Hydrol. Process. 22(25):4850-4869.

Garcia M, Raes D, Allen R, Herbas C (2004). Dynamics of reference evapotranspiration in the Bolivian highlands (Altiplano). Agric. Forest. Meteorol. 125(1-2):67-82.

Gavilán P, Lorite IJ, Tornero S, Berengena J (2006). Regional calibration of Hargreaves equation for estimating reference ET in a semiarid environment. Agric. Water. Manage. 81(3):257-281.

Granger R (1996). Comparison of surface and satellite-derived estimates of evapotranspiration using a feedback algorithm. Applications of remote sensing in hydrology. pp.71-81.

Granger R (2000). Satellite-derived estimates of evapotranspiration in the Gediz basin. J. Hydrol. 229(1-2):70-76.

Heilman J, McInnes K, Kjelgaard J, Keith Owens M, Schwinning S (2009). Energy balance and water use in a subtropical karst woodland on the Edwards Plateau, Texas. J. Hydrol. 373(3-4):426-435.

Hussey B, Odum W (1992). Evapotranspiration in tidal marshes. Estuar. Coast. 15(1):59-67.

Immerzeel W, Droogers P (2008). Calibration of a distributed hydrological model based on satellite evapotranspiration. J.

Hydrol. 349(3-4):411-424.

Jabloun M, Sahli A (2008). Evaluation of FAO-56 methodology for estimating reference evapotranspiration using limited climatic data:: Application to Tunisia. Agric. Water. Manage. 95(6):707-715.

Jacob F, Olioso A, Gu X, Su Z, Seguin B (2002). Mapping surface fluxes using airborne visible, near infrared, thermal infrared remote sensing data and a spatialized surface energy balance model. Agronomie 22(6):669-680.

Jiang D, Wang J (2003). Key hydrologiacal parameters retrieved by using remote sensing technique in the Yellow River Basin. Adv. Water Sci. 14(6):736-739.

Kannan N, White S, Worrall F, Whelan M (2007a). Sensitivity analysis and identification of the best evapotranspiration and runoff options for hydrological modelling in SWAT-2000. J. Hydrol. 332(3-4):456-466.

Kannan N, White S, Worrall F, Whelan M (2007b). Hydrological modelling of a small catchment using SWAT-2000-Ensuring correct flow partitioning for contaminant modelling. J. Hydrol. 334(1-2):64-72.

Kite G, Droogers P (2000). Comparing evapotranspiration estimates from satellites, hydrological models and field data J. Hydrol. 229(1-2):3-18.

Kustas W, Moran M, Humes K, Stannard D, Pinter P Jr, Hipps L, Swiatek E, Goodrich D (1994). Surface energy balance estimates at local and regional scales using optical remote sensing from an aircraft platform and atmospheric data collected over semiarid rangelands. Water Resour. Res. 30(5):1241-1259.

Kustas W, Norman J (1999). Evaluation of soil and vegetation heat flux predictions using a simple two-source model with radiometric temperatures for partial canopy cover. Agric. Forest. Meteorol. 94(1):13-29.

Liu R, Wen J, Wang X, Wang L, Tian H, Zhang T, Shi X, Zhang J, Lv S (2010). Actual daily evapotranspiration estimated from MERIS and AATSR data over the Chinese Loess Plateau. Hydrol. Earth. Syst. Sc. 14:47-58.

Lopez-Urrea R, Martin de Santa Olalla F, Fabeiro C, Moratalla A (2006). Testing evapotranspiration equations using lysimeter observations in a semiarid climate. Agric. Water. Manage. 85(1-2):15-26.

Mohan S (1991). Intercomparison of evapotranspiration estimates. Hydrolog. Sci. J. 36(5):447-460.

Monteith J (1973). Principles of environmental physics, Edward Arnold. London. P. 241.

Moran M, Kustas W, Vidal A, Stannard D, Blanford J, Nichols W (1994). Use of ground-based remotely sensed data for surface energy balance evaluation of a semiarid rangeland. Water. Resour. Res. 30(5):1339-1349.

Nash J, Sutcliffe J (1970). River flow forecasting through conceptual models part I--A discussion of principles. J. Hydrol. 10(3):282-290.

Neitsch S, Arnold J, Kiniry J, Williams J, King K (2001). Soil and Water Assessment Tool-Version 2000-User's Manual, Temple, Texas.

Neitsch S, Arnold J, Kiniry J, Williams J, King K (2002). Soil and water assessment tool theoretical documentation version 2000, Grassland. Soil Water Research Laboratory Temple Texas.

Overgaard J, Rosbjerg D, Butts M (2006). Land-surface modelling in hydrological perspective: a review. Biogeosciences 3(2):229-241.

Parasuraman K, Elshorbagy A, Carey S (2007). Modelling the dynamics of the evapotranspiration process using genetic programming. Hydrolog. Sci. J. 52(3):563-578.

Sun L, Song C (2008). Evapotranspiration from a freshwater marsh in the Sanjiang Plain, Northeast China. J. Hydrol. 352(1-2):202-210.

Wang J, Jiang D (2005). Study on the duality water circulation of the Yellow river basin.

Wu W, Hall C, Scatena F, Quackenbush L (2006). Spatial modelling of evapotranspiration in the Luquillo experimental forest of Puerto Rico using remotely-sensed data. J. Hydrol. 328(3-4):733-752.

Zhang B, Kang S, Li F, Zhang L (2008). Comparison of three evapotranspiration models to Bowen ratio-energy balance method for a vineyard in an arid desert region of northwest China. Agric. Forest. Meteorol. 148(10):1629-1640.

Zhou L, Zhou G (2009). Measurement and modelling of evapotranspiration over a reed (Phragmites australis) marsh in Northeast China. J. Hydrol. 372(1-4):41-47.

14

Assessment of morphometric characteristics of Shetrunji River basin using remote sensing and geographical information system (GIS)

S. S. WANDRE and H. D. RANK

Department of Soil and Water Engineering, College of Agricultural Engineering and Technology, Junagadh Agricultural University, Junagadh - 362001, India.

The study area, that is, Shetrunji basin falling in the district of Bhavanagar, Amreli and Junagadh, is a major one among 71 river basins of Saurashtra region of the Gujarat State, India. Using the remotely sensed images of the Indian Remote Sensing SatelliteP6 (IRS P6), Linear Imaging Self Scanner III (LISS III) and Carosat satellites, the maps for the theme of land use/land cover, soil, drainage, slope and contour were prepared adopting the PCI Geomatica 10.1 software. The geographical information system (GIS) analysis was made for the said themes using the ArcMap V9.2. The Shetrunji basin was found as 7th order basin. The mean bifurcation ratio was found as 4.51 for the basin and it varied from 1.8 to 4 for the 17 watersheds which indicated that the geological structures did not amply disturbed the drainage pattern. The present study aims to assess the morphometric characteristics of Shetrunji basin and the sub-watersheds for its planning and development. Various morphometric characteristics of the Shetrunji basin have been assessed by applying GIS techniques. Strahler's, method have been employed to assess the fluvial characteristics of the study region. Each morphometric characteristic is considered as a single parameter and knowledge based weight age has been assigned by considering its role in soil erosion. The morphometric properties determined for this basin as a whole and for each watershed will be useful for the efficient planning of water harvesting and groundwater recharge projects on watershed base.

Key words: Morphometry, Shetrunji river basin, watershed, thematic mapping, remote sensing and geographical information system (GIS).

INTRODUCTION

Shetrunji is a major river basin among 71 river basins of Saurashtra region of Gujarat encompassing districts of Bhavanagar, Amreli and Junagadh with 53.44, 45.21 and 1.35% of total area, respectively. The Khodiyar and Shetrunji Dams are located on Shetrunji River having catchment area of 384 and 4317 km^2, respectively. A watershed is an ideal unit for management of natural resources like land and water and for mitigation of the impact of natural disasters for achieving sustainable

development. Soil conservation is the most important measure taken to check the ravages of soil erosion in India. Land is a precious resource as it is the physical base of biomass on the earth. Conservation of such type of natural resources is important to mitigate the increasing demand of land and water resources. Morphometry is the measurement and mathematical analysis of the configuration of the earth's surface, shape and dimension of its landforms (Agarwal, 1998; Obi

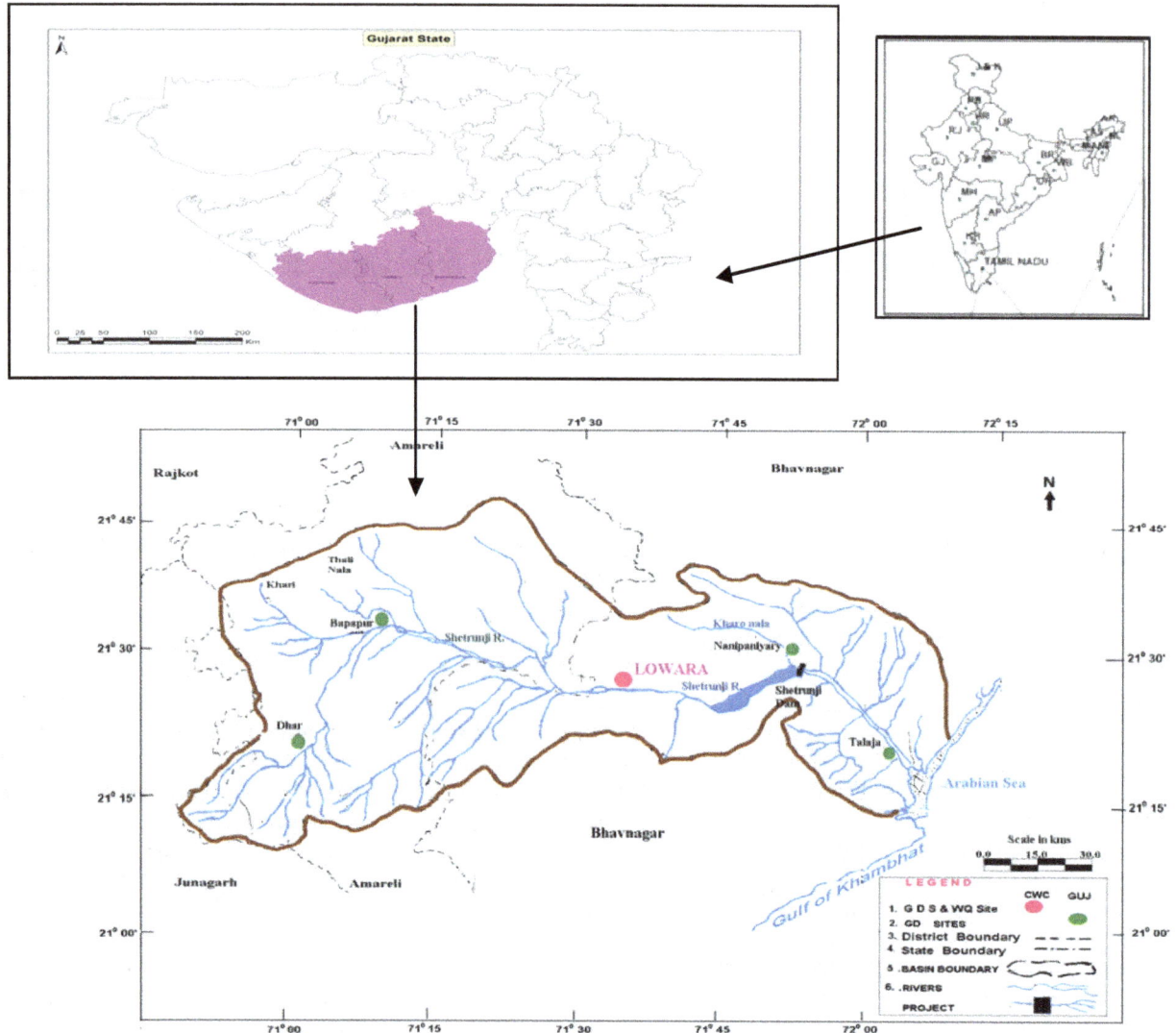

Figure 1. Location map of study area (Shetrunji river basin).

Reddy et al., 2002). The relationship between various drainage parameters and the aforesaid factors are well recognized by many workers. Recently, many researchers have used remote sensing (RS) data and analyzed them on geographical information system (GIS) platform for understanding the morphometric properties of the catchment. Analysis of various drainage parameters namely ordering of the various streams and measurement of area of basin, perimeter of basin, length of drainage channels, drainage density (Dd), drainage frequency, bifurcation ratio (Rb), texture ratio (T) and circulatory ratio (Rc) (Kumar et al., 2000). The close relationship between hydrology and geomorphology play an important role in the drainage morphometric analysis (Horton, 1932). River basins comprise a distinct morphologic region and have special relevance to drainage pattern and geomorphology (Doornkamp and King, 1971; Strahler, 1964). In a particular basin sprawl, a

drainage type is developed when a drainage network of channel lines have adjusted together with the subsurface structure. Morphometric analysis of a watershed provides a quantitative description of the drainage system which is an important aspect of the characterization of watersheds (Strahler, 1964). Various scholars have carried out morphometric analysis of river basins by using RS and GIS techniques. Shrimali et al. (2001) have worked on Sukhana lake catchment in the Shiwalik hills for the delineation and prioritization of soil erosion areas by GIS and RS.

MATERIALS AND METHODS

Study area

The study area is Shetrunji river basin in Saurashtra region of Gujarat (Figure 1). It is located between 21° 00' to 21° 47' N latitude

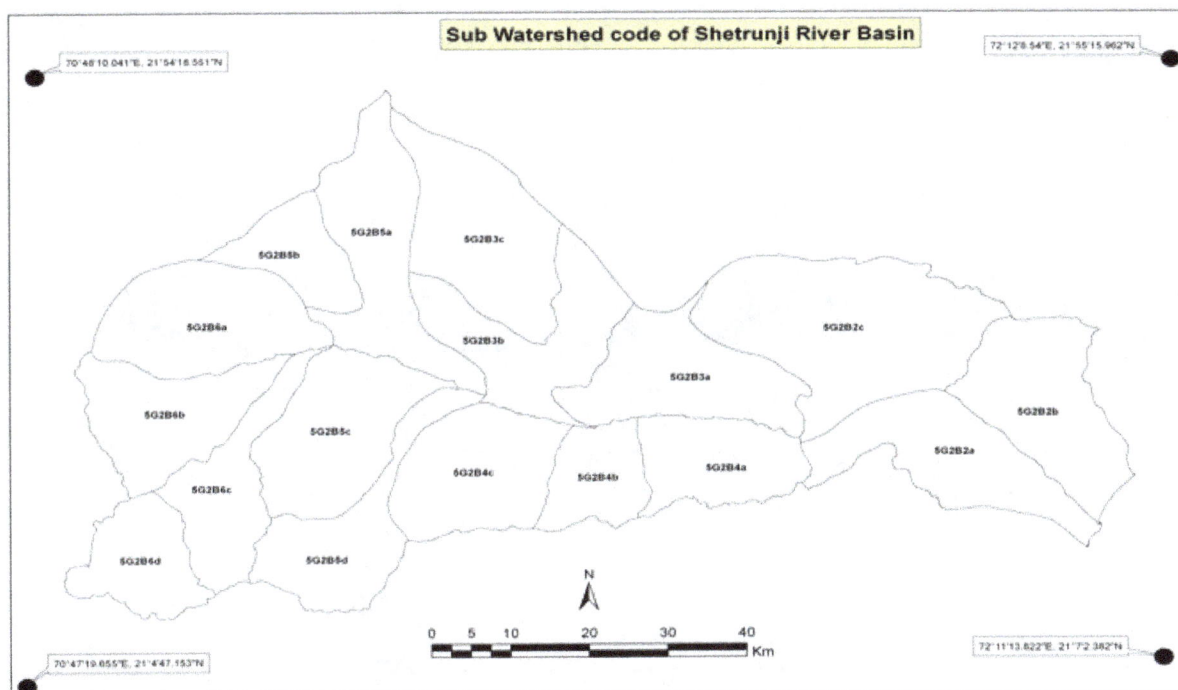

Figure 2. Subwatersheds of Shetrunji River basin.

and 70° 50' to 72° 10' E longitude. The climate of the project area can be classified as tropical and sub-tropical. Agriculture is the main occupation in the area. The river Shetrunji originates at Chchai hills in Gir forest of Junagadh district at 380 m.s.l. and flows towards east direction till it confluence with Gulf of Khambhat near Santrampur port. Its length is 227 km having 5646 km^2 catchment area with an average annual rainfall of 604.52 mm. The slope of the basin is varied from 1:1000 to 1:5000.

The RS and GIS software used for the study is PCI Geomatica V10.1 and ArcGIS-ArcMap 9.2. The satellite images of Indian Remote Sensing SatelliteP6 (IRS P6), Linear Imaging Self Scanner III (LISS III) captured in October, 2005 and February, 2008 having resolution of 23.5 × 23.5 m and images from Google Earth Pro of study area were used. Map of India with scale 1:15, 00,000, Gujarat with scale 1:37,50,000 and Watershed map of Gujarat (scale 1:37, 50,000) and soil maps of India were used for the experimental study. The thematic maps like drainage map, land use and land cover map, soil map and contour map were prepared using ArcGIS-ArcMap 9.2. Based on visual interpretation of geo-coded IRS P6-LISS III and watershed atlas of All India Land Use Survey (AILUS) the demarcation of watershed area was done in Shetrunji river basin. A total of 17 sub-watersheds were identified within this basin. Digitization work has been carried out for entire analysis of basin morphometry using GIS software (ArcGIS ver: 9.0). The order was given to each stream by following Strahler (1964) stream ordering technique. The attributes were assigned to create the digital data base for drainage layer of the river basin. The map showing drainage pattern in the study area (Figure 3) was prepared. The linear parameters like stream order, stream length, bifurcation ratio, stream length ratio and length of overland flow stream frequency, areal parameters like stream frequency, drainage density, texture ratio, elongation ratio, circularity ratio, form factor and compactness coefficient and relief parameters relief, relative relief ,relief ratio, channel slope and ground slope or watershed average slope were determined using GIS.

RESULTS AND DISCUSSION

The GIS analysis of the land use showed that area of 17.48, 17.82, 12.49, 33.32, and 19.59% were under water body, wasteland, built up, agriculture and forest, respectively. The GIS analysis showed that 25.92, 5.19, 37.80, 5.27 and 36.27% area is having soil type of clayey, skeletal clayey, fine montmorillonitic, fine loamy and loamy, respectively. It was seen that the fine soil exist in major part of the basin. It could be seen that the major land area of the basin is having the slope less than 1%. The land slope of the basin in different watershed varied from 0 to 50%. The contour values in the basin varied from 0 to 605 m. The large difference in the contour value is due to the Shetrunjay Mountain at Palitana existing in the basin. The close spacing of the contour of the basin indicated hilly ranges, while wider spacing indicates flat topography. Drainage patterns of stream network from the basin have been observed as mainly dendritic type which indicates the homogeneity in texture and lack of structural control. The basin is divided into 17 watersheds with codes viz. 5G2B2a, 5G2B2b, 5G2B2c, 5G2B3a, 5G2B3b, 5G2B3c, 5G2B4a, 5G2B4b, 5G2B4c, 5G2B5a, 5G2B5b, 5G2B5c, 5G2B5d, 5G2B6a, 5G2B6b, 5G2B6c and 5G2B6d shown in Figure 2.

Linear aspects of the watershed

Linear aspects of the 17 watersheds, related to the

Figure 3. Drainage order map of Shetrunji river basin.

channel patterns of drainage network where in the topological characteristics of the stream segments in terms of open links of the stream network system are analyzed. The parameters such as stream order, number of streams, stream length, bifurcation ratio, length of overland flow and stream length ratio are taken into account for the present study and the results have been tabulated in the Table 1 as a whole and Table 2 as watersheds.

The study area is a 7th order drainage basin covering an area of 5646.54 km^2. The total number of 8284 streams were identified of which 6285, 1512, 351, 100, 27, 8 and 1 numbers were 1st, 2nd, 3rd, 4th, 5th, 6th and 7th order streams, respectively. The higher amount of stream order indicates lesser permeability and infiltration in these sub-watersheds. The total length of the 1st order streams is highest, that is, 4861.80 km, and that of 2nd order is 1961.02 km, 3rd order is 1113.67 km, 4th order is 552.47 km, 5th order is 266.12 km, 6th order is 135.89 km and the lowest is of 7th order of 123.90 km, respectively. Generally, the higher the order, the longer the length of stream is noticed in the nature. Longer length of stream is advantageous over the shorter length, in that the former collects water from wider area and greater option for construction of a bund along the length. Lower stream lengths are likely to have lower runoff (Chitra et al., 2011).

Bifurcation ratio

Horton (1940) and Strahler (1964) defined bifurcation

ratio as the ratio of the number of streams of one order to the number of streams of the next higher order. The analysis of bifurcation value shows that the basin and its watersheds possesses well developed drainage network as the bifurcation ratio ranges between 2.8 to 4.7, that is, low value.

Stream length ratio

The value of stream length ratio ranges widely between 1.35 to 176 which shows the early stage of maturity of the watershed.

Horton's law of stream numbers

The number order relationship can be best explained by Horton's law of stream numbers which states "that the number of stream segments of successively lower orders in a given basin tend to form a geometric series beginning with the single segment of the highest order and increasing according to constant bifurcation ratio".

Horton's law of stream length

The cumulative mean lengths of stream segments of successive higher orders increase in geometrical progression starting with the mean length of the 1st order segments with constant length ratio.

Assessment of morphometric characteristics of Shetrunji River basin using remote sensing...

111

Table 1. Morphometric parameters of Shetrunji River basin.

Stream order	No. of streams	Total length of streams (km)	Mean stream length (km)	Length of overland flow (km)
		Linear aspects of basin		
1	6285	4861.80	0.77	
2	1512	1961.02	1.30	
3	351	1113.67	3.17	
4	100	552.47	5.53	0.3132
5	27	266.12	9.86	
6	8	135.89	16.99	
7	1	123.90	123.90	

$1^{st}/2^{nd}$	$2^{nd}/3^{rd}$	$3^{rd}/4^{th}$	$4^{th}/5^{th}$	$5^{th}/6^{th}$	$6^{th}/7^{th}$	Mean
			Bifurcation ratio (N_u/N_{u+1})			
4.16	4.31	3.51	3.70	3.38	8	4.51

$2^{nd}/1^{st}$	$3^{rd}/2^{nd}$	$4^{th}/3^{rd}$	$5^{th}/4^{th}$	$6^{th}/5^{th}$	$7^{th}/6^{th}$	Mean
			Stream length ratio (L_{u+1}/L_u)			
1.68	2.45	1.74	1.78	1.72	7.29	2.78

Drainage density (km/km^2)	Stream frequency (1/km)	Circularity ratio	Compactness coefficient	Form factor	Elongation ratio	Drainage texture (1/km)
			Aerial aspects of basin			
1.5965	1.4671	0.3853	1.6106	0.3023	0.6206	19.3095

Relief (km)	Relief ratio	Relative relief	Channel slope (km/km)	Ground slope (km/km)
		Relief aspects of basin		
0.605	0.004427	0.1410	0.002820	0.004427

Length of overland flow

Length of overland flow is defined as the length of flow path, projected to the horizontal, non-channel flow from point on the drainage divide to a point on the adjacent stream channel. The length of overland flow for basin 0.3132 km and for watersheds ranges from 0.2026 to 0.4419 km. The watersheds 5G2B4b, 5G2B4c, 5G2B5d, 5G2B6c and 5G2B6d are having lower values of length of overland flow that comes under the influence of high structural disturbance, low permeability, steep to very steep slopes and high surface runoff. Other remaining watersheds having length of overland flow greater than 0.25 are under very less structural disturbance, less runoff conditions and having higher overland flow. For basin, it is greater than 0.25; it comes under very less structural disturbance, less runoff conditions and having higher overland flow.

Aerial aspects of the watershed

The parameters which are governed by the area of the drainage basin are classed as area aspects of the basin. The aerial parameters include drainage density, stream frequency, elongation ratio, form factor, circularity ratio, compactness coefficient and drainage texture have been identified and results have been given in Table 2.

Drainage density

Horton has introduced drainage density (Dd) as an expression to indicate the closeness of spacing of channels. The drainage density of the basin is 1.5965 km/km^2 which comes under low drainage density. Low drainage density is more likely to occur in regions of highly permeable subsoil material under dense vegetative cover and where relief is low. The drainage density for watersheds varies from 0.1314 to 3.0857. The watersheds 5G2B4a, 5G2B4b, 5G2B5a, 5G2B5d, 5G2B6c and 5G2B6d show high drainage density (> 2 km/km^2) due to the presence of impermeable sub-surface material, sparse vegetation and high relief. Whereas remaining watersheds which fall under low drainage density indicate that the region has highly permeable

Table 2. Morphometric parameters of watersheds of Shetrunji River basin.

			Linear aspects of watersheds of Shetrunji river basin				
Watershed	Stream order	No. of streams	Total length of streams (km)	Mean stream length (km)	Length of overland flow (km)	Bifurcation ratio	Stream length ratio
5G2B2a	1	251	227.42	0.9	0.4778	4.1092	1.3586
	2	68	94.3	1.39			
	3	19	80.46	4.23			
	4	6	21.12	3.52			
	5	1	0.068	0.068			
	6	1	0.05	0.05			
	7	1	22.19	22.19			
5G2B2b	1	274	207.48	0.76	0.438	4.0736	2.0111
	2	62	120.16	1.94			
	3	16	53.98	3.37			
	4	4	28.9	7.23			
	5	1	11.58	11.58			
	6	1	0.6	0.6			
	7	1	12.97	12.97			
5G2B2c	1	580	478.75	0.81	0.3872	3.1635	1.7125
	2	148	166.67	1.13			
	3	35	118.44	3.38			
	4	10	43.12	4.31			
	5	3	20.24	6.75			
	6	1	11.4	11.4			
	7	1	15.37	15.37			
5G2B3a	1	312	231.67	0.74	0.3844	3.1647	175.54
	2	78	135.53	1.74			
	3	17	54.28	3.19			
	4	5	14.19	2.84			
	5	1	8.62	8.62			
	6	1	0.019	0.019			
	7	1	20.83	20.83			

Table 2. Contd.

5G2B3b	1	249	209.82	0.84	0.4202	2.864	1.5789
	2	63	96.83	1.54			
	3	13	31.6	2.43			
	4	5	26.76	5.35			
	5	2	13.85	6.93			
	6	2	17.39	8.7			
	7	1	11.45	11.45			
5G2B3c	1	335	273.83	0.82	0.4419	3.3716	1.8697
	2	76	96.54	1.27			
	3	16	41.99	2.62			
	4	5	37.88	7.58			
	5	2	9.56	4.78			
	6	1	10.57	10.57			
	7	0	0	0			
5G3B4a	1	474	316.08	0.67	0.1921	3.071	2.0303
	2	113	124.11	1.1			
	3	31	103.2	3.33			
	4	12	64.73	5.39			
	5	3	8.34	2.78			
	6	1	12.16	12.16			
	7	1	12.12	12.12			
5G2B4b	1	448	275.72	0.62	0.162	2.8815	1.5036
	2	113	98.02	0.87			
	3	30	59.05	1.97			
	4	13	68.68	5.28			
	5	4	28.2	7.05			
	6	2	5.11	2.56			
	7	1	2.46	2.46			
5G2B4c	1	534	477.76	0.89	0.1919	2.6554	1.6702
	2	136	185.59	1.37			

Table 2. Contd.

	3	39	115.56	2.96			
	4	12	59.61	4.97			
	5	3	39.89	13.3			
	6	1	4.01	4.01			
	7	0	0	0			
5G2B5a	1	346	277.81	0.8	0.4115	1.8461	2.2711
	2	78	82.21	1.05			
	3	18	43.63	2.42			
	4	2	26.79	13.4			
	5	3	12.12	4.04			
	6	2	9.2	4.6			
	7	1	14.02	14.02			
5G2B5b	1	172	140.3	0.82	0.3798	3.94	0.5643
	2	36	49.48	1.37			
	3	6	13.65	2.27			
	4	2	13.05	6.52			
	5	1	9.18	9.18			
	6	1	0.05	0.05			
	7	0	0	0			
5G2B5c	1	328	318.76	0.97	0.3127	3.185	20.052
	2	77	140.28	1.82			
	3	20	77.56	3.88			
	4	4	29.1	7.28			
	5	1	19.39	19.39			
	6	1	0.073	0.073			
	7	1	8.16	8.16			
5G2B5d	1	608	374	0.62	0.2026	3.7709	2.1779
	2	133	165.4	1.24			
	3	27	89.07	3.3			
	4	7	31.12	4.45			
	5	2	29.44	14.72			

Table 2. Contd.

Group	Index	Count	Value 1	Value 2			
	6	1	22.92	22.92			
	7	0	0	0			
5G2B6a	1	341	293.11	0.86			
	2	81	91	1.12			
	3	21	47.55	2.26			
	4	4	12.82	3.21	0.41	3.0528	1.249
	5	2	9.05	4.53			
	6	2	8.01	4			
	7	1	1.84	1.84			
5G2G6b	1	222	226.75	1.02			
	2	56	85.7	1.53			
	3	14	61.91	4.42			
	4	3	17.52	5.84	0.379	3.3262	1.7262
	5	1	16.68	16.68			
	6	1	1.09	1.09			
	7	0	0	0			
5G2B6c	1	410	275.31	0.67			
	2	99	116.7	1.18			
	3	24	57.59	2.4			
	4	6	18.15	3.03	0.2494	3.3778	2.0648
	5	1	6.69	6.69			
	6	1	33.79	33.79			
	7	1	2.5	2.5			
5G2B6d	1	530	267.57	0.51			
	2	122	112.46	0.92			
	3	28	64.15	2.29			
	4	8	38.93	4.87	0.2116	3.6403	1.767
	5	2	23.23	11.61			
	6	1	0.15	0.15			
	7	0	0	0			

Table 2. Contd.

| | Aerial aspects of watersheds of Shetrunji river basin | | | | | | |
Watershed	Drainage density (km/km²)	Stream frequency (1/km²)	Elongation ratio	Circularity ratio	Form factor	Compactness coefficient	Drainage texture (1/km)
5G2B2[a]	1.2262	0.9522	0.606	0.3709	0.2883	1.6416	3.1191
5G2B2[b]	1.1416	0.9393	0.7362	0.5981	0.4254	1.2927	4.0018
5G2B2[c]	1.2915	1.1905	0.8031	0.5744	0.5063	1.3191	6.5082
5G2B3[a]	1.3007	1.1605	0.6505	0.4242	0.3219	1.5349	4.0333
5G2B3[b]	1.1899	0.9777	0.6314	0.2807	0.3129	1.887	2.7056
5G2B3[c]	1.1314	1.0463	0.5702	0.5839	0.2553	1.3083	4.6002
5G2B4[a]	2.6025	2.5792	0.7456	0.6002	0.4365	1.2905	8.8469
5G2B4[b]	3.0857	3.5093	0.7043	0.6727	0.3894	1.219	10.716
5G2B4[c]	2.605	2.1403	0.7681	0.7752	0.4631	1.1354	9.7871
5G2B5[a]	1.215	1.1738	0.4672	0.3047	0.1714	1.8112	3.5797
5G2B5[b]	1.3164	1.266	0.7748	0.5669	0.4712	1.3279	3.5212
5G2B5[c]	1.5991	1.1643	0.7646	0.6475	0.4589	1.2425	5.092
5G2B5[d]	2.4678	2.6968	0.4882	0.3125	0.1871	1.7885	7.2247
5G2B6[a]	1.2195	1.1895	0.7119	0.7575	0.3978	1.1487	5.6945
5G2B6[b]	1.3191	0.9564	0.6956	0.5097	0.3798	1.4003	3.3952
5G2B6[c]	2.0047	2.1275	0.4503	0.3087	0.1592	1.7994	5.3236
5G2B6[d]	2.3628	3.2235	0.8467	0.5582	0.5628	1.3382	9.9494

| | Relief aspects of watersheds of Shetrunji river basin | | | | |
Watershed	Relief (km)	Relative relief (km/km)	Relief ratio	Channel slope (km/km)	Ground slope (km/km)
5G2B2[a]	0.34	0.3065	0.00958	0.0061	0.00958
5G2B2[b]	0.2	0.2236	0.00668	0.004256	0.00668
5G2B2[c]	0.315	0.2635	0.00876	0.005584	0.00876
5G2B3[a]	0.445	0.4325	0.01356	0.008639	0.01356
5G2B3[b]	0.13	0.105	0.00393	0.002502	0.00393
5G2B3[c]	0.11	0.1163	0.00273	0.001736	0.00273
5G2B4[a]	0.22	0.3065	0.00926	0.0059	0.00926
5G2B4[b]	0.175	0.3069	0.00828	0.005271	0.00828
5G2B4[c]	0.14	0.1889	0.00517	0.003297	0.00517
5G2B5a	0.115	0.0915	0.00243	0.001549	0.00243
5G2B5[b]	0.075	0.1217	0.00393	0.002505	0.00393

Table 2. Contd.

5G2B5c	0.15	0.1768	0.00528	0.00336	0.00528
5G2B5d	0.315	0.2925	0.00802	0.005109	0.00802
5G2B6a	0.095	0.1197	0.00307	0.001958	0.00307
5G2B6b	0.16	0.1829	0.0056	0.003564	0.0056
5G2B6c	0.275	0.2701	0.00687	0.004379	0.00687
5G2B6d	0.375	0.5399	0.0192	0.01224	0.0192

subsoil and dense vegetation cover (Sethupathi et al., 2011).

Stream frequency

The stream frequency is defined as the total number of stream segment of all order per unit area. The stream frequency for basin is 1.4671 and for watersheds varies from 0.9393 to 3.50931. It is low due to permeable rocks, the surface runoff is low and infiltration capacity is high within in the study area (Chitra et al., 2011). The stream frequency for all 17 watersheds of the study area shows direct relation with the drainage density which indicates that the stream population increases with the increase of drainage density (Rao et al., 2011).

Elongation ratio

It is the ratio of diameter of the circle of the same area in the basin to the maximum basin length. The elongation ratio for the basin is 0.6206 indicating less elongated in nature and for the watersheds, it varies from 0.45 to 0.85. The watersheds 5G2B2b, 5G2B2c, 5G2B4a, 5G2B4b, 5G2B4c, 5G2B5b, 5G2B5c, 5G2B6a and 5G2B2d are elongated in nature, while remaining are less

elongated in nature.

Form factor

The ratio of the basin area to the square of basin length is called the form factor. The form factor for basin is 0.3023 and for basin watersheds varying from 0.16 to 0.56. These low values indicate that watersheds have flatter peak flow for longer duration. The watershed 5G2B5a, 5G2B5d and 5G2B6c are circular in shape showing less side flow for shorter duration and high main flow for longer duration (Chitra et al., 2011). The remaining watersheds are elongated watershed, indicating that they will have a flatter peak flow for longer duration. Flood flows of such elongated basins are easier to manage than from the circular basin.

Circularity ratio

Circularity ratio is the ratio of the basin area to the area of a circle having the same circumference perimeter as the basin. The circularity ratio for watersheds varies from 0.37 to 0.77 and for basin it is 0.3853 indicating elongated in shape, low discharge of runoff and highly permeability of the subsoil condition (Miller, 1953).

Compactness coefficient (C_c)

The C_c is independent of size of watershed and dependent only on the shape. The compactness coefficient for watersheds ranges from 1.14 to 1.79 and for basin is 1.6106. They have elongated shape so they have enough time for discharge.

Drainage texture

It is the total number of stream segments of all orders per perimeter of the area. The texture ratio for watersheds varies from 3.3 to 10.7. For watershed 5G2B4a, 5G2B4b 5G2B4c and 5G2B6c, it is greater than 8 indicating very fine texture that is, higher runoff potential, while 5G2B5d, 5G2B5c, 5G2B6a and 5G2B6d is moderate in nature. The rest watersheds are coarser in nature that is, having less runoff potential. For basin, it is 19.3095 showing very fine nature (Smith, 1950).

Relief aspects of the watershed

The relief aspects of drainage basin are also important in water resources studies. The character of the distribution of slope, angles sampled over the whole basin depends on the height distribution within it. Relief aspects like relief, relative relief, relief ratio, channel slope and ground slope were measured.

Relief

It is defined as the elevation difference between the reference points located in the drainage basin. The relief of basin is 0.605 km. The study area is of high relief region as it is greater than 0.3 km. The high relief value indicates low gravity of water flow as well as infiltration and high runoff conditions. The relief for watersheds varies from 0.095 to 0.445 km. The watersheds 5G2B5b and 5G2B6d are of low relief region, 5G2B2c, 5G2B3a, 5G2B5d and 5G2B6a are of high relief region and remaining are of moderate relief region. The high relief value indicates low gravity of water flow as well as infiltration into the ground and high runoff conditions .

Relief ratio

It is the ratio of relief to the horizontal distance on which relief was measured. The relief ratio for basin is 0.004427 and for watersheds, it varies from 0.00273 to 0.019. It was noticed that the higher values of relief ratio indicated steep slope and high relief (5G2B6d watershed), while the lower values in case of watershed 5G2B5a indicated the presence of basement rocks that are exposed in the form of small ridges and mounds with lower degree of slope (GSI, 1981).

Relative relief

It is the ratio of relief to the perimeter of basin. It is an important morphometric variable used for the overall assessment of morphological characteristics of terrain (Suresh, 2002). The relative relief for watersheds varies from 0.0915 to 0.5399 and for basin, it is obtained as 0.1410. The watersheds having higher relative relief have higher runoff potential than others. Therefore, the watershed 5G2B5a and 5G2B5d are having the lowest and highest runoff potential.

Channel slope

For watersheds, it varies between 0.001549 to 0.01224 km/km and for basin it is 0.002820 km/km. The higher channel slopes in 5G2B6d watershed indicated less time of concentration that is, peak flow occurs in short time, while lower slope in 5G2B5a watershed indicated less peaked flow for longer duration. Therefore, while constructing the water harvesting structures on channel of watershed 5G2B6d, the outlet should be designed of higher discharge capacity and the rest components like headwall, sidewall and wing wall should also be of higher

height for the designed storage capacity (Suresh, 2002). The drop structures in series in the channels of this watershed are recommended.

Ground slope

It is the product of drainage density and relief of the basin (Suresh, 2002). For watersheds, it is obtained as 0.00307 to 0.0192 km/km and for basin 0.004427 km/km. The higher ground slopes in case of 5G2B6d lying in upper reach of the basin indicates lower time of concentration of overland flow. Also, the possibilities of soil erosion will be higher in this 5G2B6d watershed among all watersheds of this basin.

Conclusion

One of the purposes of fluvial morphometry is to derive information in quantitative form about the geometry of the fluvial system that can be correlated with hydrologic information. Usually, morphometric analysis of drainage system is a prerequisite to any hydrological study. The watersheds 5G2B4a, 5G2B4b, 5G2B5a, 5G2B5d, 5G2B6c and 5G2B6d show high drainage density due to the presence of impermeable sub-surface material, sparse vegetation and high relief, while remaining watersheds fall under low drainage density which indicate that the region has highly permeable subsoil and dense vegetation cover. The development of stream segments in the basin area is more or less affected by rainfall. The present study demonstrates the usefulness of GIS for morphometric analysis of the watersheds of Shetrunji river basin, Gujarat. Thus, the morphometric properties determined for this basin as whole and for each watershed will be useful for the sound planning of water harvesting and groundwater recharge projects on watershed base.

REFERENCES

Agarwal CS (1998). Study of drainage pattern through aerial data in Naugarh area of Varanasi district. U.P. J. Indian Soc. Rein. Sens. 24(4):169-175.

Chitra C, Alaguraja P, Ganeshkumari K, Yuvaraj D, Manivel M (2011). Watershed characteristics of Kundah sub basin using Remote Sensing and GIS techniques. Int. J. Geomatics Geosci. 2(1):311-335

Doornkamp JC, King CAM (1971). Numerical analysis in Geomorphology: An Introduction. St. Martin's Press, New York. P. 372.

GSI (1981). Geological and Mineralogical Map of Karnataka and Goa. Geological Survey of India.

Horton RE (1932). Drainage basin characteristics. Trans. Amer. Geophysics Union. 13:350-361.

Horton RE (1940). An approach toward a physical interpretation of infiltration capacity. Proc. Soil Sci. Soc. Am. 5:399-417.

Kumar R, Lohani AK, Nema RK, Singh RD (2000). Evaluation of Geomorphological characteristics of catchment using GIS. GIS India. 9(3):13-17.

Miller VC (1953). A quantitative geomorphic study of drainage basin characteristics in the Clinch Mountain area, Varginia and Tennessee, Project NR 389042, Tech. Rept. 3.,Columbia University, Department of Geology, ONR, Geography Branch. New York.

Obi Reddy GE, Maji AK, Gajbhiye KS (2002). GIS for Morphometric Analysis of drainage basins. GIS India. 4:9-14.

Rao LAK, Ansari ZR, Yusuf A (2011). Morphometric A nalysis of Drainage Basin Using Remote Sensing and GIS Techniques: A Case Study of Etmadpur Tehsil, Agra District,U.P. Int. J. Res. Chem. Environ. 1(2):36-45.

Shrimali SS, Agarwal SP, Samra JS (2001). Prioritizing Erosion Prone areas in hills using Remote Sensing and GIS: A case study of Sukhna Lake Catchment, North India. J. Appl. Geol. 3(1):54-60.

Strahler AN (1964). Quantitative geomorphology of drainage basins and channel networks. In: V.T. Chow (Ed.), Handbook of Applied Hydrology. McGraw-Hill, New York. pp. 4.39-4.76.

Suresh R (2002). Soil and water conservation engineering, Standard Publishers Distributors. Delhi. pp. 793-812.

15

Effects of seasonal change in Osinmo reservoir on arginase and rhodanese activities in *Clarias gariepinus* Burchell and *Heterotis niloticus* Cuvier

R. E. Okonji[1], O. O. Komolafe[2], M. O. Popoola[2] and A. Kuku[1]

[1]Department of Biochemistry, Obafemi Awolowo University, Ile-Ife, Nigeria.
[2]Department of Zoology, Obafemi Awolowo University, Ile-Ife, Nigeria.

This work reports the seasonal change and variation in the physico-chemical properties of the water in Osinmo reservoir in Southwestern Nigeria and the effects on tissue distribution of the metabolic enzymes, arginase and rhodanese are the two fish species (*Clarias gariepinus* and *Heterotis niloticus*) collected from the reservoir. This study was carried out to correlate the activities of the enzymes with metabolic status of the fishes and physico-chemical properties of the water. The enzyme activities varied significantly in the different tissues of the *C. gariepinus*. The liver showed the highest mean value of activity and the intestine had the least mean value. In the *H. niloticus*, they showed the highest mean value of activity while the bile showed the lowest mean value. The activities of the two enzymes in the water reservoir were determined at regular intervals over a period of ten months in 2011, spanning both dry and rainy seasons. The activities of the enzymes varied significantly through the two seasons. Arginase was at its peak in June while rhodanese was at its peak in September. The distribution of urea in the reservoir varied significantly in the period studied; the highest mean value was in July while the lowest was in February.

Key words: Water properties, urea; cyanide, enzymes, tissue distribution.

INTRODUCTION

Fish are important vertebrates which contribute as much as 17% of the world's animal protein. Inland fisheries play important role in the provision of protein to Nigerians. The downward trend in fish as food has been attributed partly to environmental, population increase, poor management practices and over-exploitation of water ways (Komolafe and Arawomo, 2008). Free cyanide is the primary toxic agent in the aquatic environment (Eisler, 1991). Numerous accidental spills of sodium cyanide or potassium cyanide (KCN) into rivers and streams have resulted in massive kills of fishes, amphibians, aquatic insects, and aquatic vegetation through discharge of substances generating free hydrogen cyanide (HCN) in the water from hydrolysis or decomposition (Leduc, 1984; Eisler, 1991).

Cyanide adversely affects fish reproduction by reducing the number of eggs spawned, and the viability of the eggs by delaying the process of secondary yolk deposition in the ovary (Lesniak and Ruby, 1982; Ruby et al., 1986). Other adverse effects of cyanide on fish include delayed mortality, pathology, impaired swimming ability and relative performance, susceptibility to predation, disrupted respiration, osmoregulatory disturbances, and altered growth patterns. Cyanide acts rapidly in aquatic environments, does not persist for extended periods, and is highly species selective; organisms usually recover quickly on removal to clean water. The critical sites for cyanide toxicity in freshwater organisms include the gills, egg capsules and other sites where gaseous exchange and osmoregulatory processes occur (Eisler, 1991).

In aquatic organisms the most common form of nitrogen waste is ammonia, while land-dwelling organisms convert the toxic ammonia to either urea or uric acid. Urea is found in the urine of mammals and amphibians, as well as some fish. It is noteworthy that tadpoles excrete ammonia but shift to urea production during metamorphosis. Despite the generalization above, the urea pathway has been documented not only in mammals and amphibians but in many other organisms such as birds, invertebrates, insects, plants, yeast, fungi and microorganisms. Environmental conditions have also been reported to be the common stimulus for urea synthesis in some fishes (Campbell and Anderson, 1991; Okonji et al., 2011). Anthropogenic impacts on fish populations in Osinmo and other reservoirs in the Southwestern Nigeria include not only water pollution, but also the extensive fishery, as well as habitat destruction by dam building and river modifications (Lenhardt et al., 2004; Atobatele, 2008; Komolafe and Arawomo, 2008; Okonji et al., 2011). In addition, temperature and water levels are both crucial factors affecting enzyme activities in aquatic environment. Therefore, the present study investigates the effects of seasonal variation in the water properties of Osinmo reservoir and its effect on arginase and rhodanese distribution in the tissues of *Clarias gariepinus* and *Heterotis niloticus*.

MATERIALS AND METHODS

Study area and collection of fish samples

Osinmo reservoir was created in 2005 by the impoundment of Ataro river and other streams. The reservoir lies between latitude 07° 52.8' N to 07° 53.2' N and Longitude 04° 21.2' E to 04° 21.7' E. The catchment area is about 102 Km^2. The surface area of the reservoir is about 0.78 Km^2 with a mean maximum depth of 3.2 m, cast-net was used once a month to collect fish samples. Water samples were collected forthnight between April 2010 and February, 2011 in Osinmo reservoir. The fish samples were stored in an ice-chest covered with ice before transporting to the laboratory where they were stored at temperature below 0°C until ready for use. The fish species, *C. gariepinus* and *H. niloticus* were identified using the keys by Reed (1967), Paugu et al. (2003) and Adesulu and Sydenham (2007). All reagents used were of analytical grades.

Measurement of physico-chemical parameters

Mercury-in-glass thermometer was used to take water temperature. Dissolved oxygen (DO) was fixed on the field between 8.00 and 9.00 am. The DO was determined by titrating fixed water samples with sodium thiosulphate using starch indicator (N/40) until the sample changed from blue-black to colourless solution (Golterman et al., 1978). The pH of water was measured using the pH meter (Mettler MP200). Total alkalinity of the water was determined by titrating water samples with sulphuric acid standard solution, using a drop of phenolphthalein solution and one sachet of bromocresol green-methyl red as indicator until the sample changed from blue green to pink (Golterman et al., 1978). Water transparency of the

reservoir was determined with a Sechi-disc measuring 15 cm in diameter (Quayle, 1988).

Preparation of tissue extract

Prior to extraction, the *C. gariepinus* and *H. niloticus* fish species were slit open and the various tissues of interest (liver, intestine, bile and stomach) were removed and stored at 4°C until required. Tissue extracts were prepared by homogenising 10 g (w/v) of each tissue in 3 volume of homogenisation buffer (phosphate buffer, pH 7.2). The suspensions were centrifuged for 20 min at 4,000 rpm in a Microfield Centrifuge Model 800 D. The supernatants were used as the source of enzyme.

Arginase assays

Arginase activity was determined by the measurement of urea produced by the reaction of Ehrlich's reagent according to the modified method of Kaysen and Strecker (1973). The reaction mixture contained, in final concentration, 1.0 mM Tris-HCl buffer, pH 9.5 containing 1.0 mM $MnCl_2$, 0.1 M arginine solution and 50 µl of the enzyme preparation in a final volume of 1.0 ml. The mixture was incubated for 10 min at 37°C. The reaction was terminated by the addition of 2.5 ml Erhlich reagent (2.0 g of p-dimethylaminobenzaldelyde in 20.0 ml of concentrated hydrochloric acid and made up to 100 ml with distilled water). The optical density reading was taken after 20 min at 450 nm. The urea produced was estimated from the urea curve (graph of optical density against urea concentration). The unit of activity of arginase is defined as the amount of enzyme that will produce one µmol of urea per min at 37°C.

Rhodanese and protein assay

Rhodanese activity was measured during purification and routinely according to the method employed by Agboola and Okonji (2004) using KCN and $Na_2S_2O_3$ as substrates. The activity of the enzyme is expressed in Rhodanese unit (RU). One Rhodanese unit is taken as the amount of enzyme which under the given conditions will produce an optical density reading of 1.08 at 460 nm. Bradford (1976) method was used to measure the protein concentration of the enzyme using Bovine Serum Albumin (BSA) as standard.

Urea concentration of Osinmo reservoir

Urea concentration was determined using Erlich spectrophotometric method. The mixture contained 0.2 mM Tris-HCl buffer pH 7.5, water sample and 20% Erlich reagent. Optical density was read after 25 min at 450 nm. A calibration standard was prepared with 50 µM urea.

Statistical analysis

The results are presented as means ± SD. Data were analyzed by one-way ANOVA using SAS/PC soft ware. Duncan multiple range test was used for paired comparisons. A p-value < 0.05 was considered statistically significant.

RESULTS

The physicochemical parameters of Osinmo reservoir is

Table 1. Physicochemical parameters of Osinmo reservoir (May, 2010-February, 2011).

Water parameter (Mean)	Rainy season	Dry season
Water temperature (°C)	24.83±0.75	27.0±0.56
Dissolved O_2 content (mg/L)	1.97±0.17	2.0±0.33
Transparency (cm)	78.17±3.27	89.83±1.56
pH	7.05±0.18	7.43±0.05
Total alkalinity (mg/L)	77.33±3.29	96.5±3.68

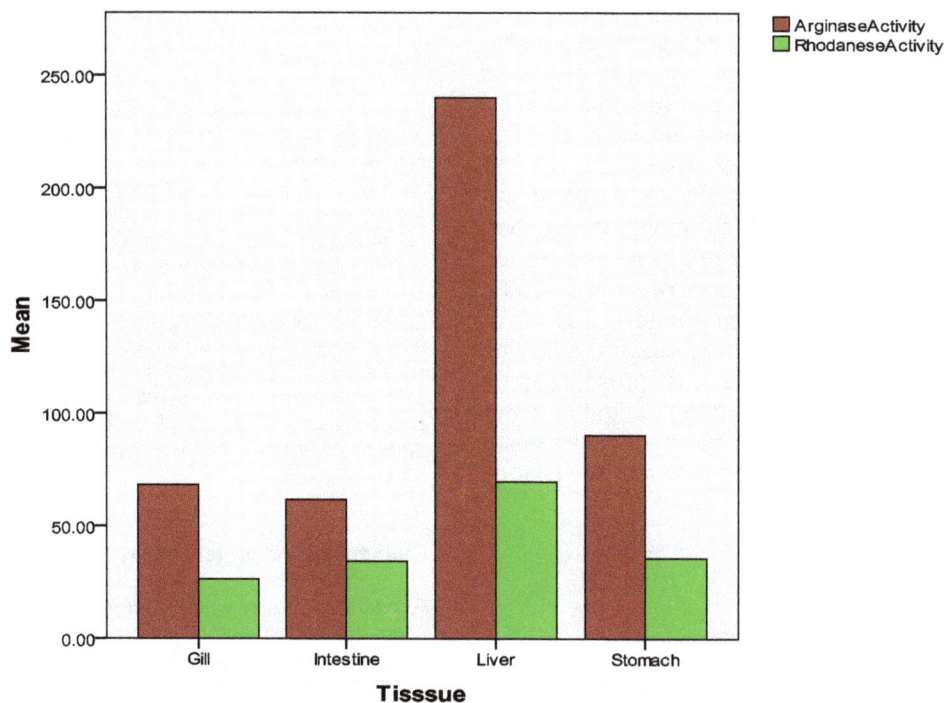

Figure 1. Tissue distribution of arginase and rhodanese enzymes in the tissues of *C. gariepinus*.

presented in Table 1. The dissolved oxygen concentration was fairly low in the rainy season (1.97 ± 0.17) as compared to the dry season (2.0 ± 0.33), while total alkalinity was lower in the rainy season (77.33 ± 3.29) than the dry season (96.5 ± 3.68). The variation in water temperature was high in the dry season as compared to rainy season (27.0 ± 0.56°C and 24.83 ± 0.75 respectively). Similarly, hydrogen ion concentration was moderately high in dry season (7.43 ± 0.05) than in the rainy season (7.05 ± 0.18). High water temperature in the dry season was also observed when compared to the rainy season.

Two different species of fish used in the study were *C. gariepinus* and *H. niloticus*. In Osinmo reservoir, *C. gariepinus* with 20.1% of the population was well represented (Komolafe and Arawomo, 2008) while *H. niloticus* was not observed until the present study. Distribution of the enzyme activities varies significantly in

both species of fish. The liver of *C. gariepinus* showed the highest mean value as compared to the intestine with lowest activity (Figure 1). The distribution of the enzyme activities in the *H. niloticus* tissues also showed the liver to have the highest mean value while the bile showed lowest mean value (Figure 2). The enzymes, arginase and rhodanese varied significantly in their distribution throughout the season. Arginase peak was in June, 2010, while Rhodanese activity was at its peak in September, 2010 (Figure 3). The distribution of urea varies significantly in the different months. With the highest mean value in July, 2010 and lowest mean value in February, 2011 (Figure 4).

DISCUSSION

It is well documented that humans have greatly altered

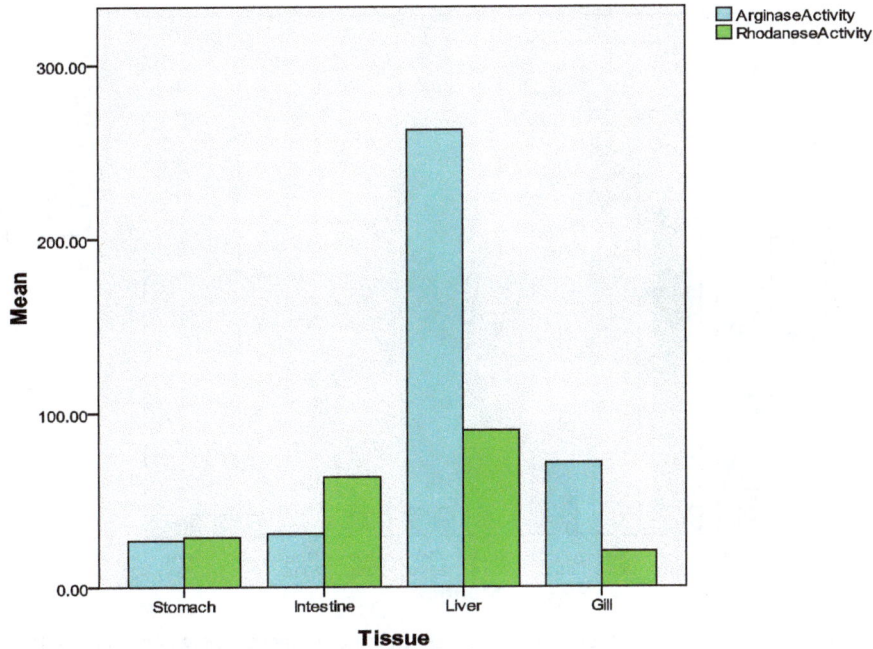

Figure 2. Tissue distribution of arginase and rhodanese enzymes in the tissues of *H. niloticus*.

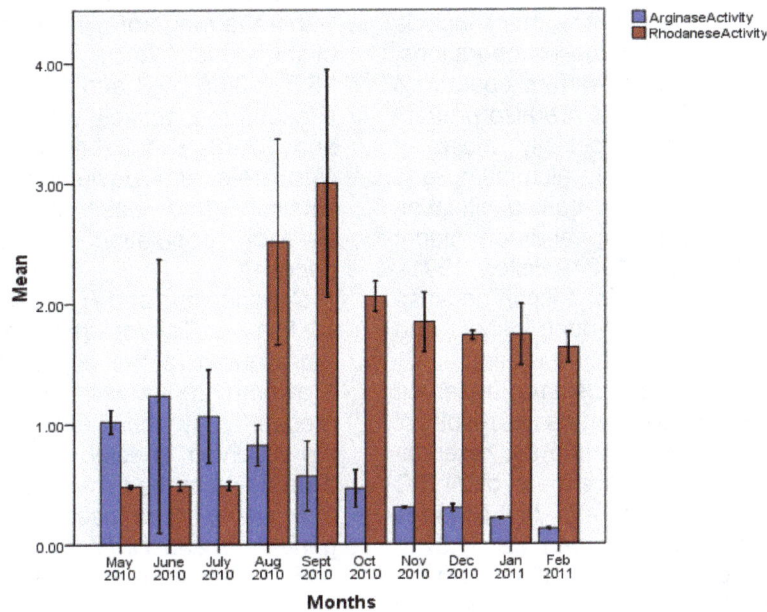

Figure 3. Distribution of arginase and rhodanese in water of Osinmo reservoir between May, 2010 and March, 2011.

predatory fish communities worldwide, especially through industrialized commercial and recreational fisheries (Christensen et al., 2003; Worm et al., 2005; Stallings, 2009). Coastal regions and areas that are home to large and growing proportion of the world's population are undergoing environmental decline. The problem is particularly acute in developing countries. The reasons for environmental decline are complex, but population factors play a significant role. Contaminants and activities that destroy habitats and ecosystems also contribute to

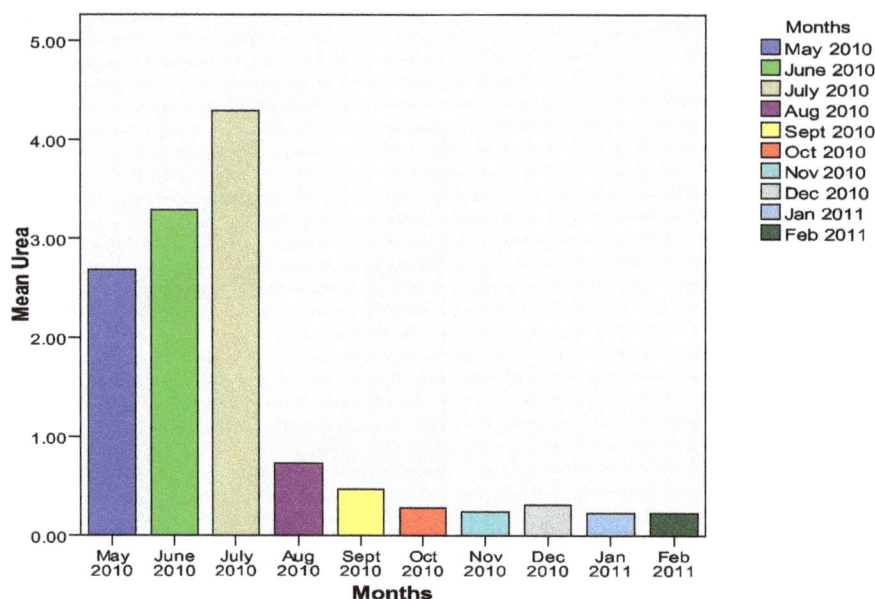

Figure 4. Distribution of urea in water of Osinmo reservoir between May, 2010 and February, 2011.

the loss of fresh water and marine fishes on which many people rely for food and income. Human anthropogenic activities relating to cyanide in an environment include industrial processes, laboratories, fumigation operations, cyanogenic drugs, fires and chemical warfare operations (Marrs and Ballantyne, 1987). Cyanides are also present in many industrial wastewaters, especially those of electroplaters; manufacturers of paint, aluminum, and plastics; metal finishers; metallurgists; coal gasification processes; certain mine operations; and petroleum refiners (Towill et al., 1978; Way, 1984; Eisler, 1991). Maintaining a healthy environment is critical because most of the world's fish produce their young inshore and feed on organisms in both fresh and marine waters.

Physicochemical parameters of Osinmo reservoir showed variations during the season as a result of suspended particulate matters brought into the reservoir by flood. Water temperature varied with a mean of 24.0°C ± 0.75 and 27.0°C ± 0.56 in rainy and dry seasons respectively. The means dissolved oxygen was 1.97 ± 0.17 mg/L during rainy season and 2.0 ± 0.33 mg/L in the dry season. Akinbuwa (2008) and Komolafe and Arawomo (2008) also observed variation in Osinmo and Opa reservoir. Oke (1998) also recorded a dissolved oxygen concentration below 5 mg/ml in Owena reservoir. Low dissolved oxygen production in the present study could be due to phytoplankton bloom and decomposition of allochthonou organic materials in the reservoir as reported by Okayi (2003). Water transparency was higher in dry season (89.83 ± 1.56 cm) than in the rainy season. Low transparency of the reservoir in rainy season might be attributed to flood resulting to increase in turbidity. The

pH of 7.05 ± 0.18 and 7.43 ± 0.05 observed in rainy and dry seasons respectively were moderately alkaline and within the range of pH known for most lakes and streams of the world (Welch, 1952). The mean total alkalinity of 96.5 ± 3.68 mg/l and 77.33 ± 3.29 mg/l in rain and dry seasons respectively were high compared to 66.15 ± 15 and 63.68 ± 1.29 mg/ml reported to Aiba reservoir (Atobatele and Ugwumba, 2008). Wide range in total alkalinity had been attributed to season, location, plankton population and the nature of the bottom deposits.

C. gariepinus and *H. niloticus* represents two of species of fish in Osinmo reservoir. *C. gariepinus* was well represented in the population by 5.8% (Komolafe and Arawomo, 2008), while *H. niloticus* was no observed until present study. Influx of water from adjoining streams into the reservoir affected the quality of water and enzyme distribution in the fishes during the season. The presence of arginase in the reservoir water and its distribution in *C. gariepinus* and *H. niloticus* is an indication of the acidity and alkalinity of the reservoir. Arginase, an enzyme known to catalyse the conversion of arginine to urea becomes more effective in aquatic organisms, especially fresh water fishes when their environment becomes polluted and made more alkaline (Campbell and Anderson, 1991; Mommsen and Walsh, 1992; Wood, 1993). During the rainy season, allochotonous materials flushed into the reservoir and affected its status (Eisler, 1991; Okonji et al., 2010, 2011). Saha and Ratha (2007) had also reported *Heteropneustes fossilis* and *C. batrachus* to be hardy and capable of living in derelict water bodies and tolerating temporary water deprivation.

Several studies have been made on their ureogenic adaptations, ureogenic metabolic machinery and regulation under different physiological and environmental conditions. Both species (*H. fossilis* and *C. batrachus*) are potentially ureogenic teleosts expressing the complete repertoire of ornithine-urea cycle (OUC) enzymes, not only in hepatic tissue but also in certain non-hepatic tissues. They have developed peculiar ureogenic machinery and the induction of ureogenesis during adaptation to deal with the various stressful conditions such as exposure to high environmental ammonia, water deprivation and highly alkaline environment (Saha and Ratha, 2007).

Rhodanese also known as thiosulphate-cyanide sulphurtransferase is widely distributed in the body (Westley, 1980; Agboola and Okonji, 2004), but activity levels in mammals are highest in the mitochondrial fraction of liver. Rhodanese detoxifies cyanide to a less toxic thiocyanate. Its activity was found to be high in rainy season at Osinmo reservoir. This high activity could be explained based on the premise that the enzyme can be induced in the presence of cyanide (Nakajima et al., 2008). Leduc (1984) reported the increase in the concentration of cyanide in larger rivers with peaks more frequent during the summer because of cyanide production by plants and low in winter owing to dilution by high runoff.

The dissolved oxygen concentration of Osinmo reservoir obtained in the rain season was lower compared to the dry season. Reports have shown that cyanide is more toxic to freshwater fish under conditions of low dissolved oxygen (Doudoroff, 1976; Towill et al., 1978; Smith et al., 1979; EPA, 1980; Leduc, 1984). The pH levels within the range 6.8 to 8.3 had little effect on cyanide toxicity but enhanced toxicity at acidic pH (Smith et al., 1979; EPA, 1980; Leduc et al., 1982; Leduc, 1984). The present pH result throughout the season stable. However, changes in the dissolved oxygen, high concentration of cyanide affects survival of different fish species at different developmental stages. This could explain the low population of *H. niloticus* in the Osinmo reservoir. This is also supported by the report that juveniles and adults fishes were the most sensitive life stages tested and embryos and sac fry where the most resistant (Smith et al., 1978; Leduc, 1984; Eisler, 1991). On the distribution of arginase and rhodanese enzymes in different tissues of the fish species (*C. gariepinus* and *H. niloticus*), the activities of these enzymes were found to be more in the liver. This is not uncommon going by the function of liver in metabolism and in particular, detoxification. These results provide further evidence of the importance of the two enzymes in the survival of the fishes. Raymond (1998) has shown the importance of trimethylamine Oxide (TMAO) and urea in some cold water fishes, his result demonstrated that the synthetic machinery for these osmolytes are present in the liver. Similarly, Wai et al. (2003) reported that marine elasmobranchs (sharks, skates and rays) which are common in tropical waters, exhibit osmoconforming hypoionic regulation with body fluid osmolalities equal to or slightly higher than the environment. They also found that marine elasmobranchs possess an active ornithine-urea cycle synthesizing urea through carbamoylphosphate synthetase III primarily for osmoregulation (Ballantyne, 1997; Anderson, 2001; Yancey, 2001; Wai et al., 2003).

In conclusion, the results from this investigation on urea, arginase and rhodanese activities could be used as good indicators of fish population in the water systems. This could be supported by Lenhardt et al. (2004) report that urea and creatinine were used as indicators of state of fish population in some water systems. They were also reported to be indicators of gill and kidney dysfunction, respectively. As human population density increases, presence of large-bodied fishes declines, and fish communities become dominated by a few smaller-bodied species with complete disappearance of several large-bodied fishes suggesting ecological and local extinctions (Christensen et al., 2003). Fish populations and other aquatic resources are also likely to be affected by changes in seasonal flow regimes, total flows, water levels and water quality. These changes affect the health of aquatic ecosystems, with impacts on productivity, species diversity, and species distribution.

REFERENCES

Adesulu EA, Sydenham DHJ (2007). The fresh water fishes and fisheries of Nigeria. Ibadan: Macmillan Nigeria Publishers Ltd. Lagos, Ibadan. P. 397.

Agboola FK, Okonji RE (2004). Presence of rhodanese in the cytosolic fraction of the Fruit Bat (*Eidolon helvum*) liver. Int. J. Biochem. Mol. Biol. 37:275-281.

Akinbuwa O (2008). The studies of the physico-chemical factors and the rotifer fauna of opa reservoir. (M.Sc Thesis). Obafemi Awolowo University, Ile-Ife, Nigeria. P. 162.

Anderson PM (2001). Urea and glutamine synthesis: environmental influences on nitrogenexcretion. In: Wright PA, Anderson PM (eds), Fish Physiology and Nitrogen Excretion, New York Academic Press. pp. 239-277.

Atobatele OE (2008). Physicochemical parameters, plankton, macrozoobenthos and some aspects of the biology of two species of chrisichthys (Bagridae) of Aiba reservoir, Iwo, Osun State Nigeria. PhD thesis, University of Ibadan, Nigeria. P. 390.

Atobatele OE, Ugwumba OA (2008). Seasonal variation in the physico-chemistry of a small tropical reservoir (Aiba reservoir, Iwo, Osun, Nigeria). Afr. J. Biotech. 7(12):1962-1971.

Ballantyne JS (1997). Jaws: the inside story. The metabolism of elasmobranch fishes. Comp. Biochem. Physiol. 118:703-742.

Bradford KM (1976). A rapid and sensitive method for the quantification of microgram quantities of protein, utilizing the principle of protein dye binding. Anal. Biochem. 72:248-254.

Campbell JW, Anderson PM (1991). Evolution of mitochondrial enzyme system in fish: the mitochondrial synthesis of glutamine and citrulline. In Hochachka PW, Mommsen TP, (eds) Biochemistry and Molecular Biology of Fishes. Elsevier, Amsterdam I:43-76.

Christensen V, Guenette S, Heymans JJ, Walters CJ, Watson R (2003). Hundred-year decline of North Atlantic predatory fishes. Fish Fish. 4:1-24.

Doudoroff P (1976). Toxicity to fish of cyanides and related compounds-a review. United State Environmental Protection Agency Report No. 600/3-76-038. P. 161.

Eisler R (1991). Cyanide Hazards to Fish, Wildlife, and Invertebrates: A Synoptic Review. United State Fish and Wildlife Services. Biol. Report. 85:1-23.

EPA (Environmental Protection Agency). (1980). Ambient water quality criteria for cyanides. United State Environmental Protection Agency Report No. 440/5-80-037. P. 72.

Golterman HL, Clymo RS, Ohnstad MAM (1978). Methods for physical and chemical analysis of fresh waters.IBP Handbook n°8, Blackwell Sci. Pub. Oxford, U.K.

Kaysen GA, Strecker HJ (1973). Purification and Properties of Arginase of Rat Kidney. Biochem. J. 133:779-807.

Komolafe OO, Arawomo GAO (2008). Preliminary observations on fish species in a newly impounded Osinmo Reservoir. Turkey J. Fish Aqu. Sci. 8:288-292.

Leduc G (1984). Cyanides in water: toxicological significance. In Weber IJ (ed), Aquatic toxicology, Raven Press, New York. pp. 153-224.

Leduc G, Pierce RC, McCracken IR (1982). The effects of cyanides on aquatic organisms with emphasis upon freshwater fishes. NRCC Series No. 19246. National Research Council of Canada.

Lenhardt M, Cakić P, Kolarević J (2004). Influence of the HEPS Djerdap I and Djerdap II dam construction on catch of economically important fish species in the Danube River. Ecohydro. Hydro. 4:499-502.

Lesniak JA, Ruby SM (1982). Histological and quantitative effects of sublethal cyanide exposure on oocyte development in rainbow trout. Arch. Environ. Contam. Toxi. 11:343-352.

Marrs TC, Ballantyne B (1987). Clinical and experimental toxicology of cyanides: an overview. In Ballantyne B, Marrs TC (eds), Clinical and experimental toxicology of cyanides. John Wright, Bristol, England. pp. 473-495.

Mommsen TP, Walsh PJ (1992). Biochemical and environmental perspectives on nitrogen metabolism. Experientia 48:583-593.

Nakajima T, Taki K, Wang B, Ono T, Matsumoto T, Oghiso Y, Tanaka K, Ichinohe K, Nakamura S, Tanaka S, Nenoi M (2008). Induction of rhodanese, a detoxification enzyme in livers from mice after long-term irradiation with low-dose rate gamma-rays. J. Rad. Res. 49:661-666.

Okayi RG (2003). Effect of effluent discharge on water quality, distribution and abundance of plankton and fish species of River Benue. PhD Thesis, University of Ibadan.

Oke OO (1998). Plankton diversity, abundance and productivity in the Owena reservoir, Southwestern Nigeria. PhD Thesis, University of Ibadan.

Okonji RE, Popoola MO, Komolafe OO, Kuku A (2011). The distribution of the enzyme arginase in the tissues of selected Cichlidae Species: *Tilapia zillii, Sarotherodon galilaeus* and *Oreochromis niloticus*. West Afr. J. Appl. Ecol. 18:47-52.

Okonji RE, Popoola MO, Kuku A, Aladesanmi OT (2010). The distribution of cyanide detoxifying enzymes (rhodanese and 3-mercaptopyruvate sulphurtransferase) in different species of the family Cichlidae (*Tilapia zillii, Sarotherodon galilaeus* and *Oreochromis niloticus*). Afr. J. Biochem. Res. 4:163-166.

Paugu D, Levegue C, Teugels GG (2003). The fresh and brackish water fishes of West Africa. Vol I & II. IRD Editions. Publications Scientifiques du Museum. P. 457.

Quayle DB (1988). Pacific oyster culture in British Columbia. Can. Bull. Fish Aqua. Sci. pp. 218-241.

Raymond JA (1998). Trimethylamine Oxide and Urea Synthesis in Rainbow Smelt and Some Other Northern Fishes. Physiol. Zool. 71:515-523.

Reed W, Burchard J, Hopson AJ, Jenness J, Ibrahim Y (1967). Fish and fisheries of Northern Nigeria. Ministry of Agriculture, Northern Nigeria. Gaskiya Cooperation, Zaria.

Ruby SM, Idler DR, So YP (1986). The effect of sublethal cyanide exposure on plasma vitellogenin levels in rainbow trout (*Salmo gairdneri*) during early vitellogenesis. Arch. Environ. Contam. Toxicol. 15:603-607.

Saha N, Ratha BK (2007). Functional ureogenesis and adaptation to ammonia metabolism in Indian freshwater air-breathing catfishes. Fish Physiol. Biochem. 33: DOI: 10.1007/s10695-007-9172-3

Smith LL, Broderius SJ, Oseid DM, Kimball GH, Koenst WM, Lind DT (1979). Acute and chronic toxicity of HCN to fish and invertebrates. USEPA Serial NO. 600/3-79-009. United State Environmental Protection Agency.

Smith LL, Broderius SJ, Oseid DM, Kimball GL, Koenst WM (1978). Acute toxicity of hydrogen cyanide to freshwater fishes. Arch. Environ. Contam. Toxicol. 7:325-337.

Stallings CD (2009). Fishery-Independent Data Reveal Negative Effect of Human Population Density on Caribbean Predatory Fish Communities. PLoS ONE 4(5): e5333. doi:10.1371/journal.pone.0005333.

Towill LE, Drury JS, Whitfield BL, Lewis EB, Galyan EL, Hammons AS. (1978). Reviews of theenvironmental effects of pollutants: v. cyanide. USEPA Serial NO. 600/1-78-027. United State Environmental Protection Agency.

Wai LT, Wong WP, Loong AM, Hiong KC, Chew SH, Ballantyne JS, Yuen KI (2003). The osmotic response of the Asian freshwater stingray (*Himantura signifer*) to increased salinity: A comparison with marine (*Taeniura lymma*) and Amazonian freshwater (*Potamotrygon motoro*) stingrays. J. Experim. Biol. 206:2931-2940.

Way JL (1984). Cyanide intoxication and its mechanism of antagonism. Annu. Rev. Pharm. Toxicol. 24: 451-481.

Welch PS (1952). Limnology. 2nd Edition. Mc-Graw Hill Book Company Inc. London. P. 538.

Wood CM (1993). Ammonia and urea excretion. In Evans DH (ed), Physiology of fishes pp. 379- 425. CRC Press.

Worm B, Sandow M, Oschlies A, Lotze HK, Myers RA (2005). Global patterns of predator diversity in the open oceans. Sci. 309:1365-1369.

Yancey PH (2001). Nitrogen compounds as osmolytes. In Wright PA, Anderson PM (eds), Fish Physiology and Nitrogen Excretion. New York. Academic Press. pp. 309-341.

16

Dry matter accumulation studies at different stages of crop growth in mesta (*Hibiscus cannabinus*)

K. Pushpa[1]*, N. Krishna Murthy[1] and R. Krishna Murthy[2]

[1]Department of Agronomy, College of Agriculture, GKVK, Bangalore- 560 065.
[2]Soil and Water Management, Zonal Agricultural Research Station, V.C. Farm, Mandya -571 405, Karnataka, India.

In crop like mesta, stem is the main source of fibre and therefore, fibre yield depends greatly on fresh biomass yield as well as on dry stalk yields of a particular variety at the time of harvesting. At 30 DAS (days after sowing), dry matter accumulation in stem varied significantly with respect to different plant spacing, varieties and nutrient sources. Between the varieties, variety HS-108 had recorded significantly higher dry matter accumulation in stem (0.46 g/plant) than variety AMV-4 (0.29 g/plant). Significantly higher dry matter accumulation in stem was achieved under 45 cm × 10 cm spacing (0.42 g/plant) than 30 cm × 10 cm (0.34 g/plant). Further, application of 5 t of FYM per ha along with 40:20:20 kg NPK per ha fertilizer induced higher dry matter accumulation in stem (0.44 g/plant) compared to 100% N equivalent through FYM (0.32 g/plant). The interaction effects between varieties, plant spacing and nutrient sources were found to be significant. Significantly higher dry matter accumulation in stem (0.71 g/plant) was obtained with variety HS-108 planted at 45 cm × 10 cm compared to variety AMV-4 planted at 30 cm × 10 cm (0.36 g/plant). Significantly highest dry matter accumulation in stem (0.75 g/plant) was observed in the variety HS-108 supplied with 5 t of FYM per ha along with 40:20:20 kg NPK per ha fertilizer than 100% N equivalent through FYM alone in the variety AMV-4 (0.31 g/plant). Application of 5 t of FYM per ha along with 40:20:20 kg NPK per ha fertilizer registered higher dry matter accumulation in stem (0.69 g/plant) compared to 100% N equivalent through FYM (0.37 g/plant).

Key words: Mesta, uptake, nutrient, fibre, biomass, India.

INTRODUCTION

Mesta (*Hibiscus sabdariffa)* is being successfully grown as a commercial crop in many tropical and subtropical countries of the world including India. FAO (2007) estimates shows that the total annual production of jute and jute like fibres including mesta is 3.68 million tonnes with developing countries accounting for production that is six times more than developed countries. World production of *H. sabdariffa var. altissima* far exceeds its other related species (Krishna Murthy et.al.,1992). In India, both *H. sabdariffa var. altissima* and *H. cannabinus* are grown. However, *H. sabdariffa var. altissima* accounts for more than 75% of the area under mesta in India (Krishna Murthy et al., 1992). The importance of varieties with high yield potential combined with its wider adoptability for boosting up production of mesta fibre need hardly be emphasized. A variety found superior in a particular locality may not exhibit similar performance under certain other environmental conditions. It is therefore, essential to evaluate the varieties in different agro-climatic conditions of a region to assess their performance and selecting the best one's to exploit their yield potentiality through various agro-techniques.

MATERIALS AND METHODS

The experiments were conducted in 19E block at field unit Gandi Krishi Vignana Kendra (GKVK), University of Agricultural Sciences, Bangalore which is located at latitude of 12°58' north, longitude of

77°3' east and at an altitude of 930 m above mean sea level in Eastern dry zone (zone 5) of Karnataka, India. The soil of the experimental site was Red Sandy Loam. The soil was near neutral in pH (5.8) with low organic carbon content. The soil was also found to be medium in available nitrogen (273.18 kg/ha), available phosphorus (39.00kg/ha), and available potassium content (231.80 kg/ha).

The experiment conducted in 2007 and 2008 comprised 16 treatment combinations consisting of two varieties AMV-4 and HS-108, two spacing trails 30 × 10 cm and 45 × 10 cm and four nutrient treatments, that is, 40:20:20 kg NPK/ha, 40:20:20 kg NPK/ha + 5 t/ha FYM, 30:20:20 kg NPK/ha + 7.5 t/ha FYM and 100% N equivalent through FYM. The experiment was laid out in Split-Split plot design.

Five plants were uprooted at random from the adjacent net plot rows excluding border rows at 30, 60, 90, 120 DAS and at harvest (135-140DAS). The root portion of the plant was discarded. The above ground portion of the plant samples were separated into leaves, stem and reproductive parts. The separated plant parts were air dried and kept it in hot air oven at 65°C to get a constant weight. After thorough drying, the weight was recorded for all the parts separately and expressed in grams. The mean of five plants was taken as the dry matter produced per plant. The mean dry weight of all the plant parts was added and taken as total dry matter accumulation per plant.

RESULTS

Dry matter accumulation in stem

Averages values on dry matter accumulation in stem of mesta varieties as influenced by the plant spacing and nutrient sources at different stages of crop growth are presented in Table 1. At 30 DAS, dry matter accumulation in stem varied significantly with respect to different plant spacing, varieties and nutrient sources. Between the varieties, variety HS-108 achieved significantly higher dry matter accumulation in stem (0.46 g/plant) than variety AMV-4 (0.29 g/plant).

Significantly higher dry matter accumulation in stem was recorded under 45 cm × 10 cm spacing (0.42 g/plant) than 30 cm × 10 cm (0.34 g/plant). Further, application of 5 t of FYM per ha along with 40:20:20 kg NPK per ha fertilizer recorded higher dry matter accumulation in stem (0.44 g/plant) compared to 100% N equivalent through FYM (0.32 g/plant).

The interaction effects between varieties, plant spacing and nutrient sources were found to be significant but the interaction effects between varieties, plant spacing and nutrient sources were found to be significant.

Dry matter accumulation in leaves

Pooled data on dry matter accumulation in leaves of mesta varieties as influenced by the plant spacing and nutrient sources at different stages of crop growth are presented in Table 2. At 30 DAS, dry matter accumulation in leaves varied significantly with respect to different plant spacing, varieties and nutrient sources.

Between the varieties, variety AMV-4 recorded significantly higher dry matter accumulation in leaves (0.18 g/plant) than variety HS-108 (0.12 g/plant). Significantly higher dry matter accumulation in leaves was recorded under 45 cm × 10 cm spacing (0.15 g/plant) than 30 cm × 10 cm (0.14 g/plant). Further, application of 5 t of FYM per ha along with 40:20:20 kg NPK per ha of fertilizer induced higher dry matter accumulation in leaves (0.20 g/plant) compared to 100% N equivalent through FYM (0.09 g/plant).

Among the interactions, dry matter accumulation in leaves was found to be significant between varieties and plant spacing as well as varieties with nutrient.

Total dry matter accumulation

Average data on total dry matter accumulation of mesta varieties as influenced by the plant spacing and nutrient sources at different stages of crop growth are presented in Table 3. Total dry matter accumulation differed significantly due to different plant spacing, varieties and nutrient sources at all the stages of crop growth. At 30 DAS, significantly largest total dry matter accumulation was noticed in variety HS-108 (0.58 g/plant) than, variety AMV-4 (0.47 g/plant). Significantly higher total dry matter accumulation was recorded under 45 cm × 10 cm spacing (0.58 g/plant) than 30 cm × 10 cm (0.47 g/plant). Further, application of 5 t of FYM per ha along with 40:20:20 kg NPK per ha fertilizer recorded higher total dry matter accumulation (0.62 g/plant) compared to 100% N equivalent through FYM (0.41 g/plant).

The interaction effects between varieties, plant spacing and nutrient sources were found to be significant.

DISCUSSION

Dry matter accumulation in stem

In mesta, stem is the main source of fibre and therefore, fibre yield depends greatly on fresh biomass yield as well as dry stalk yields of a particular variety at the time of harvesting. Dry stalk yield is the ultimate factor which decides the final fibre yield. While in fresh biomass yield, stalk portion constitutes nearly 70% and remaining 30% is constituted by leaf, petiole and reproductive parts (Krishna Murthy et.al.,1992). Bhattacharjee et al. (1987) observed higher fibre yield in variety which was superior in total green biomass. Results of the present investigations are in conformity with the findings of Naidu et al. (1996b). In the present study also, HS-108 belonging to 'roselle' showed superiority over AMV- 4 in biomass production. Higher biomass production with 'roselle' could be due to its superior genetic potential (Naidu et al., 1996a). Similar results were also reported by Lakshminarayana et al. (1980).

Varietal differences in total fresh biomass and dry stalk

Table 1. Dry matter accumulation in stem (g) of mesta varieties as influenced by plant spacing and nutrient sources at different stages of crop growth.

Treatments	30 DAS			60 DAS			90 DAS			120 DAS			At Harvest		
	2007	2008	Pooled	2007	2008	Pooled	2007	2008	Pooled	2007	2008	Pooled	2007	2008	Pooled
Variety (V)															
AMV-4 (V_1)	0.28	0.29	0.29	0.63	0.66	0.65	10.07	10.36	10.21	14.69	15.18	14.94	16.53	17.12	16.83
HS-108 (V_2)	0.45	0.47	0.46	0.97	0.99	0.98	14.26	14.65	14.46	18.67	18.86	18.76	19.47	19.64	19.56
S.Em ±	0.01	0.01	0.01	0.01	0.02	0.02	0.04	0.04	0.04	0.03	0.01	0.02	0.21	0.01	0.12
C.D. (0.05)	0.03	0.03	0.03	0.02	0.06	0.04	0.12	0.10	0.11	0.09	0.03	0.06	0.48	0.03	0.28
Plant spacing (S)															
30 cm × 10 cm (S_1)	0.33	0.35	0.34	0.74	0.76	0.75	11.27	11.45	11.36	16.22	16.33	16.28	17.29	17.59	17.44
45 cm × 10 cm (S_2)	0.40	0.43	0.42	0.85	0.90	0.88	13.06	13.56	13.31	17.15	17.72	17.42	18.72	19.17	18.95
S.Em ±	0.01	0.02	0.02	0.02	0.03	0.03	0.04	0.05	0.05	0.02	0.02	0.02	0.22	0.02	0.12
C.D. (0.05)	0.02	0.05	0.04	0.06	0.09	0.08	0.12	0.13	0.10	0.06	0.04	0.06	0.56	0.06	0.32
Nutrient source (N)															
40:20:20 kg/ha (N_1)	0.40	0.39	0.40	0.84	0.86	0.85	12.54	12.87	12.71	17.19	17.21	17.20	18.69	18.45	18.57
40:20:20 kg/ha + 5 t FYM/ha (N_2)	0.43	0.45	0.44	0.88	0.90	0.89	13.09	13.19	13.14	17.59	17.76	17.68	18.42	19.11	18.76
30:20:20 kg/ha + 7.5 t FYM/ha (N_3)	0.35	0.37	0.36	0.79	0.81	0.80	12.13	12.44	12.29	16.67	16.76	16.72	17.97	18.19	18.08
100% N equivalent through FYM (N_4)	0.29	0.34	0.32	0.68	0.76	0.72	10.90	11.51	11.21	15.28	16.37	15.82	16.95	17.77	17.36
S.Em ±	0.02	0.02	0.02	0.02	0.01	0.01	0.06	0.07	0.07	0.03	0.02	0.03	0.29	0.01	0.15
C.D.(0.05)	0.05	0.06	0.06	0.04	0.03	0.03	0.18	0.20	0.21	0.09	0.06	0.09	0.64	0.03	0.40
Interactions															
V×S															
S.Em ±	0.02	0.01	0.02	0.01	0.01	0.01	0.12	0.13	0.14	0.04	0.03	0.04	0.60	0.05	0.32
C.D.(0.05)	0.04	0.03	0.04	NS	NS	0.09	0.33	0.35	0.38	0.12	0.09	0.11	NS	0.13	0.84
V×N															
S.Em ±	0.02	0.02	0.03	0.02	0.03	0.02	0.17	0.19	0.18	0.06	0.05	0.04	0.83	0.04	0.42
C.D.(0.05)	NS	NS	0.06	0.05	0.09	0.06	0.38	0.40	0.36	0.18	0.15	0.14	NS	0.12	NS
S×N															
S.Em ±	0.02	0.03	0.03	0.03	0.04	0.04	0.16	0.18	0.18	0.08	0.05	0.06	0.84	0.05	0.42
C.D.(0.05)	0.04	NS	0.09	0.06	NS	0.10	0.34	0.39	0.42	0.21	0.13	0.16	1.78	0.12	0.99
V×S×N															
S.Em ±	0.01	0.02	0.02	0.01	0.02	0.02	0.08	0.09	0.09	0.04	0.03	0.03	0.41	0.01	0.23
C.D.(0.05)	0.03	NS	NS	0.03	NS	NS	0.23	0.26	0.28	0.12	0.09	0.09	NS	0.03	NS

DAS, Days after sowing; FYM, farm yard manure; NS, non-significant.

Table 2. Dry matter accumulation in leaves (g) of mesta varieties as influenced by plant spacings and nutrient source at different stages of crop growth.

Treatments	30 DAS			60 DAS			90 DAS			120 DAS			At Harvest		
	2007	2008	Pooled	2007	2008	Pooled	2007	2008	Pooled	2007	2008	Pooled	2007	2008	Pooled
Variety (V)															
AMV-4 (V_1)	0.17	0.19	0.18	1.14	1.28	1.21	2.26	2.34	2.30	3.70	3.74	3.77	0.10	0.14	0.12
HS-108 (V_2)	0.10	0.13	0.12	0.84	0.85	0.85	1.72	1.78	1.75	2.87	2.84	2.80	0.04	0.06	0.05
S.Em ±	0.01	0.01	0.01	0.02	0.01	0.02	0.01	0.05	0.03	0.08	0.03	0.06	0.01	0.01	0.01
C.D. (0.05)	0.03	0.02	0.03	0.06	0.03	0.05	0.03	0.12	0.09	0.20	0.09	0.14	0.03	0.03	0.03
Plant spacing (S)															
30 cm x 10 cm (S_1)	0.13	0.15	0.14	0.90	0.95	0.93	1.95	2.07	2.01	3.27	3.25	3.22	0.06	0.09	0.08
45 cm x 10 cm (S_2)	0.14	0.17	0.15	1.08	1.18	1.13	2.02	2.05	2.04	3.29	3.32	3.35	0.08	0.11	0.10
S.Em ±	0.02	0.01	0.02	0.03	0.02	0.03	0.02	0.05	0.04	0.08	0.04	0.06	0.02	0.01	0.02
C.D. (0.05)	0.04	0.03	0.04	0.09	0.06	0.09	0.06	NS	0.10	NS	0.12	0.14	0.05	0.03	0.04
Nutrient source (N)															
40:20:20 kg/ha (N_1)	0.15	0.17	0.16	1.08	1.19	1.14	2.08	2.16	2.12	3.34	3.37	3.39	0.07	0.12	0.09
40:20:20 kg/ha + 5 t FYM/ha (N_2)	0.17	0.22	0.20	1.20	1.26	1.23	2.19	2.34	2.27	3.52	3.51	3.52	0.11	0.14	0.12
30:20:20 kg/ha + 7.5 t FYM/ha (N_3)	0.14	0.15	0.15	0.98	1.03	1.01	2.02	2.04	2.03	3.41	3.38	3.40	0.07	0.09	0.08
100 per cent N equivalent through FYM (N_4)	0.08	0.10	0.09	0.70	0.78	0.74	1.66	1.71	1.69	2.85	2.88	2.91	0.04	0.05	0.05
S.Em ±	0.01	0.01	0.01	0.01	0.02	0.03	0.01	0.06	0.04	0.07	0.03	0.05	0.02	0.01	0.02
C.D.(0.05)	0.03	0.03	0.03	0.02	0.06	0.06	0.02	0.14	0.08	0.14	0.09	0.15	0.04	0.03	0.06
Interactions															
V×S															
S.Em ±	0.02	0.01	0.01	0.02	0.02	0.02	0.01	0.13	0.06	0.23	0.03	0.12	0.02	0.02	0.02
C.D.(0.05)	NS	NS	NS	0.05	0.05	0.05	0.02	NS	NS	NS	0.09	0.08	0.05	0.04	0.04
V×N															
S.Em ±	0.03	0.03	0.02	0.02	0.03	0.03	0.02	0.18	0.12	0.24	0.05	0.15	0.03	0.02	0.03
C.D.(0.05)	NS	NS	NS	0.06	0.05	0.06	0.05	0.40	0.23	NS	0.14	0.14	0.06	0.06	0.08
S×N															
S.Em ±	0.03	0.03	0.02	0.02	0.03	0.03	0.02	0.18	0.10	0.23	0.04	0.13	0.02	0.02	0.03
C.D.(0.05)	NS	NS	NS	0.05	0.06	0.06	0.04	NS	0.05	NS	0.12	0.10	NS	NS	NS
V×S×N															
S.Em ±	0.02	0.02	0.01	0.01	0.02	0.02	0.01	0.09	0.06	0.10	0.02	0.06	0.01	0.01	0.01
C.D.(0.05)	NS	NS	NS	0.03	0.06	0.04	0.03	NS	NS	NS	NS	NS	NS	NS	NS

DAS, Days after sowing; FYM, farm yard manure; NS, non-significant.

Table 3. Total dry matter accumulation (g) of mesta varieties as influenced by plant spacings and nutrient source at different stages of crop growth.

Treatments	30 DAS			60 DAS			90 DAS			120 DAS			At Harvest		
	2007	2008	Pooled	2007	2008	Pooled	2007	2008	Pooled	2007	2008	Pooled	2007	2008	Pooled
Variety (V)															
AMV-4 (V_1)	0.45	0.48	0.47	1.77	1.94	1.86	12.33	12.70	12.51	19.48	20.04	19.77	22.79	23.47	23.14
HS-108 (V_2)	0.55	0.60	0.58	1.81	1.84	1.83	15.98	16.43	16.21	22.10	22.82	22.46	24.88	25.12	25.01
S.Em ±	0.01	0.01	0.01	0.01	0.02	0.01	0.02	0.01	0.02	0.02	0.03	0.03	0.02	0.02	0.02
C.D. (0.05)	0.02	0.03	0.03	0.03	0.05	0.02	0.04	0.03	0.04	0.05	0.08	0.08	0.04	0.05	0.05
Plant spacing (S)															
30 cm × 10 cm (S_1)	0.46	0.47	0.47	1.65	1.71	1.68	13.18	13.46	13.33	20.24	20.46	20.35	23.09	23.47	23.28
45 cm × 10 cm (S_2)	0.54	0.65	0.58	1.93	2.08	2.00	15.18	15.58	15.38	21.22	22.45	21.84	24.38	25.12	24.75
S.Em ±	0.01	0.04	0.03	0.01	0.05	0.03	0.01	0.03	0.02	0.01	0.05	0.03	0.01	0.02	0.01
C.D. (0.05)	0.03	0.12	0.07	0.03	0.14	0.09	0.02	0.07	0.06	0.02	0.15	0.09	0.03	0.05	0.03
Nutrient source (N)															
40:20:20 kg/ha (N_1)	0.54	0.57	0.55	1.91	2.05	1.98	14.62	15.02	14.82	21.32	21.80	21.58	24.14	24.45	24.30
40:20:20 kg/ha + 5 t FYM/ha (N_2)	0.58	0.65	0.62	2.08	2.16	2.12	15.28	15.47	15.38	21.91	22.41	22.20	24.43	25.25	24.84
30:20:20 kg/ha + 7.5 t FYM/ha (N_3)	0.50	0.52	0.51	1.76	1.83	1.80	14.26	14.41	14.33	20.83	19.09	19.96	23.82	24.13	23.98
100 per cent N equivalent through FYM (N_4)	0.37	0.43	0.41	1.38	1.54	1.46	12.56	13.18	12.87	18.86	17.80	18.33	22.55	23.34	22.95
S.Em ±	0.01	0.02	0.02	0.01	0.02	0.02	0.01	0.03	0.02	0.02	0.10	0.06	0.01	0.02	0.02
C.D.(0.05)	0.03	0.04	0.04	0.03	0.04	0.04	0.03	0.06	0.05	0.04	0.28	0.19	0.03	0.04	0.04
Interactions															
V×S															
S.Em ±	0.02	0.01	0.02	0.02	0.01	0.02	0.02	0.05	0.04	0.03	0.11	0.07	0.02	0.03	0.03
C.D.(0.05)	0.06	0.03	0.04	0.05	0.03	0.04	0.04	0.15	0.10	0.07	0.30	0.18	0.06	0.10	0.08
V×N															
S.Em ±	0.03	0.04	0.04	0.02	0.03	0.03	0.03	0.08	0.05	0.05	0.24	0.15	0.03	0.06	0.05
C.D.(0.05)	0.08	NS	0.08	0.06	0.06	0.06	0.06	0.16	0.10	0.10	0.50	0.30	0.6	0.13	0.10
S×N															
S.Em ±	0.03	0.04	0.03	0.03	0.03	0.03	0.03	0.09	0.06	0.04	0.26	0.15	0.03	0.06	0.05
C.D.(0.05)	NS	NS	0.06	0.06	0.07	0.07	0.06	0.19	0.13	0.09	0.55	0.32	0.07	0.13	0.10
V×S×N															
S.Em ±	0.02	0.02	0.02	0.01	0.02	0.02	0.02	0.04	0.03	0.02	0.14	0.08	0.02	0.03	0.03
C.D.(0.05)	NS	NS	NS	0.03	0.06	0.05	0.06	0.12	0.09	0.06	NS	NS	0.06	0.09	0.08

DAS, Days after sowing; FYM, Farm yard manure; NS, Non-significant.

production might be attributed to their variability in total dry matter production and its accumulation in different plant parts and stem in particular. During the initial period (at 60 DAS), total dry matter production (pooled) was significantly superior with AMV- 4 compared to HS-108 but in later stages (90 DAS, 120 DAS and harvest stage) total dry matter production (pooled) of variety HS-108 was significantly superior to AMV-4. Dry matter accumulation in stem of variety HS-108 was significantly superior over AMV- 4 at all the growth stages (pooled) (Table 1). Higher dry matter accumulation in stem of HS-108 variety had favourable influence on production of significantly higher fresh biomass as well as dry stalk yields which in turn resulted in significantly higher fibre yield per hectare.

Dry matter accumulation in leaves

Total dry matter production and its greater partitioning into stem depends upon photosynthetic capacity of the plant during its vegetative period (up to 130 to 140 DAS) and translocation of photosynthates from source (leaf and petiole) to ultimate sink (stem). Photosynthetic ability of a plant at any stage depends upon dry weight of leaf, leaf area index and photosynthetic efficiency of leaves (Donald, 1963). Dry weight of leaves in variety AMV- 4 was significantly higher than variety HS-108 at all the growth stages. At harvest, AMV- 4 variety recorded significantly higher dry weight of leaf per plant (Table 2). Variety AMV- 4 showed lower translocation of photosynthates from leaf to stem resulting in poor dry matter accumulation in stem.

Higher dry matter accumulation in leaf of variety AMV-4 could be attributed to production of more number of leaves per plant at all the stages of crop growth. Leaf area, leaf area index and leaf area duration also followed similar trend at all the growth stages (Pushpa, 2009).

Total dry matter production

Higher total dry matter production and its partitioning into stem may be attributed to higher rate of dry matter production and crop growth rate. From 60 to 90 DAS, rate of dry matter production in variety HS-108 was 0.48 g per plant per day as against 0.36 g per plant per day in AMV-4 variety. Similarly, crop growth rate (pooled) in AMV- 4 variety during the same period was 1.60 g per m^2 per day when compared to 1.18 g per m^2 per day of HS-108 variety. In both the varieties, the rate of dry matter production and growth rate declined towards maturity (120 to 160 DAS) as compared to 60 to 120 DAS. But the decline was more in AMV- 4 variety when compared to HS-108 variety which indicates that dry matter accumulation and its diversion to different plant parts in AMV- 4 variety takes place for a shorter period when

compared to HS-108 resulting in lower dry matter accumulation in stem. Higher dry matter accumulation in reproductive parts, depends on photosynthetic capacity of the plant during capsule developing period (after flowering to maturity) and translocation of assimilates from other plant parts to developing capsules (Krishna Murthy et al.,1992)

Photosynthetic ability of a plant in turn depends on photosynthetic efficiency of leaves. In the present study, dry weight of leaves during the capsule development period (120 DAS to harvest) was significantly higher with AMV-4 (3.77 g/plant and 0.12 g/plant, respectively) compared to HS-108 variety (2.80 and 0.05 g/plant, respectively). Leaf area, leaf area index and leaf area duration also followed similar trend, this might have resulted in increased rate of dry matter production and accumulation in reproductive parts compared to HS-108 variety. In variety HS-108, lower rate of dry matter production and crop growth rate was observed during the capsule formation period (100 to 120 DAS). This might be due to infestation of crop by top shoot borer which had restricted the growth of plants preventing formation of capsules and further dry matter accumulation. Premature shedding and drying of capsules was also noticed. This might have reduced the number of capsules and dry matter accumulation in capsules as a consequence; lower number of seeds per capsules was observed which in turn resulted in lower seed yield per hectare.

Effect of plant spacing

Plant spacing adjustment is an important agronomic manipulation for attaining higher yields. Maintenance of optimum plant population helps to utilize available moisture, nutrients, solar radiation efficiently and enable the crop to produce higher yields.

Dry matter accumulation in leaves and reproductive parts also decreased significantly with increase in plant population from 0.25 million per ha to 0.38 million per ha. Higher dry matter accumulation in leaf at 45 cm × 10 cm plant spacing might have helped in increased total dry matter production per plant and its greater accumulation in stem due to increased photosynthesis and greater translocation of assimilates from source to sink on account of higher leaf area per plant and greater availability of growth resources, have been reported by Bhangoo et al. (1986), Krishnamurthy et al. (1992), Sarma and Bordoloi (1995) and Guggari (2002).

Total fresh biomass production per hectare increased from 16.26 to 17.99 t per ha with increase in inter row spacing from 30 to 45 cm and it may depends upon photosynthetic activity of a crop. There was a significant difference in total fresh biomass production between 30 and 45 cm row spacing (16.26 and 17.99 t/ha, respectively). This may be attributed to significant difference in leaf area per plant and crop growth rate between the plant populations at different plant spacing.

Higher total dry mater production per plant and its distribution in stem, leaf, and reproductive parts under 45 ×10 cm plant spacing might be attributed to higher rate of dry matter production, relative growth rate, crop growth rate and net assimilation rate and they decreased with increase in plant population. This might be due to mutual shading of leaves and increased respiratory losses, resulting in decreased net photosynthesis per unit leaf area as the inter row spacing decreased.

Effect of nutrient sources

The production efficiency of a crop though depends on its genetic potential, its yield could be improved to a perceptible magnitude through proper nutrient management, and further the nutrient requirement of a crop varies with the variety. Therefore, study was conducted to find out the optimum requirement of the nutrients for higher yields.

Total dry matter production per plant and its greater accumulation in stem is vital for higher fibre yield. At harvest, significantly higher average dry matter accumulation (pooled) was observed with the application of 5 t FYM per ha along with 40:20:20 kg NPK per ha (24.43 g/plant) when compared to application of 40:20:20 kg NPK per ha alone (24.30 g/plant), application of 7.5 t FYM along with 30:20:20 kg NPK per ha (23.98 g/plant) and application of 100% N equivalent through FYM alone (22.95 g/plant). Similar trend was observed in respect of dry matter accumulation in stem (Table 3). This increased total dry matter production per plant and its accumulation in stem at higher levels of nitrogen was attributed to increased plant height coupled with higher stem diameter (Krishnamurthy et al., 1992).

Dry matter accumulation in leaves and reproductive parts increased significantly with the combined application of both organic and inorganic sources at all the growth stages (pooled). Dry matter accumulation in leaves per plant was maximum at 120 DAS and there after it declined drastically towards maturity. At 120 DAS (pooled), application of 5 t FYM per ha along with 40:20:20 kg NPK per ha accumulated significantly higher dry matter in leaves (3.52 g/plant) compared to application of 100% N equivalent through FYM alone (2.91 g/plant). At 30, 60, 90 DAS and at harvest a similar trend was noticed. Dry matter accumulation in reproductive parts also followed similar trend. This increased dry matter accumulation in leaves and reproductive parts with combined application of N might be the reason for higher total dry matter production per

plant and its accumulation in stem. Higher dry matter accumulation in leaf at combined application of N may be due to higher number of leaves and increased leaf area per plant.

Higher total dry matter production per plant and its distribution in different plant parts with increasing rate of nitrogen with the combined application of both organic and inorganic source of N might be due to higher leaf area, leaf area index, leaf area duration, rate of dry matter production, crop growth rate and net assimilation rate at all the growth stages. Krishanmurthy et al. (1992) opined that increased biomass / dry matter yield due to increased N application was attributed to higher leaf area index. Higher biomass production with the application of 5 t FYM per ha along with 40:20:20 kg NPK per ha may be attributed to increased crop growth rate at all the growth stages.

REFERENCES

Bhangoo MS, Tehrani HS, Henderson J (1986). Effect of planting date, nitrogen levels, row spacing and plant population on Kenaf performance in the San Joaquin Valley, California. Agron. J. 78:600-604.

Bhattacharjee AK, Mukerjee N, Datta AN, Goswami KK (1987). Suitability of rainfedkenaf(Hibiscus cannabinus L.) as the source of raw materials for newsprints. Jute Dev. J. 2:27-29.

Donald CM (1963). Competition among crop and pasture plants. Adv. Agron., 10: 435-473.

FAO (2007). Statistics:9-10

Guggari AK (2002) Studies on production technologies for mesta in medium black soils of northern dry zone of Karnataka. Ph.D. Thesis, University of Agricultural Sciences, Dharwad.

Krishna murthy N, Hunsagi G, Singlachar MA, Rudraradhya M (1992). Influence of plant density and nitrogen levels on mesta: Growth components, dry matter and biomass yields. Mysore J. Agric. Sci. 26:239-243.

Lakshminarayana A, Murthy R K, Rao MR, Rao PA (1980). Efficiency of nitrogen utilization by roselle and kenaf. Indian J. Agric. Sci. 50(3):224-248.

Naidu MV, Prasad PR, Lakshmi MB (1996a). Influence of species, fertility level and stage of harvest on biomass production and pulp yield in mesta (Hibiscus sp.). J. Res. Angrau 24(3):1-4, 1996.

Naidu MV, Reddy AT, Prasad PR, Lakshmi MB (1996b). Response of mesta(Hibiscus sabdariffa L.) to NPK under rainfed conditions. J. Research, Angrau 24(2):79-81.

Pushpa K (2009), Studies on production potentials of mesta varieties as influenced by spacing, nutrient sources and intercropping systems in alfisols, Ph.D. Thesis, University of Agricultural Sciences, Bangalore.

Sarma TC, Bordoloi DN (1995) Yield of pulpable biomass of Kenaf(Hibiscus cannabinus) varieties under various row spacing and nitrogen. Indian J. Agron. 40 (4):722-724.

17

Influence of drip fertigation on water productivity and profitability of maize

S. Anitta Fanish

Department of Agronomy, Tamil Nadu Agricultural University, Coimbatore, Tamil Nadu– 641 003, Indian.

Field experiments were conducted during the 2008 and 2009 cropping season at Tamil Nadu Agricultural University, Coimbatore to study the effect of drip fertigation on water use efficiency in intensive maize (*Zea mays* L.) based intercropping system. In 2008, drip fertigated maize at 150% recommended dose of fertilizer (RDF) recorded significantly higher grain yield of 7.3 t ha^{-1}. Whereas in 2009, higher grain yields of 7.5 t ha^{-1} was recorded under drip fertigation of 100% RDF with 50% P and K as water soluble fertilizer (WSF). Drip irrigation helps to save the water up to 43% compared to surface irrigation besides enhancing the water use efficiency. The highest net return (Rs 56858) and B:C ratio (3.24) was obtained under drip fertigation of 150% RDF + radish as intercrop combination. It is inferred that drip fertigation once in three days at 100% RDF with 50% P and K as water soluble fertilizer enhanced the productivity of maize-based intercropping system. Considering the high cost of water soluble fertilizers, drip fertigation of 150% RDF with radish as intercrop could be an alternative option to realize a reasonably good yield and income in maize-based intercropping system.

Key words: Drip fertigation, intercropping, maize grain equivalent yield (MEY), water use efficiency (WUE).

INTRODUCTION

Water is the most important and critical input in man's life especially in agriculture. The pressure for the most efficient use of water for agriculture is intensifying with the increased competition for water resources among various sectors with mushrooming population. In spite of having the largest irrigated area in the world, India too has started facing severe water scarcity in different regions. Efficient utilization of available water resources is crucial for India, which shares 17% of the global population with only 2.4% of land and 4% of the water resources. Improper management of water and nutrient has contributed extensively to the current water scarcity and pollution problems in many parts of the world, and is also a serious challenge to future food security and environmental sustainability. Addressing these issues requires an integrated approach to soil-water-plant-nutrient management at the plant-rooting zone.

Fertigation, a latest technology wherein nutrients are applied along with the irrigation water and opens new possibilities for controlling water and nutrient supplies to crops besides maintaining the desired concentration and distribution of water and nutrients to the soil (Bar-Yosef, 1999). Specialty fertilizers are high in analysis and totally water soluble and are available in double and multi nutrient combinations. One of the important strategies to increase agricultural output is development of new high intensity cropping system including intercropping system. Diversification of cropping pattern particularly in favour of vegetable crops is becoming popular among farmers because in a balanced diet, vegetables are most important component and also it gives more income.

Maize or Indian corn (*Zea mays* L.) is one of the most important cereal crops in the global agricultural economy both as a food for man and feed for animal and the crop of

Table 1. Fertigation schedule for maize.

Crop stages	Quantity (%)		
	N	P	K
Vegetative stage (6 - 30 days)	25	25	25
Reproductive stage (30 - 60 days)	50	50	50
Maturity stage (60 - 75 days)	25	25	25
Total	100	100	100

immense potentiality. By 2020 AD, the requirement of maize for various sectors will be around 100 million tonnes, of which the poultry sector demand alone will be around 31 million tonnes (Seshaiah, 2000). Research works on drip irrigation under intercropping is very limited. Hence, the present field research was initiated to assess the feasibility of drip fertigation in maize-based inter cropping system.

MATERIALS AND METHODS

Study site

The experiment was conducted during of 2008 and 2009 cropping seasons at Tamil Nadu Agricultural University, Coimbatore, India. The farm is situated in North Western Agro-Climatic Zone of Tamil Nadu at 11°N latitude and 77°E longitude with an altitude of 426.7 m above MSL. The soil was sandy clay loam with pH 7.53; EC, 0.76 dS m^{-1}; organic carbon, 0.32%; available N, 220 kg ha^{-1}; available P, 17 kg ha^{-1} and available K, 425 kg ha^{-1}.

Experimental design, treatments and field management

The experiment was laid out in split plot design with three replications. The treatments comprised of nine fertigation levels in main plot, M_1, surface irrigation with soil application of 100% RDF (150:75:75 kg NPK per ha); M_2, drip irrigation with soil application of 100% RDF; M_3, drip fertigation of 75% RDF; M_4, drip fertigation with 100% RDF; M_5, drip fertigation of 125% RDF; M_6, drip fertigation of 150% RDF; M_7, drip fertigation of 50% RDF (50% P and K as water soluble fertilizer (WSF)); M_8, drip fertigation of 75% RDF (50% P and K as WSF); M_9, drip fertigation of 100% RDF (50% P and K as WSF) and four intercrops in sub plots consisting S_1, Vegetable coriander; S_2, Radish; S_3, Beet root and S_4, Onion. Fertilizer was given based on base crop (maize) requirement. RDF for maize is 150:75:75 kg NPK ha^{-1}. Maize hybrid COH (M) 5 was sown with spacing of 75 × 20 cm.

In the surface irrigated plots, ridges and furrows were formed at 60 cm apart and maize was sown at a spacing of 60×20 cm. In between the two row of maize, one row of inter crops was sown. Paired row planting system was adopted under drip irrigation with spacing of 75 × 20 cm. In between the two rows of maize two rows of inter crops were sown by adopting a spacing of 30 cm between rows. One lateral with in-line drippers (discharge rate, 4 L/h) was laid at the centre of the raised flat bed (1.2 m width and 20 m length) and it covered the two row of maize and two rows of intercrops. The lateral spacing between two raised flat beds was 1.5 m with furrow in-between of 30 cm width and 15 cm depth. For radish, beetroot and coriander, seeds were sown and for onion, bulbs were used for sowing.

The recommended doses of inorganic fertilizers were applied directly to soil for the treatments M_1 and M_2. Fertilizer sources used for supplying NPK were urea (46% N), di-ammonium phosphate (18% N and 46% P_2O_5) and muriate of potash (60% K_2O), respectively. The entire quantity of phosphorus was applied as basal in treatments M_1 and M_6 in the form of di-ammonium phosphate one day before sowing. In the treatments M_1 and M_2 involving soil application of fertilizers, recommended dose of nitrogen and potassium were applied in the form of urea in three splits doses (25% N as basal, 50% N on 25 DAS and 25% N on 45 DAS as top dressing) and muriate of potash in two splits doses (50% as basal and 50 as top dressing on 45 DAS). For treatments M_3 to M_9 fertilizers were given through drip fertigation.

In treatments M_3 to M_6 normal fertilizer was used as sources for supplying NPK through drip irrigation. Normal fertilizers viz., urea and muriate of potash were used to supply N and K respectively. For the treatments M_7 to M_9 50% P and K were supplied through water soluble fertilizer and the remaining through normal fertilizer. Mono-ammonium phosphate (12: 61: 0) and multi-K (13: 0: 46) were used as water soluble fertilizer for supplying P and K, respectively. The fertilizer solution was prepared by dissolving the required quantity of fertilizer with water in 1:5 ratio and injected into the irrigation system through ventury assembly. Considering the nutrient uptake pattern at phenological growth phases of maize, the fertigation schedule was worked out and presented in Table 1.

Fertigation was given in once in three days. The quantity of irrigation water supplied through drip was 173 and 198 mm during the 2008 and 2009 cropping season, respectively. The effective rainfall received during the cropping period was 158 mm (2008) and 130 mm (2009). The total water used under the drip irrigation treatments was 331 and 328 mm for each year, respectively. Under surface irrigation method, irrigation was given immediately after sowing followed by life irrigation at 5 cm depth thereafter irrigation was given as per the IW/CPE ratio of 0.8. The quantity of water applied was 300 and 350 mm in 2008 and 2009, respectively. An effective rainfall of 192 and 161 mm was received during crop growth period, and totally 492 and 511 mm of water was consumed by surface irrigated crop during each year.

Economic analysis

Generally the life span of a well maintained drip system would be 10 years. But here in the intercropping system situation, the drip system may be used for three seasons per year, so on an average, 7 years was considered for calculating the economics. The interest rate was fixed at 8% and the depreciation cost of 15% on the drip system was considered. The fixed cost towards the installation of drip fertigation system was worked out to be Rs 72,510. A seasonal cost of Rs 3,453 was included in the cost of cultivation for the annual maintenance and repairs, interest rate and depreciation of the drip system.

RESULTS AND DISCUSSION

Growth parameters

The growth of maize was influenced by various fertigation treatments as shown through the positive response on root characters and dry matter production. Roots are the main component in the absorption of water and minerals, which are essential in plant physiological processes. The results on rooting depth revealed that there was a significant variation in rooting depth of maize due to irrigation methods, fertigation levels and different intercrops (Table 2). Among the fertigation treatments, 100% RDF with 50% P and K as WSF and 150% RDF resulted in higher root parameters. Adequate quantity of nutrients coupled with adequate moisture might have resulted in higher root proliferation. With the application of readily available form of fertilizer, particularly in frequent intervals (once in three days) by reducing the quantity of nutrients at one application, the crops were able to utilize maximum quantity of nutrients, thereby reducing leaching and volatilization loss and increasing the nutrient use efficiency which might have resulted in higher root growth. The biological efficiency of any crop species could be reflected in the amount of dry matter it produces. The results of this study (Table 2) clearly showed significant increase in dry matter production of maize. Application of fertilizer (100% RDF with 50% P and K as WSF) through fertigation resulted in higher growth characters which were followed by 150% RDF. The crops responded to higher dose of fertilizers which were applied as water soluble fertilizers through fertigation which resulted in higher uptake and lead to higher dry matter production over drip irrigation. The increased P uptake and greater P mobility through frequent or continuous low-volume irrigation can maintain three-dimensional distribution patterns of water and nutrients and provide improved conditions for growth, water and nutrient uptake (Ben-Gal and Dudley, 2003).

Dry matter production and root growth of maize was highest under maize + vegetable coriander intercropping system. Increased dry matter accumulation and root growth under this system may be due to less competition for moisture and nutrient as compared to other intercropping system. Vegetable coriander was harvested at 25 DAS. Afterwards, maize crop was grown as sole maize. So availability of moisture and nutrients are higher under this intercropping system. This might be one of the reasons for higher growth parameter of maize in maize + vegetable coriander intercropping system. Tiwari et al. (2003) reported that leafy vegetables like coriander did not show any adverse effect on growth and development of main crop which may be attributed to the fact that coriander is shallow rooted and with short stature and short duration. Application of fertilizer dose of 150% RDF resulted in higher crude protein content and 75% RDF with drip fertigation resulted in lower crude protein content in maize (Table 2). The protein content of the grain was decided by the nitrogen content of the grain. Since the fertigation dose was higher and sufficient under

150% RDF, it might have resulted in higher uptake of nitrogen and ultimately resulted in higher protein content in maize.

Maize grain yield

Generally the maize grain yield increased with increase in fertilizer level (Table 3). In 2008, drip fertigated maize at 150% RDF recorded significantly higher grain yield of 7.3 t ha^{-1}. The yield increase observed under 150% RDF over drip irrigation with conventional method of fertilizer application was 39%. In 2009, higher maize grain yield (7.5 t ha^{-1}) was recorded under drip fertigation of 100% RDF with 50% P and K through WSF. The yield increase over drip irrigation with soil application of fertilizer was 35% in 2009. Application of water soluble fertilizer also influenced the grain yield of maize compared to straight fertilizer. In this present investigation, drip fertigation with 100% RDF in which 50% P and K as WSF increased the grain yield to the tune of 14 and 17% during 2008 and 2009, respectively, compared to drip fertigation of 100% RDF with normal fertilizer. The pooled data revealed that higher grain yield of maize was observed under fertigation of 100% RDF with 50% P and K as WSF.

However, it was on par with fertigation of 150% RDF through normal fertilizer. Different intercrops also influenced the grain yield of maize significantly. Among the four intercrops, vegetable coriander intercrop recorded a highest yield of 6467, 6576 and 6522 kg ha^{-1} in 2008, 2009 and pooled data, respectively. The increase in yield under 100% RDF with P and K as WSF might be due to the fact that fertigation with more readily available form of fertilizer obviously resulted in higher availability of all the three (NPK) major nutrients in the soil solution which led to higher uptake and better translocation of assimilates from source to sink thus in turn increased the yield. The highest number of fruits per plant under liquid fertilizer treatments could be due to continuous supply of NPK from the liquid fertilizers as reported by Kadam and Karthikeyan (2006) in tomato. Hebbar et al. (2004) reported that fertigation with normal fertilizer gave significantly lower yield compared to fertigation with water soluble fertilizers. This was attributed to complete solubility and availability of the water soluble fertilizer as compared to normal fertilizer. Water soluble fertilizer had higher concentration of available plant nutrient in the top layer of soils. Intercrops also had a significant impact on yield components and yield of maize. Yield components were significantly higher in maize + vegetable coriander intercropping system. This could be explained by easy access of resources like moisture and nutrient by maize in this cropping system compared to those in other intercropping

Table 2. Effect of drip fertigation on root character, DMP, stover yield and crude protein content of maize in intensive maize based intercropping system (pooled data).

Treatments	Root volume (cm³)	DMP (t ha⁻¹)	Stover yield (t ha⁻¹)	Crude protein content (%)
M$_1$- SI+ SA of 100% RDF	74.4	8.1	8.2	9.1
M$_2$- DI+ SA of 100% RDF	75.9	8.9	8.8	9.3
M$_3$- DF + 75% RDF (NF)	77.3	9.7	9.2	9.8
M$_4$- DF + 100% RDF (NF)	80.4	10.77	9.4	10.1
M$_5$- DF + 125% RDF (NF)	81.8	11.87	10.1	11.7
M$_6$- DF + 150% RDF (NF)	85.3	13.4	10.4	12.2
M$_7$- DF + 50% RDF (50% P and K- WSF)	78.7	99.5	9.3	9.7
M$_8$- DF + 75% RDF (50% P and K- WSF)	83.3	12.6	9.7	10.6
M$_9$- DF + 100% RDF (50% P and K-WSF)	86.9	14.2	10.6	11.3
S$_1$- Coriander	83.0	12.0	9.8	10.8
S$_2$- Radish	79.6	11.4	9.6	10.6
S$_3$- Beet root	78.2	10.9	9.3	10.2
S$_4$- Onion	81.0	10.4	9.5	10.0

	SEm±	CD(p-0.05)	SEm±	CD(p-0.05)	SEm±	CD(p-0.05)	SEm±	CD(p-0.05)
M	0.56	1.40	315.7	791.1	107.8	269.6	0.08	0.22
S	0.42	1.06	182.6	454.7	81.2	203.0	0.04	0.10
M x S	0.79	NS	480.6	NS	155.1	387.8	0.12	0.30
S x M	0.67	NS	289.3	NS	116.8	291.9	0.06	NS

system (Kumar and Bangarwa, 1997). The increased trend in yield component might be due to the increased supply of nutrients under this cropping system. The stover yield was higher with drip fertigated maize compared to surface irrigated crop (Table 2). The same trend was observed with that of grain yield.

Yield of intercrops

In this study, fertigation at 150% RDF produced significantly more leaf yield of coriander (2825 kg ha⁻¹) and radish tuber yield (5101 kg ha⁻¹). Root yield of beet root was significantly highest under 100% with 50% P and K as WSF (4826 kg ha⁻¹) followed by 75% RDF in which 50% P and K as WSF (Table 4). In case of onion, fertigation of 125% RDF recorded significantly higher bulb yield (4990 kg ha⁻¹) followed by fertigation of 150% RDF. In all intercrops, lower yield was recorded under surface irrigation with soil application of fertilizer compared to drip ferigation. Moisture stress and less availability of nutrient might be the reasons for yield reduction under the surface irrigation method.

Maize grain equivalent yield

In pooled analysis, fertigation with 100% RDF with 50% P and K as WSF produced a highest MEY in all intercropping system (Table 4). In general, among the different system, the MEY was lowest under maize + vegetable coriander system while radish intercropped with maize had a higher MEY of 11153 kg ha⁻¹.

Water use efficiency (WUE)

The water saving under drip irrigation was due to

Table 3. Effect of drip fertigation on grain yield of maize in intensive maize based intercropping system.

Treatments	Pooled data				
	S_1- Coriander	S_2- Radish	S_3-Beet root	S_4- Onion	Mean
M_1- SI+ SA of 100% RDF	5065	4968	4607	4841	4720
M_2- DI+ SA of 100% RDF	5601	5493	5095	5353	5386
M_3- DF + 75% RDF (NF)	6120	6003	5567	5849	5885
M_4- DF + 100% RDF (NF)	6574	6447	5979	6283	6321
M_5- DF + 125% RDF (NF)	7081	6945	6441	6768	6809
M_6- DF + 150% RDF (NF)	7501	7357	6823	7169	7212
M_7- DF + 50% RDF (50% P and K- WSF)	6310	6189	5740	6031	6068
M_8- DF + 75% RDF (50% P and K- WSF)	6842	6710	6223	6539	6578
M_9- DF + 100% RDF (50% P and K -WSF)	7602	7455	6915	7265	7309
Mean	6522	6396	5932	6233	
	M	**S**	**M at S**	**S at M**	
SEm±	100	72	152	117	
CD (P=0.05)	235	181	380	288	

low application rate at frequent intervals matching the actual crop water needs at various stages. Under drip irrigation, only a portion of the soil surface around the crop was wetted whereas under surface irrigation the entire field was wetted. Under drip irrigation, irrigation was practiced frequently once in three days during which the soil moisture was always maintained near to the field capacity. Hence much of the rainfall received during the crop period has gone as ineffective rainfall under drip irrigation.

WUE was higher under drip fertigation treatments compared to surface irrigation method. The increase in WUE in all drip irrigated treatments over surface irrigation was mainly due to considerable saving of irrigation water, greater increase in yield of crops and higher nutrient use efficiency (Ramah, 2008). Increase in irrigation amount did not increase the marketable yield of crops but reduced the irrigation production efficiency significantly (Imtiyaz et al., 2000). Ardell (2006) reported that application of N and P fertilizer will frequently increase crop yields, thus increasing crop water use efficiency. Adequate levels of essential plant nutrients are needed to optimize crop yields and WUE.

Economics

The computed data on the economics of drip fertigation in maize-based intercropping system were presented in Table 5. Though the initial capital investment was high for the drip fertigation system, however, the benefits obtained would be greater considering the longer life of the system. The cost of cultivation was generally higher in the fertigated treatments than the soil applied treatments due to high cost of water soluble fertilizers. Higher cost of cultivation (Rs.38, 157/ha) was observed in fertigation of 100% RDF with 50% of P and K as WSF and onion as intercrop (M_9S_4). The economic analysis of the fertilizer application methods revealed that the cost of cultivation under drip irrigation and fertigation was higher than the surface irrigation with soil application of fertilizers.

The data on the economics of drip fertigation in maize based intercropping system indicated higher net return (Rs. 61, 343 and Rs. 52, 372 and Rs 56858 /ha in 2008, 2009 and pooled data, respectively) and B:C ratio (3.42, 3.06 and 3.24) was obtained under drip fertigation at 150% RDF + radish as intercrop. Drip fertigation at 100% RDF with 50% P and K as WSF resulted in highest productivity in maize-based intercropping system (Table 5). However the net return and benefit cost ratio were lower due to high cost of water soluble fertilizers. Drip fertigation technique aims to achieve water saving and efficient use of applied nutrients for higher productivity. Drip irrigation is the need of the hour especially in areas with water deficit problems. The adoption of drip system should not be merely viewed on the economic point of view. In the context of shrinking land availability for cultivation and diversion of available water for non-agricultural activities, it is paramount important that the available water for agriculture purpose needs to be efficiently utilized by adopting modern irrigation techniques like drip system. The benefits of drip system in terms of water and nutrient savings and enhancement in cropping intensity and productivity of crops are to be taken into consideration. Above all in the water scarcity

Table 4. Effect of drip fertigation on yield of intercrops (kg ha⁻¹) and MEY in maize based intercropping system (Pooled data).

Treatments	Yield of intercrops (t ha⁻¹)				Maize grain equivalent yield (t ha⁻¹)			
	Coriander	Radish	Beetroot	Onion	Coriander	Radish	Beetroot	Onion
M₁- SI+ SA of 100% RDF	1.6	3.0	3.0	3.5	6.2	7.1	6.7	7.3
M₂- DI+ SA of 100% RDF	1.9	3.9	3.6	4.1	7.0	8.4	7.7	8.3
M₃- DF + 75% RDF (NF)	2.1	4.1	3.8	4.2	7.6	9.0	8.3	8.9
M₄- DF + 100% RDF (NF)	2.3	4.3	4.2	4.5	8.2	9.7	9.0	9.5
M₅- DF + 125% RDF (NF)	2.6	4.7	4.2	5.0	8.9	10.3	9.4	10.1
M₆- DF + 150% RDF (NF)	2.8	5.1	4.4	4.9	9.5	11.0	10.0	10.7
M₇- DF + 50% RDF (50% P and K- WSF)	2.2	4.0	4.0	4.4	7.9	9.1	8.6	9.2
M₈- DF + 75% RDF (50% P and K- WSF)	2.6	5.0	4.6	4.7	8.7	10.3	9.6	10.0
M₉- DF + 100% RDF (50% P and K -WSF)	2.7	5.2	4.8	4.8	9.6	11.2	10.4	10.8
Mean	-	-	-	-	8.2	9.6	8.8	9.4
SEm±	44.8	41.4	41.4	32.8	-	-	-	-
CD (P=0.05)	111.8	103.6	104.1	82.2	-	-	-	-

Table 5. Effect of drip fertigation levels on economics and water use efficiency (WUE) of maize based intercropping system.

Treatments	Cost of cultivation (Rs.ha⁻¹)	Net income (Rs.ha⁻¹)	B:C ratio	WUE (kg/ha/mm)
M₁- SI+ SA of 100% RDF	27406	24265	1.90	9.72
M₂- DI+ SA of 100% RDF	26359	32605	2.26	16.35
M₃- DF + 75% RDF (NF)	24305	39434	2.65	17.86
M₄- DF + 100% RDF (NF)	25159	43114	2.74	19.19
M₅- DF + 125% RDF (NF)	26013	46985	2.83	20.67
M₆- DF + 150% RDF (NF)	27653	37545	2.38	18.42
M₇- DF + 50% RDF (50% P and K- WSF)	30606	41578	2.38	19.97
M₈- DF + 75% RDF (50% P and K- WSF)	33562	45118	2.36	22.19
M₉- DF + 100% RDF (50% P and K -WSF)	27406	24265	1.90	9.72
S₁- Coriander	24543	37632	2.55	19.21
S₂- Radish	24543	37632	2.55	19.21
S₃- Beet root	27443	39043	2.44	17.47
S₄- Onion	32143	38627	2.21	18.36

areas, drip system is the only answer to enhance the productivity of crops.

Conclusion

Drip fertigation once in three days at 100% RDF with 50% P and K as water soluble fertilizer enhanced the productivity of maize-based intercropping system. Considering the high cost of water soluble fertilizers, drip fertigation of 150% RDF with radish as intercrop could be an alternative option to realize a reasonably good yield and income in maize based intercropping system.

REFERENCES

Ardell DH (2006). Water use efficiency under different cropping situation. Ann. Agric. Res. 27(5):115-118.

Bar-Yosef B (1999). Advances in fertigation. Adv. Agron. 65:2-67.

Ben-Gal A, Dudley LM (2003). Phosphorus availability under continuous point source irrigation. Soil Sci. Soc. Am. J. 67:1449-1456.

Hebbar SS, Ramachandrappa BK, Nanjappa HV, Prabhakar M (2004). Studies on NPK drip fertigation in field grown tomato (*Lycopersicon esculentum* Mill.). Eur. J. Agron. 21:117-127.

Imtiyaz M, Mgadla NP, Manase SK, Chendo K, Mothobi EO (2000). Yield and economic return of vegetable crops under variable irrigation. Irrig. Sci. 19:87-93.

Kadam JR, Karthikeyan S (2006). Effect of soluble NPK fertilizers on the nutrient balance, water use efficiency, fertilizer use efficiency of drip system in a Tomato. Int. J. Plant Sci. 1:92-94.

Kumar S, Bangarwa AS (1997). Yield and yield components of winter maize (*Zea mays* L.) as influenced by plant density and nitrogen levels. Agric. Sci. Digest. 17(3):354-359.

Ramah K (2008). Study on drip irrigation in maize (*Zea mays* L) based cropping system. Ph.D., Thesis, TNAU, Coimbatore. Seshaiah, M. P. 2000. Sorghum grain in poultry feed. In: Proc. of International Conference on Technical and Institutional options for sorghum grain mould management. ICRISAT, Patencheru, Andhra Pradesh, India. 18 – 19 May. pp. 240-241.

Tiwari KN, Ajai S, Mal PK (2003). Effect of drip irrigation on yield of cabbage (*Brassica oleraceae* L. var. capitata) under mulch and non mulch conditions. Agric. Water Manage. 58:19-28.

18

Effect of irrigation with wastewater from swine in the chemical properties of a latossol

José Antonio Rodrigues De Souza[1] , Débora Astoni Moreira[1], Antonio Texeira De Matos[2] and Aline Sueli De Lima Rodrigues[3]

[1]Department of Agricultural Engineering, Instituto Federal Goiano – Câmpus Urutaí, Urutaí – GO, Brazil.
[2]Department of Agricultural Engineering, Universidade Federal de Viçosa, Viçosa - MG, Brazil.
[3]Department of Environmental Geology, Instituto Federal Goiano – Câmpus Urutaí, Urutaí – GO, Brazil.

The present study aimed to evaluate the variation chemical (electrical conductivity, phosphorus and nitrogen) of an dystrophic red-yellow latossol fertilized with wastewater from swine after filtering process (FSW). For this, tomato plants of the variety Fanny TY were cultivated in drainage lisymeters in a greenhouse, and fertirrigated with different doses of FSW, with and without added chemical fertilizer. The results showed that the c treatments that received lower FSW doses and higher quantities of chemical fertilization presented higher values of electrical conductivity of the saturation extract of soil (ECs); there were an increase in the concentration of available phosphorus, mainly in the superficial layers; the FSW addition resulted in increments in the nitrogen concentration in the superficial layers, while the chemical fertilizer application resulted in larger displacement in the soil profile.

Key words: Reuse, chemical alteration, wastewater, fertirrigation, electrical conductivity.

INTRODUCTION

Nowadays, one of the great global concerns has been the production of wastewater from diverse activities, primarily due to the impacts it causes to the environment, especially with regard to contamination of soil, surface and underground water sources for various processes (Silva Junior et al., 2012). The care with environmental preservation has increased in parallel with the increase of waste production generated by agribusiness. Thus, aware of environmental degradation caused by the launching of wastewater in collections of water and due to the inspection action undertaken by public agencies responsible for environmental quality, swine farmer seek specific solutions towards, to treat, dispose of or reuse the waste (Souza et al., 2010). An alternative that has been presented to reduce environmental degradation

resulting from inadequate disposal of these wastewaters is the use of these effluents in the agriculture, which has long been practiced around the world, earning nowadays importance to the reduction of availability of good quality water resources (Caovilla et al., 2010). The disposition of wastewater in the soil-plant system, when done without agronomic and environmental criteria, may cause problems of contamination of soil, surface water and groundwater and toxicity to plants (Silva Junior et al., 2012). On the other hand, if well planned, this application may bring benefits such as source of nutrients and water for plants, reduction in the use of fertilizers and their pollution potential (Erthal et al., 2010; Souza et al., 2010). The reuse of water for irrigation is a practice widely studied and recommended by many researchers as a

Table 1. Results of physical and chemical analyses of the soil used to fill the lisymeters.

Characteristic	Value	Characteristic	Value
Texture class	Very clayey	Clay (%)	75
Coarse sand (%)	10	Soil specific mass (kg dm^{-3})	0.98
Fine sand (%)	10	Specific mass of the particles (kg dm^{-3})	2.64
Silte (%)	5	Total porosity ($dm^3\,dm^{-3}$)	0.63
pH	7.01	H+Al ($cmol_c\,dm^{-3}$)[d]	0.80
P (mg dm^{-3})[a]	0.90	SB ($cmol_c\,dm^{-3}$)	2.64
K (mg dm^{-3})[a]	9.00	t ($cmol_c\,dm^{-3}$)	2.64
Na (mg dm^{-3})[a]	5.50	T ($cmol_c\,dm^{-3}$)	3.44
P-rem (mg dm^{-3})[e]	11.80	V (%)	76.72
Ca^{2+} ($cmol_c\,dm^{-3}$)[c]	2.02	m (%)	0.00
Mg^{2+} ($cmol_c\,dm^{-3}$)[c]	0.57	ISNa (%)	0.91
Al^{3+} ($cmol_c\,dm^{-3}$)[c]	0.00	OC (dag kg^{-1})[b]	0.52
N_T (mg kg^{-1})[f]	817.00	OM (dag kg^{-1})[b]	0.90

a - Mehlich-1 method; b - Walkley and Black method; c - KCl 1 mol L^{-1} method; d - Ca(OAc)$_2$ 0.5 mol L^{-1} method and - concentration of phosphorus in balance after agitation for 1 h of the TFSA with CaCl$_2$ 10 mmol L^{-1} solution, containing 60 mg L^{-1} of P, in the relation 1:10; f – salicylic acid method. pH - Hydrogenionic potential in water 1:2.5; P - phosphorus available; K – exchangeable potassium; Na – exchangeable sodium; P-rem – remaining phosphorus; Ca^{2+} - exchangeable calcium; Mg^{2+} -exchangeable magnesium; Al^{3+} - exchangeable acidity; H+Al - potential acidity; SB – sum of bases; t -capacity of effective cation exchange; T – cation exchange capacity at pH 7.0; V –index of saturation by bases; m – index of saturation by aluminum; ISNa – index of saturation by sodium; OM – organic matter; OC – organic carbon; N_T – total nitrogen.

viable alternative to meet the water needs and, largely, the nutritional needs of the plants (Grants et al., 2012; Souza and Moreira, 2010; Scheierling et al., 2010; WHO, 2006).

Despite the comproved advantages of the use of swine wastewater as fertilizer of soil, and studies that aim to understand the effects of chemicals in the soil disposal, most not based on agronomic criteria for the calculation of the doses to be applied. Considering that the plants have a fundamental role in the technical viability and sustainability of the treatment system, the objective of this study was to evaluate the effects of fertirrigation with effluent in the chemical characteristics (electrical conductivity, phosphorus and nitrogen) of an dystrophic red-yellow latossol cultivated with tomato (*Lycopersicon esculentum* Mill).

MATERIALS AND METHODS

The experiment was carried out at the Lysimeter Station of the Experimental Area of Hydraulics, Irrigation and Drainage), in the campus of the Universidade Federal de Viçosa (UFV), in Viçosa, MG, from September 2010 to May 2011. Twenty-one drainage lysimeters were utilized under a greenhouse, filled with dystrophic red-yellow latossol previously air-dried, harrowed, sieved in a 0.004 m mesh, with adjusted acidity, and homogenized up to the formation of the profile of 0.60 m. Table 1 presents the physical and chemical characteristics of the soil used to fill the lysimeters. In these lysimeters, after the formation of four definite leaves, the saplings of tomato plants (*L. esculentum* Mill), hybrid Fanny TY were transplanted for furrows with 0.15 m of depths, and a 1.00 × 0.50 m spacing, totaling four plants per lysimeter. The tomato plants were conducted with a single stem, without tip pruning, without the removal of the first inflorescence, with the maintenance

of only six inflorescences per plant, which were vertically staked with polypropylene cord, starting the fastening ten days after the transplanting (DAT), as recommended by Guimarães (2004). The treatments consisted the control (T1 – irrigation and fertilization recommended for the tomato plant) and fertirrigation with filtered swine wastewater providing 100, 150 and 200% of the nitrogen dose recommended for tomato plants, without addition of chemical fertilization (T2, T3 and T4) and with addition of chemical fertilization (T5, T6 and T7), respectively, and the experiment was carried out in a completely randomized design with seven treatments and three replications. The fertirrigations were carried out with swine wastewater (SW) of the Swine Sector of the Department of Animal Science of the UFV, which was conducted to a treatment tank with average hydraulic detention time of 339 h, whose effluent was submitted to a sequence of filtering procedures, passing through two 10 mesh stainless steel screens and one 25 mesh screen. The filtered swine wastewater (FSW) was pumped into the wastewater reservoir of the lysimeter station to be used in fertigation.

Table 2 presents the physical, chemical and microbiological characteristics of the FSW, while Table 3 presents the chemical characteristics of the irrigation water. For the calculation of the doses of the FSW, nitrogen was taken as the reference nutrient, whose doses, necessary to the application of the different percentages of nitrogen, were calculated by means of Equation 1, recommended by reference EPA (1981).

$$L_W = \frac{C_p\,(PR - ET) + 10\,U}{(1 - f)\,C_n - C_p} \tag{1}$$

Where Lw - lamina of annual application, cm $year^{-1}$; Cp - nitrogen concentration in the percolation water, mg L^{-1}; PR - local precipitation, cm $year^{-1}$; ET - evapotranspiration of the local culture, cm $year^{-1}$; U - absorption of nitrogen by the culture, kg ha^{-1} $year^{-1}$; Cn - concentration of nitrogen in the wastewater, mg L^{-1}; and F - fraction of the nitrogen that is removed by denitrification and volatilization, dimensionless.

Table 2. Average values of the physical, chemical and microbiological characteristics of the filtered swine wastewater (FSW) used in the fertirrigation.

Characteristics	Values	Characteristics	Values
pH	7.43	K_T (mg L^{-1})	162
EC (μS cm^{-1})	3.403	Na (mg L^{-1})	40
N_T (mg L^{-1})	480	TOC (dag kg^{-1})	0.12
N-NO_3^- (mg L^{-1})	0.44	OM (dag kg^{-1})	0.20
N-NH_4^+ (mg L^{-1})	0.30	Ca + Mg ($mmol_c L^{-1}$)	4.40
Cl (mg L^{-1})	181.40	BDO (mg L^{-1})	89
Alkalinity (mg L^{-1} de $CaCO_3$)	1954	CDO (mg L^{-1})	370
P_T (mg L^{-1})	139	RAP (($mmolL^{-1}$)$^{-1/2}$)	2.81
TS (mg L^{-1})	1067	RAS (($mmolL^{-1}$)$^{-1/2}$)	1.18
STS (mg L^{-1})	126	TC (NMP/100 mL)	13.4×10^5
TVS (mg L^{-1})	381	FC(NMP/100 mL)	4.1×10^5

pH – Hidrogenionic potential; EC – electrical conductivity; N_T – total nitrogen; N-NO_3^- - nitrogen in the nitrate form; N-NH_4^+ - nitrogen in the ammoniacal form; Cl - chloride; P_T – total phosphorus; TS – total solids; STS - solids in total suspension; TVS – total volatile solids; K_T – total potassium; Na - sodium; TOC – total organic carbon; OM – organic matter; Ca+Mg – calcium plus magnesium; BDO - biochemical demand of oxygen; CDO – chemical demand of oxygen; RAP - ratio of potassium adsorption; RAS - Ratio of sodium adsorption; TC - total coliforms; FC – thermotolerant coliforms; NMP most probable number.

Table 3. Chemical characteristics of the irrigation water.

pH	EC	CDO	N_T	K_T	Na	Cl	Alc	Ca+Mg	RAS	RAP
	μS cm^{-1}			mg L^{-1}			mg L^{-1} de $CaCO_3$	$mmol_c L^{-1}$	($mmol_c L^{-1}$)$^{-1/2}$	
7.44	70.40	9.80	3.47	2.63	3.83	1.00	26.00	0.58	0.31	0.13

Where: pH - hidrogenionic potential; EC – electrical conductivity; CDO – chemical demand of oxygen, N_T – total nitrogen l; K_T – total potassium; Na - sodium; Cl - chloride; Alc – total alkalinity, Ca+Mg – calcium plus magnesium, RAS - relation of sodium adsorption; RAP - relation of potassium adsorption.

This method considered Cp as 10 mg L^{-1} (CONAMA 357/2005; COPAM/CERH n° 01/2008), null PR-ET (handling in a greenhouse and evapotranspiration reposition), U equals to 400 kg ha^{-1} (tomato plant cultivated in a greenhouse, vertically staked, according to CFSEMG (1999), f equals to 20% (Matos, 2007) and Cn achieved in bimonthly evaluations. The complementary chemical fertilization was calculated by subtracting from the values of P and K recommended by CFSEMG (1999), the quantity of these nutrients comes from the different doses of the FSW applied. Therefore, 261.10, 229.80 and 181.4 g $furrow^{-1}$ of super-simple and, 49.70, 40.90 and 32.70 g $furrow^{-1}$ of potassium chloride were added to the soils under treatments 5, 6 and 7, respectively. In the soils submitted to the control treatment, 100 g $furrow^{-1}$ of ammonium sulfate, 375 g $furrow^{-1}$ of super-simple and 69 g $furrow^{-1}$ of potassium chloride were added. The meteorological variables, necessary to the determination of the evapotranspirometric demand, were achieved by means of a Davis automatic station, installed in a greenhouse. The reposition of the evapotranspirometric demand of tomato plants was determined considering the culture evapotranspiration (ETc) achieved by the multiplication of the reference evapotranspiration (ET0) by the coefficients of cultivation (Kc) of tomato plants suggested by Moreira (2002), the percentage of shaded area, the coefficient of localization proposed by Keller and Bliesner (1990), and the efficiency of the application system.

The applications of the irrigation water and fertirrigation were carried out by dripping, with the use of 0.016 m diameter polyethylene hose, whose emitters were integrated in the spacing of 0.50 m (one emitter per plant) and presented a flow of 1.90 L h^{-1} for an operating pressure of 10 MPa. The fertirrigations were carried out with the reposition of 100, 150 and 200% of the daily ETc for the treatments that received, respectively, 100, 150 and 200% of the nitrogen by means of the FSW doses, thus, making the most needed nutrients available for the plants in due time. The fertirrigation started after the transplanting of the saplings by means of daily applications of FSW doses, which were concluded 68 days after the transplanting (DAT), then totaling 114.29, 171.43 and 228.58 mm, corresponding to 100, 150 and 200% of the nitrogen required by the culture, calculated by Equation 1. After this period, only water was applied to replace the evapotranspirometric demand by tomato plants. Thus, as observed by Batista (2007), when clean water is prevented from passing through polyethylene lines during the period of the FSW application, biofilm formation and, consequently, the clogging of drippers are reduced.

During transplanting (0 DAT), in the middle (60 DAT) and end (120 DAT) of the tomato plant cycle, samples of soil were collected, with the use of a Dutch auger, 0.10 m far from the stem of a plant, in each lysimeter, in the layers of depths of 0.18 to 0.22, 0.38 to 0.42 and 0.56 to 0.60 m, except for the samples for the determination of the electrical conductivity of the saturation extract of soil (electrical conductivity of the saturation extract of soil (ECs)), which were collected in the layer of 0.20 m, in the period of 44, 77 and 112 DAT, corresponding to the formation of the first and sixth inflorescences and final phase of the tomato plant cycle. These samples were identified and sent to be analyzed as for the CEes, phosphorus (P) and total nitrogen, at the Laboratory of

Table 4. Results of the evaluations of the electrical conductivity of the saturation extract of soil (ECs) (dS m^{-1}) and respective average tests, in different evaluation periods, for the 0 to 0.20 m layer.

TRAT	DAT		
	44	77	112
1	4.42^{Aa}	4.79^{Aa}	2.20^{Ab}
2	2.52^{Db}	3.90^{Ba}	1.76^{Ac}
3	2.64^{Db}	4.03^{Ba}	1.72^{Aa}
4	3.21^{Cb}	4.42^{ABa}	1.87^{Ac}
5	3.94^{ABb}	4.13^{Ba}	2.01^{Ac}
6	3.70^{BCa}	4.43^{ABa}	2.13^{Ab}
7	3.45^{BCb}	4.33^{ABa}	1.85^{Ac}

Averages followed by at least one same lower case letter in the lines indicate that, for the treatment (TRAT), the evaluations at the time (DAT) do not differ, at 5% of probability, by the Tukey test.

Soil Fertility and Laboratory of Soil Physics of the Department of Soils of the UFV, according to methodologies described in EMBRAPA (1997).

RESULTS AND DISCUSSION

Effects on the electrical conductivity

The doses of FWS needed to supply 100, 150 and 200% of the nitrogen required by tomato plants, calculated by Equation 1 were, respectively, 114.29, 171.43 and 228.58 mm, while the ETc during the period was 211.62 mm. Finalized the fertirrigation with FWS, 68 DAT, were applied only doses of irrigation water in order to replenish the daily ETc, totaling 97.33 mm. It was verified that, even with the application of 200% of the daily ETc, the daily doses applied were not enough to produce effluents in the lysimeters and ensured that the all FSW was available to the plants. According to the classification proposed by Ayers and Westcot (1991), due to the low electrical conductivity (EC) and the sodium adsorption ratio (SAR), the water used in the irrigations present high risk of sodicity and no risk of soil salinization, while the FSW present a high risk of salinization. However, with regard to the potential to cause reduction in the soil infiltration capacity, these guidelines should not be used for FSW, because they do not include the solid organic elements contained in the wastewaters. Table 4 presents the results of the evaluations of the soil electrical conductivity of the saturation extract of soil (electrical conductivity of the saturation extract of soil (ECs)), in different periods, in the 0 to 0.20 m layer, for the different treatments. In Table 4, it can be observed that the electrical conductivity of the saturation extract of soil (ECs) has increased with increments of doses of FWS applied and, when added chemical fertilizers, the opposite behavior occurred, presented higher electrical conductivity of the saturation extract of soil (ECs), the treatments had the smallest FWS doses, however, with larger quantities of supplementation by chemical fertilizer. Because the swine wastewater has a slower release of nutrients, in relation to chemical fertilizer, it can be verified that the treatments that received a larger input of chemical fertilizer showed highest electrical conductivity of the saturation extract of soil (ECs), indicating a higher concentration of salts in the soil at depths monitored. Therefore, it is evident that chemical fertilizer generally was more effective in increasing the electrical conductivity of the saturation extract of soil (ECs) of the soil than the FWS. This fact may be associated to the presence of ions participants organic chains or are complexed/chelated who thus are not detected by the conductivimeter electrode.

Silva et al. (2012) studying the effects of application of cattle farm wastewater in EC of soil founded that the chemical fertilizer was more effective in the ionization of the soil, and when made the supplementation of the fertilization, higher EC were observed in soils that received larger quantities of fertilizers, smaller doses of chemicals and wastewater. Similarly, Freitas et al. (2005) evaluating the effect of swine wastewater on soil chemical characteristics, found an increase of 2.75% from the electrical conductivity of the saturation extract of soil (ECs) regarding the treatment managed with clean water. These authors also verify that the application of swine wastewater made the saline soil. Lo Monaco et al. (2009) found an increase in the electrical conductivity of the saturation extract of soil (ECs) with increased doses wastewater from the processing of the coffee fruit. These authors claim that such behavior is associated with increased ions in the soil solution when high doses of wastewater from processing the coffee fruit were applied. The application of FSW doses during transplanting 68 DAT and their suppression after this period, when only irrigation water started to be applied, and the end of the chemical fertilization 90 DAT, carried out in the treatment 1, were responsible for the salinity reduction observed in the evaluation carried out 112 DAT.

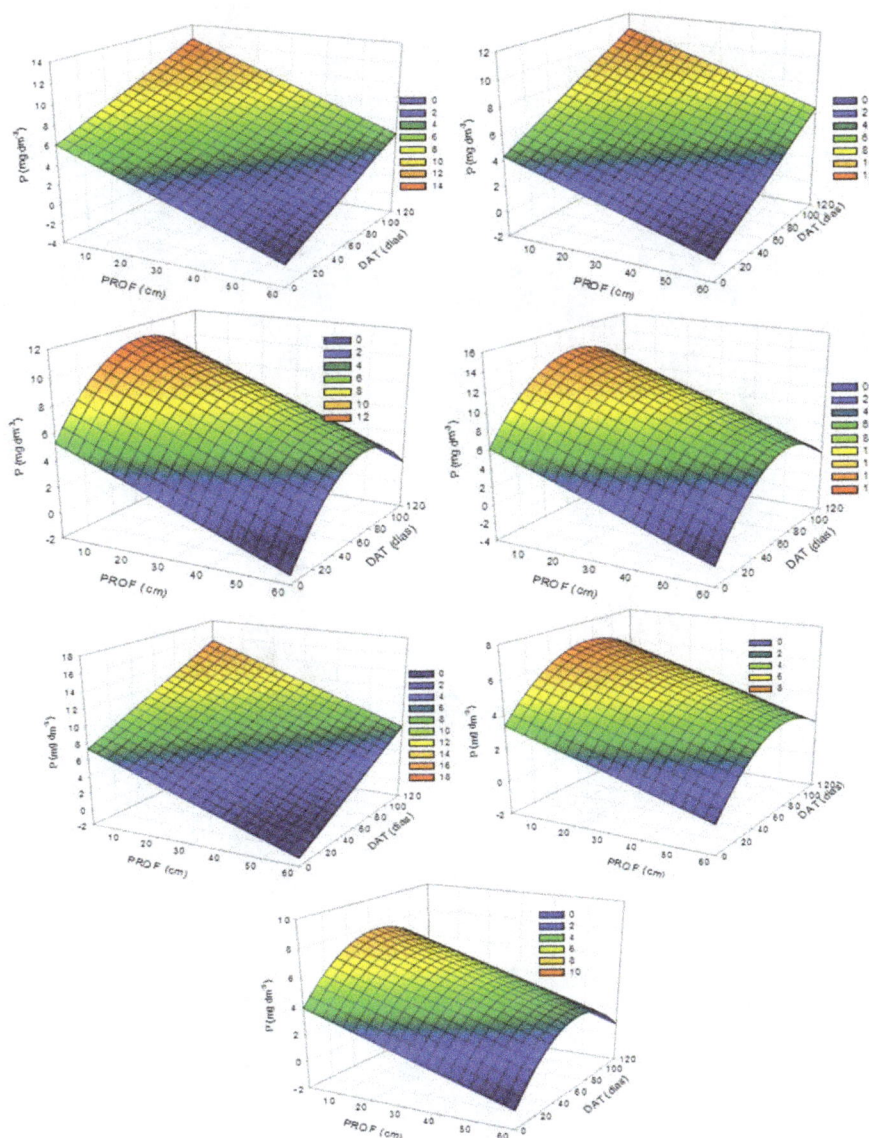

Figure 1. Variation in the P concentration available in the soil profile, according to the depth (PROF) and days after transplanting (DAT), in the soils submitted to the treatments 1 (A), 2 (B), 3 (C), 4 (D), 5 (E), 6 (F) and 7(G).

Phosphorus

Figure 1 presents the phosphorus variation available with depth and time, in the soils submitted to the different treatments. It is possible to observe that the phosphorus concentration presented a negative linear relation with depth and a quadratic relation with time, except for the soils submitted to the treatments 1, 2 and 5, whose relation was positive linear. It is also observed that, compared to the initial conditions, there was an increase in the phosphorus concentration available, mainly in the superficial layers. According to Klein and Agne (2012) and Ceretta et al. (2005), the low phosphorus concentrations in the lower layers are due to the low mobility of this nutrient on soil, which is probably

adsorbed by the soil particles and absorbed by the plants, while the remaining are precipitated. For José et al. (2009), the content of available phosphorus normally tends to decrease with depth, following the content of soil organic matter. The application of daily doses of FSW during transplanting until 68 DAT and its suppression after this period may have been responsible for the quadratic effect on time, while the positive linear behavior, observed in the soils submitted to the treatments 2 and 5, may have been a consequence of the virus symptoms presented by the tomato plants cultivated in these soils, which hindered their development and yield, thus causing lower absorption of this nutrient. Similarly, several authors (Roveda et al., 2011; José et al., 2009; Erthal et al., 2010; Garcia et al.,

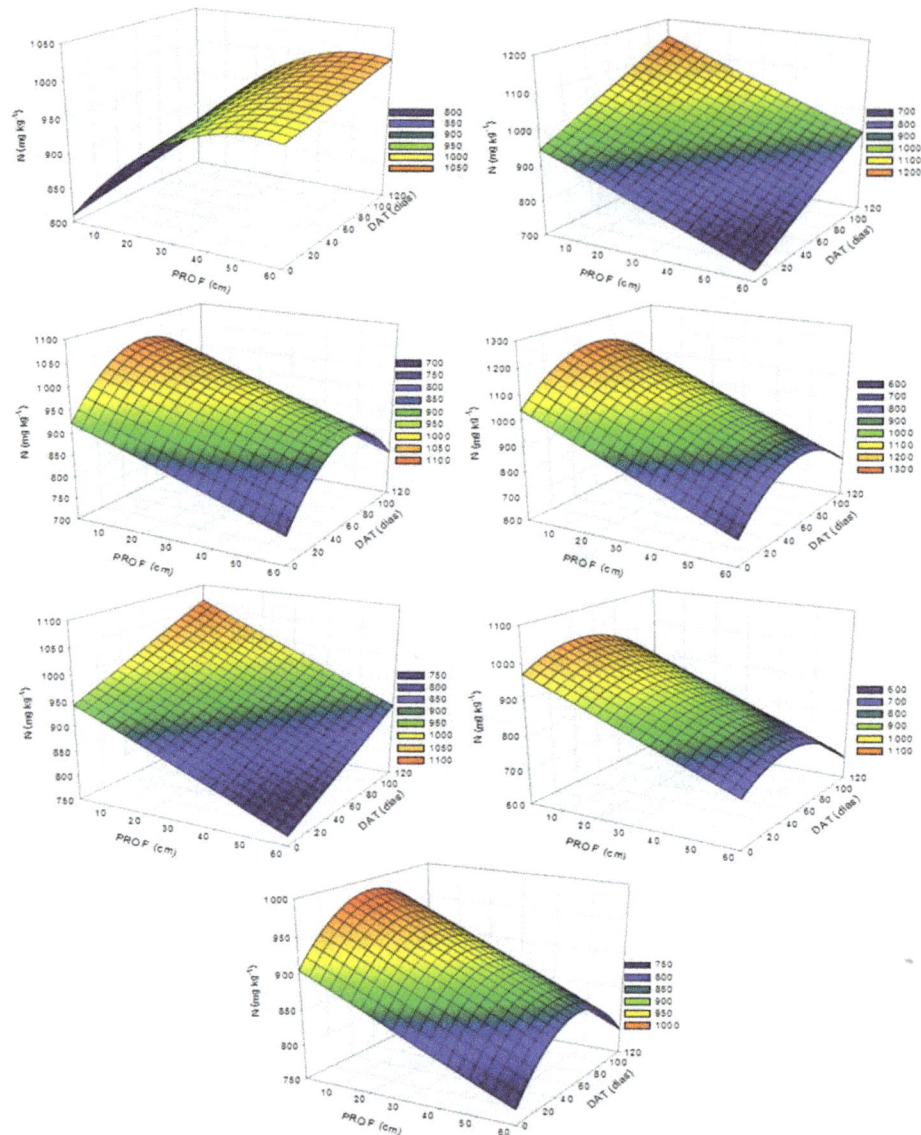

Figure 2. Variation in the N concentration in the profile, according to depth (PROF) and days after transplanting (DAT) in the soils submitted to the treatments 1 (A), 2 (B), 3 (C), 4 (D), 5 (E), 6 (F) and 7(G).

2008; Queiroz et al., 2004; Oliveira, 2006; Berwanger, 2006) studying the effects of wastewater application, the soil also found increases in the concentration of available phosphorus in the surface layers, obtaining higher values when applied to the highest doses of wastewater.

At the end of the experimental period, at a depth of 0.10 m, reductions were observed in the concentrations of available phosphorus in relation to the witness, having been obtained reductions of 10.85, 30.98, 17.05, 54.20 and 59, 20% on soils submitted to treatments 2, 3, 4, 6 and 7, respectively, and increase of 25.63% on soils submitted to treatment 5. Thus, except for the soils submitted to the treatments 2 and 5, because of disease symptoms, it was observed that the highest FSW doses provided increments in the absorption of phosphorus by

the culture, that were intensified by the nutrient balance through the fertilization addition. In relation to the classes of interpretation of phosphorus availability suggested by CFSEMG (1999), before the experimental period, the soil of all the experimental plots presented very low availability of phosphorus and, after this period, at the depth of 0.10 m, the soils presented low (treatments 6 and 7), average (treatment 3), good (treatments 1, 2 and 4) and very good (treatment 5) phosphorous availability.

Nitrogen

Figure 2 presents the variation of the total nitrogen concentration with soil depth and time. It is possible to observe that, in the soils of the treatments that received the application of FSW, the nitrogen concentration presented a negative linear relation with depth and quadratic relation with time, except for the soils submitted to the treatments 2 and 5, whose relation was positive linear. For the soil submitted to the treatment 1, it was observed a quadratic relation with depth and positive linear relation with time. The predominance of the organic form of nitrogen (99%), added to the treatments by means of the FSW application, may have been responsible for the increase in the concentration of this nutrient in the superficial layers, while the quadratic effect on time may be related to the FSW application until 68 DAT and its suppression after this period. The positive linear behavior in time, observed in the soils submitted to the treatments 2 and 5, is probably related to the virus symptoms presented by the tomato plants cultivated in these lysimeters, which resulted in a lower development of the plants and, consequently, lower values for plant growth, dry matter production, nutrient concentration in the fruits and yields. In the soils submitted to the treatment 1, the addition of ammonium sulfate as a source of nitrogen, clearly presenting great mobility on the soil, the liming and the application of irrigation laminas may have caused the quadratic effect observed with the depth in the soil profile, favoring the displacement of NH_4^+ and NO_3^-. The variation in time may have been caused by the split application of nitrogen, according to recommendations for the tomato culture suggested by CFSEMG (1999).

Fortes et al. (2012) applied treated sewage in the production of oats, Ceretta et al. (2003) analyzing the alterations caused by the WS application in soil cultivated with natural pasture and Dal Bosco et al. (2008), by applying WS in arable soil for eight consecutive years, also observed low nitrogen mobility in the soil profile, and achieved higher values in the superficial layers, which were increased with the addition of WS. According to these authors, this behavior could be associated with the nitrogen immobilization reaction by microorganisms to decompose organic matter in the soil and also the absorption by the roots of the crop, presenting thus the same tendency of organic matter. It is also possible to observe that the maximum values of the nitrogen concentration in the soil, in the soil receptors of FSW, occurred in the higher layers, following the application of all the laminas, except for the soils submitted to the treatments 2 and 5, in which the maximum values occurred in the end of the experimental period. In the soils submitted to the treatment 1, the maximum value was also observed in the end of the experimental period, but in the lower layers, which indicates higher tendency to groundwater contamination. Similar results were observed by Fortes et al. (2012), Fagundes et al. (2007), Mohammad and Mazahreh (2003) and Wang et al. (2003), when studied the effect of reuse of wastewater on the soil. In the end of the experimental period, it was verified that, in relation to the initial conditions, at the depth of 0.10 m, there were increments in the nitrogen concentration of 11.00, 36.17, 13.83, 26.00, 27.21, 4.41 and 9.77%, in the soils submitted to the treatments 1, 2, 3, 4, 5, 6 and 7, respectively. Therefore, it is possible to observe that, except for the soils submitted to the treatments 2 and 5; higher doses provided higher increases in the concentration of nitrogen in the soil, with lower values observed when complementary fertilization was carried out, which, due to nutrient balance, favored a higher absorption by the culture.

According to Lopes (1998), one of the problems in culture fertilization is the unbalanced use of nitrogen and potassium, which causes much damage to the agricultural production.

Conclusions

Under the experimental conditions and according to the results, it can be concluded that: a) the chemical fertilization was more effective in the ionization of the soil solution than the filtrate swine wastewater (FSW). The treatments that received lower FSW doses and higher quantities of chemical fertilization presented higher values of electrical conductivity of the saturation extract of soil (ECs); b) in comparison to the initial conditions, there was an increase in the concentration of phosphorus available, mainly in the superficial layers; c) the addition of FSW resulted in increases in the nitrogen concentration in the superficial layers, while the chemical fertilization resulted in higher displacement in the soil profile.

REFERENCES

Ayers RS, Westcot DW (1991). A qualidade da água na agricultura: UFPB. P. 218.

Batista RO (2007). Desempenho de sistema de irrigação por gotejamento utilizado na aplicação de água residuária de suinocultura. Tese (Doutorado em Engenharia Agrícola). Universidade Federal de Viçosa, Viçosa. P. 146.

Berwanger AL (2006). Alterações e transferências de fósforo do solo para o meio aquático com o uso de dejeto líquido de suínos. UFSM: P. 99.

Caovilla FA, Sampaio SC, Smanhotto A, Nóbrega LHP, Queiroz MMF, Gomes BM (2010). Características químicas de solo cultivado com soja e irrigado com água residuária da suinocultura. Revista Brasileira de Engenharia Agrícola e Ambiental Campina Grande. 14(7):692-697.

Ceretta CA, Basso CJ, Vieira FCB, Herbes MG, Moreira ICL, Berwanger AL (2005). Dejeto líquido de suínos: I – perdas de nitrogênio e fósforo na solução escoada na superfície do solo, sob plantio direto. Ciência Rural 3:1296-1304.

Ceretta CA, Durigon R, Basso CJ, Barcellos LAR, Vieira LAR (2003). Características químicas de solo sob aplicação de esterco líquido de suínos em pastagem natural. Pesquisa Agropecuária Brasileira. Brasília. 6(38):729-735.

CFSEMG (1999). Comissão de Fertilidade do Solo do Estado de Minas Gerais. Recomendações para o uso de corretivos e fertilizantes em Minas Gerais – 5ª aproximação. Ribeiro AC, Guimarães PTG, Alvarez VH (ed)., editores. Viçosa, MG. P. 359.

Conselho Estadual De Política Ambiental/Conselho Estadual De Recursos Hidricos – Copam/Cerh (2008). Deliberação Normativa nº 01 de 05 de maio de 2008. Dispõe sobre a classificação dos corpos de água e diretrizes ambientais para o seu enquadramento, bem como estabelece as condições e padrões de lançamentos de efluentes, e da outras providencias. Belo Horizonte. Available at: <http://www.siam.mg.gov.br/sla/download.pdf?idNorma=8151>. Accessed on 11 February 2012

Conselho Nacional Do Meio Ambiente – Conama (2005). Resolução Nº 357, De 17 de março de 2005. Dispõe sobre a classificação dos corpos de água e diretrizes ambientais para o seu enquadramento, bem como estabelece as condições e padrões de lançamento de efluentes. Brasília. Available at: <http://www.siam.mg.gov.br/sla/download.pdf?   idNorma=2747>. Accessed on 04 October. 2012.

Dal Bosco TC, Sampaio SC, Iost C; Silva LN, Carnellosi, CF, Ebert DC, Schreiner JS (2008). Utilização de água residuária de suinocultura em propriedade agrícola – estudo de caso. Irriga, Botucatu. 13(1):139-144.

EMBRAPA (1997). Manual methods of soil testing . 2. ed. Brasília: Embrapa-SPI, p. 212.

Environmental Protection Agency (1981). Process design manual – land treatment of municipal wastewater. Washington, D.C.: Department of the interior. P. 625.

Erthal VJT, Ferreira PA, Matos AT, Pereira OG (2010). Alterações físicas e químicas de um Argissolo pela aplicação de água residuária de bovinocultura. Revista Brasileira de Engenharia Agrícola e Ambiental. 14(5):467–477.

Fagundes JD, Santiago G, Mello AM, Bellé RA, Streck NA (2007). Crescimento, desenvolvimento e retardamento da senescência foliar em girassol de vaso (Helianthus annuus L.): fontes e doses de nitrogênio. Ciência Rural, Santa Maria. 37(4):987-993.

Fortes Neto P, Veiga PGA, Fortes NLP, Targa MS, Gadioli JL, Peixoto PHM (2012). Alterações químicas do solo e produção de aveia fertilizada com água residuária do tratamento de esgoto sanitário. In: The 4th International Congress on University-Industry Cooperation – Taubate, SP – Brazil – December 5th through 7th.

Freitas WS, Oliveira RA, Pinto FA, Cecon PR, Galvao JCC (2005). Efeito da aplicação de aguas residuais da suinocultura na produção do milho para silagem. Revista Brasileira de Engenharia Agrícola e Ambiental, Campina Grande. 8(1):120-125.

Garcia GO, Ferreira PA, Matos AT, Ruiz HA, Martins Filho S (2008). Alterações químicas em três solos decorrentes da aplicação de águas residuárias da lavagem e despolpa de frutos do cafeeiro conilon. Engenharia na Agricultura. 16(4):416-427.

Grant SB, Saphores J, Feldman DL, Hamilton AJ, Fletcher TD, Cook PLM, Stewardson M, Sanders BF, Levin LA, Ambrose RF, Deletic A, Brown R., Jiang SC, Rosso D, Cooper J, Marusic I (2012). Taking the "Waste" Out of "Wastewater" for Human Water Security and Ecosystem Sustainability. Science 337:681-685.

Guimarães MA (2004). Influência da poda apical e da posição do cacho do tomateiro no crescimento da planta e na qualidade dos frutos. Dissertação (Mestrado em Fitotecnia) – Universidade Federal de Viçosa, Viçosa. P. 93.

Keller J, Bliesner RD (1990). Sprinkle and trickle irrigation. New York: Van Nostrand Reinold. P. 652.

Klein C, Agne SAA (2012). Fósforo: de nutriente à poluente. Revista Eletrônica em Gestão, Educação e Tecnologia Ambiental. 8(8):1713-1721.

Lo Monaco PA, Matos AT, Martinez HEP, Ferreira P, Mota MM (2009). Características químicas do solo após a fertirrigação do cafeeiro com aguas residuarias da lavagem e descascamento de seus frutos. Irriga, Botucatu. 14(3):348-364.

Lopes AS (1998). Manual internacional de fertilidade do solo. 2° ed. Instituto da Potassa & Fosfato. Piracicaba. P. 177.

Matos AT (2007). Disposição de águas residuárias no solo. Viçosa, MG: AEAGRI. (Caderno didático n. 38):142.

Mohammad MJ, Mazahreh N (2003). Changes in soil fertility paramenters in response to irrigation of forage crops with secondary treated wastewater. Commun. Soil Science Plant Anal. New York 34(9-10):1281-1294.

Moreira HM (2002). Desempenho de métodos de manejo de irrigação para a cultura do tomateiro cultivado em campo e em casa de vegetação. Dissertação (Mestrado em Engenharia Agrícola) – Universidade Federal de Viçosa, Viçosa. P. 111.

Oliveira W (2006). Uso de agua residuária da suinocultura em pastagens da Brachiária Decumbens e Grama Estrela Cynodom Plesctostachyum. Dissertação (Mestrado em Agronomia). Escola Superior de Agricultura "Luiz de Queiroz", Universidade de São Paulo, Piracicaba. P. 104.

Queiroz FM, Matos AT, Pereira OG, Oliveira RA (2004). Características químicas de solo submetido ao tratamento com esterco líquido de suínos e cultivado com gramíneas forrageiras. Ciência Rural, Santa Maria. 34(5):1487-1492.

Roveda LF, Motta ACV, Dionísio JÁ, Brondani GE; Gabardo J, Pimentel IC, Cuquel FL (2011). Fertilidade, produção e nutrição de plantas submetidas à água residuária da indústria de enzimas Revista Brasileira Agrociência, Pelotas. 17(3-4):338-347.

Scheierling SM, Bartone C, Duncan DM, Drechsel P (2010). Improving wastewater use in agriculture: an emerging priority. World Bank Policy Research Working. P. 5412.

Silva JBG, MArtinez MA, Pires C S, Andrade IPS, Silva GT (2012). Avaliação da condutividade elétrica e ph da solução do solo em uma área fertirrigada com água residuária de bovinocultura de leite. Irriga, Botucatu, Edição Especial, pp. 250-263.

Silva Junior JJ, Coelho EF, Sant'ana JAV, Accioly AMA (2012). Physical, chemical and microbiological properties of a Dystrophic yellow latosol using manipueira. Engenharia Agrícola., Jaboticabal. 32(4):736-744.

Souza JAR, Moreira DA (2010). Efeitos do uso da água residuária da suinocultura na condutividade elétrica e hidráulica do solo. Engenharia Ambiental, UNIPINHAL. 7:134– 143.

Souza JAR, Moreira DA, Coelho DF (2010). Crescimento e desenvolvimento de tomateiro fertirrigado com água residuária da suinocultura. Revista Ambiente. Água. 5:144 –157.

Wang Z, Chang AC, Wu L, Crowley D (2003). Assessing the soil quality of longterm reclaimed wastewater-irrigated cropland. Geoderma 114:261-278.

WHO (World Health Organization), WHO Guidelines for the safe use of wastewater (2006). Excreta and Greywater. Wastewater Use Agriculture 2:182.

19

Decision support system for soil and water analysis and fertiliser recommendation for flue-cured Virginia (FCV) tobacco

H. Ravi Sankar, C. Chandrasekhararao and K. Sivaraju

Central Tobacco Research Institute, Rajahmundry, Andhra Pradesh – 533 105, India.

Soil fertility and water quality play an important role on the productivity of any crop. Soil and water test based fertilizer recommendation will help the farmers to optimize the resources and to improve the productivity. Decision Support System (DSS) was developed for calculating the instrument reading into final value of the test parameters of soil (pH, EC, organic carbon, available nitrogen, phosphorous, potassium and chloride) and water (pH, EC, chloride), to suggest the recommended doses of fertilizers for Flue-Cured Virginia (FCV) tobacco crop and to test their suitability for tobacco cultivation. DSS on soil and water analysis for tobacco crop was developed with Visual Basic. Net as Front-end and MS-Access as Back-end. In this system, by entering instrument reading, it will be transformed into soil / water testing value of the concerned parameter which will be compared with recommended values and finally suitability of soil and water for tobacco cultivation will be judged and recommended fertilizer doses for different tobacco zones will be prepared and printed in the prescribed format. This software will be helpful to the researchers / technical personnel in reducing time for preparation of reports, minimizing the errors in manual calculation and improves the precision. This system can be applicable to any soil testing laboratories, where recommendations are given to different crops based on soil and water test values with suitable modifications.

Key words: Decision support system, flue cured tobacco, irrigation water quality, soil testing.

INTRODUCTION

In India, tobacco is an important commercial crop grown in an area of 0.45 million ha with 750 million kg production (Anonymous, 2010). There are ten tobacco types grown in India, out of which flue-cured Virginia (FCV) tobacco is an important type used for manufacturing cigarettes. In India, FCV tobacco is grown in an area of 0.24 million ha in Andhra Pradesh and Karnataka states with an annual production of 300 million kg leaf out of which nearly 200 million kg is exported to different countries (Krishnamurthy and Anuradha, 2011). Tobacco is a quality conscious commercial crop. Soil, water and climatic factors play a predominant role on

tobacco quality and yield (Davis and Nielsen, 2007). The chemistry and fertility of soils greatly influences the tobacco plant growth, leaf size, yield and physical, chemical and manufacturing properties of tobacco leaf. Evaluation of fertility status of the soil before tobacco planting is a pre-requisite for optimum NPK fertilizer recommendations to get quality tobacco. Among the several other factors influencing tobacco productivity, soil fertility and fertilizer use contributes nearly 50% of yield and quality improvement of tobacco crop (Krishnamurthy and Deosingh, 2002).

Irrigation is defined as an artificial application of

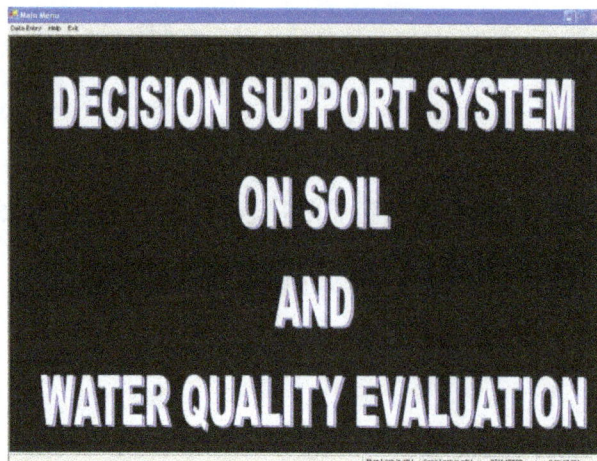

Figure 1. Main menu.

water to the soil for the purpose of crop production. Quality of irrigation water plays an important role on the quality and quantity of tobacco production. Fertilizer recommendation based on soil test values will help in improving the fertilizer use efficiency thereby increase in yield and quality of tobacco. Soil and water testing laboratories for tobacco farmers are available at Central Tobacco Research Institute (CTRI), Rajahmundry and also in other parts of the country under the management of Tobacco Board, Ministry of Commerce, Government of India. In addition to that soil testing laboratories is available all over India to serve different crops. In these laboratories a large number of soil and water samples are being analyzed regularly and test reports along with fertilizer recommendations are being prepared manually, which consumes a lot of time and there is a possibility of committing errors. The information collected from technical persons involved in soil and water analysis laboratories indicated the difficulties involved in the calculation of soil and water test parameters and recommendation of fertilizer doses for a particular zone. As the laboratories are short of human resources and number of samples to be analyzed is more, development of an integrated system for calculation of soil and water parameter and to suggest the test based recommendations for a crop will help not only technical / research personnel but also the end users that is farmers.

Information Technology has become an integral part of our day-to-day life. In agriculture, computers are extensively used to disseminate the information on soils, water and climatic conditions. Decision Support Systems (DSS) are defined as computerized systems, which include models and databases and they are used in decision-making. They are "tools" that help farmers and everyone who makes decisions and in choosing the best (economic, social or environmental) alternative solution (Manos et al., 2004). Several scientific disciplines support

the development of DSSs and constitute the necessary background for their effective planning.

In Agriculture, DSS with varying capabilities have been developed viz., cropping systems simulation model (Stockle et al., 1994), crop rotation planning systems (Stone et al., 1992), Model of an Integrated dry-land agricultural system (Pannell, 1996). A few expert systems on tobacco cultivation were developed at CTRI including nutrient deficiency symptoms and disease management (Ravisankar et al., 2009, 2010). In the present study, efforts were made to develop a decision support system which will help technical personnel and researchers on preparation of soil and water test reports along with fertilizer recommendations instantaneously, which improves the precision, saves time and energy there by more number of soil and water samples can be analyzed.

MATERIALS AND METHODS

The DSS was developed using Visual Basic.Net (Balena, 2005; Gaddis et al., 2003) as front-end application and MS-Access (Nelson and Kelly, 2002) as back-end with user-friendly menus consisting 10 parameters as the attributes. The parameters 'Name of the farmer, soil type, source of water, pH, electrical conductivity, organic carbon, available nitrogen, phosphorus, potassium and chlorides' were selected for inclusion in the package. These fields are created with text-boxes for data and label-boxes for title of the text. The knowledge base for this system was soil and water testing methodologies developed by different scientists which are widely being followed in tobacco and other crops (Krishnamurthy and Nagarajan, 2001; Ghosh et al., 1983). Based on the knowledge base, this system has been developed (coding) which consists of 22 modules. Crystal reports 9.0 are used for designing and generating reports in an effective way.

The multiple document interface (MDI) form of the software consists of three options. The first option in the menu was "data entry" allows the user to enter the input data of various parameters for soil fertility and water quality evaluation. The second option is 'Help' which consists of a pull down menu and two sub options viz., 'About the Project' and 'Execution of the project'. The last option 'Exit' allows the user to quit from the software.

After the coding phase, the testing phase is performed by connecting the database to the developed modules with Open Database Connectivity (ODBC). Debugging is performed to correct the syntax and semantic errors in the developed program. Finally, a 'setup' program was prepared for easy loading and execution of the software.

RESULTS AND DISCUSSION

For executing this package, a PC with preloaded software of Visual Basic .Net and MS ACCESS are required. The procedure includes: Select the 'Soil Project' from the files and 'open' the software and then press F5 to execute. The first screen of the package 'Main Menu' gets displayed (Figure 1). Click 'Data entry' in the MDI form. A new screen with 'Farmer Details Form' as title bar gets displayed. Enter the 'Farmer Name', 'Type of Soil' and 'Source of Water' to enter a new record. Click 'Soil analysis / water analysis' option at the bottom of the

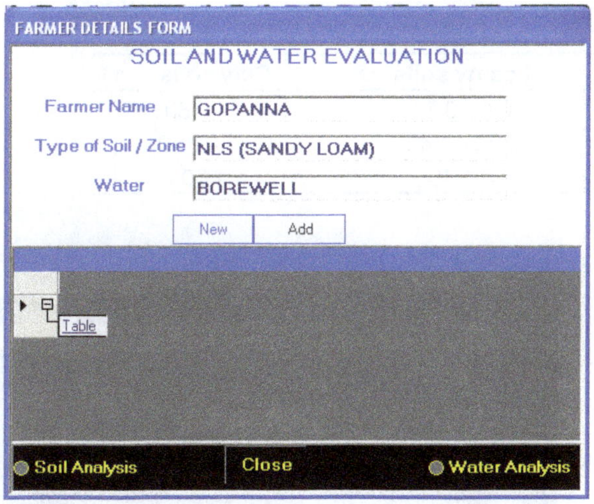

Figure 2. Farmer details form.

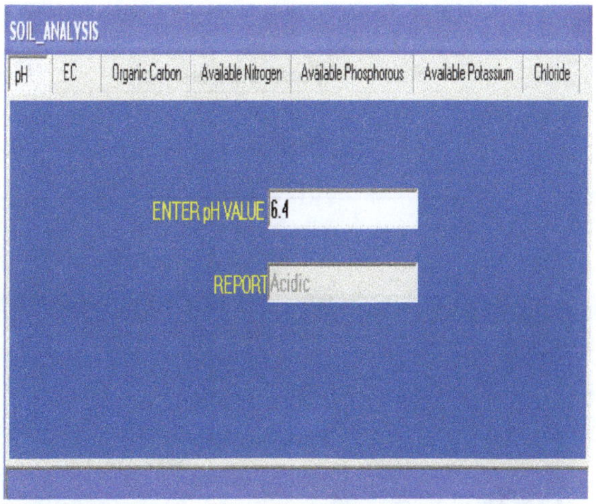

Figure 3. Data entry sheet for pH.

menu to enter the data (Figure 2). New screen with SOIL_ANALYSIS as title bar gets displayed. Various input parameters viz., pH, EC, organic carbon, available nitrogen, available phosphorous, available potassium and chloride gets displayed. Click 'Add' button to add a new record. Click on the first parameter pH (soil reaction).

Soil analysis

Soil reaction (pH) is the key to plant nutrient availability in the soil. It controls fixation, release and availability of nutrients to the plants. Soil pH is an excellent single indicator of the general soil conditions. The ideal soil reaction for tobacco plant growth falls in the range of pH 5 to 6. Enter the pH value obtained from the pH meter, which will be compared with table values of soil pH viz. Acidic: < 6.5; Neutral: 6.5 to 7.5; moderately alkaline: 7.5 to 8.8, Alkaline: > 8.8 (Figure 3). The values were compared with table values. The electrical conductivity (EC) measurement is based on the principle that ions being the carriers of electricity, the EC of a solution increases with the soluble salt concentration. EC is measured by electrical conductivity solubridge, where the instrument reading is multiplied by K (cell constant) value to get the electrical conductivity in dS/m. EC will be calculated using the formulae embedded in the software. In case of sandy soils, the limits of EC were normal: < 0.40; critical: 0.41 to 0.80; injurious: > 0.80 (Table 1). Then Next parameter on the menu is organic carbon content in soil which is estimated by Walkley and Black (1934) titration method. Blank titre value and sample titre values are fed to the computer. The organic carbon content (%) will be calculated by using the formula:

Organic carbon (%): $10 (B-T) / B * 0.003 * 100 / 0.5$.

Where B = Blank titre value, and T = sample titre value.

These values are compared with table values, that is, low: < 0.5%, medium: 0.5 to 0.75% and high: > 0.75% and the recommendations are prepared and stored.

Phosphorous is the life generating element and a good supply of phosphorus in the early stages promotes root development, growth and establishment of the seedlings and hastens crop maturity in the later stages. For estimation of available soil phosphorous, select the parameter 'available phosphorous' in the soil which is estimated by Brays method (Bray and Kurtz, 1945) in acid soil and Olsen method (Olsen et al., 1954) in alkaline soil. By feeding the instrument reading (R), soil available 'P' is calculated by using the formulae.

Olsen's P (kg/ha): $R * (50 / 5) * (1/2.5) * 2.24$; Brays P (kg/ha): $R * (50 / 5) * (1/5) * 2.24$;

The results will be compared with the table value, that is, low: < 10 kg/ ha, medium: 10 to 25 kg /ha and high: > 25 kg/ha, and the suitable recommendation will be given from the stored data. Nitrogen is the "Master Nutrient" controlling the plant growth, yield and quality of tobacco leaf. Although, potassium is absorbed in the greatest quantity, nitrogen is the key nutrient in tobacco fertilization and leaf chemistry. Too low or too high N application, adversely affects the quality of tobacco. Excessive soil nitrogen will generally produce cured leaves with dark-brown to black in colour, dry and chaffy and have a strong and pungent smoke. In the field, deficiency of nitrogen causes premature yellowing of leaves, which when cured are generally pale in colour, close-grained and thin-bodied and their smoke is flat and insipid. 'Soil available nitrogen' is estimated by alkaline permanganate method (Subbaiah and Asija, 1956). Available nitrogen (kg/ha) in soil will be obtained by using the formula $(R * N * (1/20) * 0.014 * 2.24)$, where R: volume of H_2SO_4 consumed. N: normality of H_2SO_4. By entering the input values of R and N, available nitrogen is

Table 1. Acceptable limits for salts (ds/m) in soil for FCV tobacco crop.

S/N	Total soluble salts	Sandy soils	Loamy soils	Clay soils
1	Normal	0.0 - 0.40	0.0 - 0.70	0.0 - 0.80
2	Critical	0.41 - 0.80	0.71 - 1.40	0.81 - 1.60
3	Injurious	> 0.80	> 1.40	> 1.60

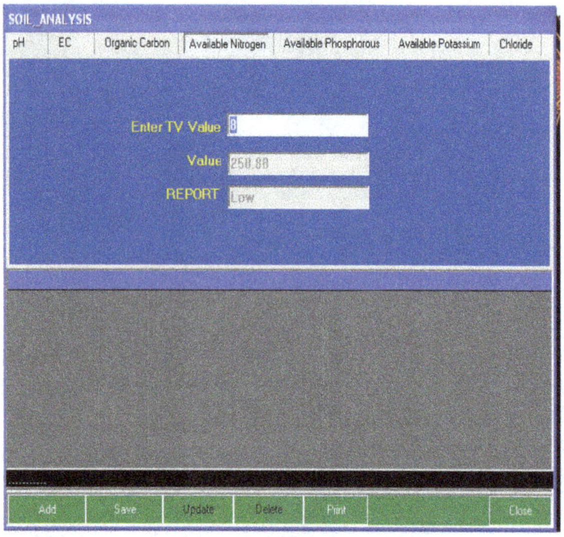

Figure 4. Data entry sheet for available nitrogen.

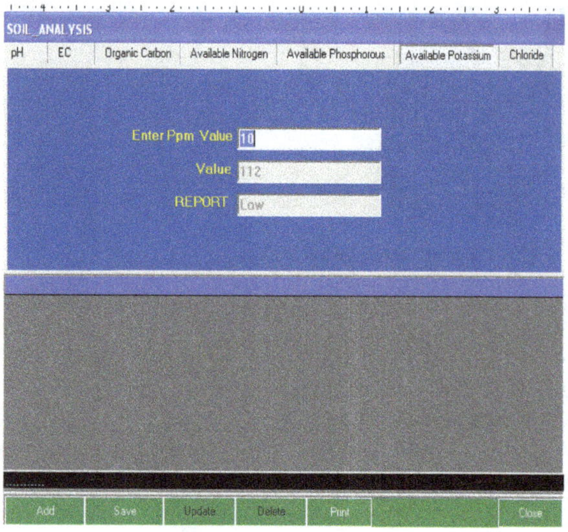

Figure 5. Data entry sheet for available potassium.

calculated and compared to table value that is, low: <250 kg, medium: 250 to 500 kg/ha and high: > 500 kg/ha, and nitrogen recommendation is provided from the stored data (Figure 4).

Potassium is the "Element of Quality" in tobacco. Tobacco is known to be a luxury consumer of potassium and the crop removes about 160 kg K_2O/ha from the soil (Krishnamurthy and Nagarajan 2001). K-deficiency is noticed in tobacco crop grown in light soils due to inherent low K status. K-deficiency may be aggravated by excess nitrogen fertilization. Leaf colour, texture, combustibility and hygroscopic properties of cured leaves are believed to be enhanced by potash fertilization. 'Soil available potassium' is estimated by neutral normal ammonium acetate using flame photometer (Hanway and Heidal, 1952). By feeding the instrument reading (R), available potassium in kg / ha will be calculated by the formula R × 11.2, where R = ppm of potassium in the extract. Results will be compared with table values, that is, low: <118 kg / ha, medium: 118 to 280 kg/ha and high: >280 kg/ha, and suitable suggestions will be made for the FCV tobacco for the particular zone (Figure 5).

Chloride is an essential micronutrient for tobacco. It plays an important role in influencing the leaf quality and leaf burn. When present in small quantities (< 1.0%), it improves the yield and certain quality factors like colour, moisture content and keeping quality. Larger amount of

chloride (> 1.5%) produces cured leaves of muddy and uneven colour with excessive hygroscopic and poor burn. Soils containing chlorides <80 ppm: highly suitable, 80 to 100 ppm: suitable, >100 ppm: unsuitable. Chlorides in soil and water are estimated by titrating with N / 35.5 silver nitrate using potassium chromate as indicator (Krishnamurthy and Nagarajan, 2001). Soil chlorides (ppm) will be calculated by R* 200, where R is the titre value. The results are compared with table values for recommendation. Once the soil samples are analyzed for different parameters, depending upon the tobacco zone from where samples are collected, site specific fertilizer doses are recommended automatically from the data already fed to computer. Then Click 'Save' option at the bottom of the menu to store the record into the database. To modify the input value of any parameter click 'Modify' option at the bottom of the screen. To take the hardcopy of the report for all parameters, click 'Print' option at the bottom of the screen (Table 2).

Water analysis

In Northern Light Soils (NLS) of Andhra Pradesh, the flue-cured tobacco crop is given 8 to 10 irrigations while in Southeren Light Soils, one life-saving irrigation is given to the crop. The quality of water used for irrigation in

Table 2. Final report for soil analysis and recommendation of fertilizers.

Soil analysis		
Parameter	**Value**	**Report**
pH	6.40	Acidic
EC	0.42	Critical
Organic carbon	0.55	Medium
Available nitrogen	250.88	Low
Available phosphorus	8.96	Low
Available potassium	112.00	Low
Chloride	80	Suitable

An amount of 120 kg N/ha – 60 kg P_2O_5 and 120 kg; K_2O is recommended in the form of diammonium phosphate, calcium ammonium nitrate and potassium sulphate.

Table 3. Final Report for Water Analysis and Recommendation of fertilizers.

Water analysis		
Parameter	**Value**	**Report**
pH	6.40	Acidic
EC	0.42	Unsuitable
Chloride	16.00	Suitable

The water is suitable for irrigating FCV tobacco.

different agro-climatic zones varies widely, particularly the ground water. Continued use of high chloride waters can lead to salinization of soils and consequently the production of saline leaf (Cl > 1.5%). The quality of irrigation water generally depends on the total concentration of dissolved salts, particularly ions like Na^+ and Cl^-. Therefore, analysis of the water samples in these areas is as important as the soil analysis. For evaluating the quality of irrigation waters, electrical conductivity (taken as a measure of total soluble salt content), chlorides and pH of the water are determined. The pH of water is determined by pH meter. In case of water, the limits were Acidic: < 6.5, Neutral: 6.5 to 7.5, Alkaline: > 7.5. After comparing, the inference will be stored.

EC is measured by electrical conductivity solubridge as it was in soil. When the irrigation water contains EC more than 0.4 (ds / m) at 25°C, it is unsuitable for irrigation. The obtained EC of water will be compared with recommended values. Chloride content in irrigation water is estimated as it was in soil analysis. Chlorides in water is calculated by using the formula R*40, where R is the titre value. If the chloride content is greater than 40 ppm, such water is rated as saline and is unsuitable for irrigating flue-cured tobacco crop in light soils. Once the water samples are analyzed for different parameters, then Click 'Save' option at the bottom of the menu to store the record into the database. To modify the input value of any parameter click 'Modify' option at the bottom of the screen. To take the hardcopy of the report for all

parameters, click 'Print' option at the bottom of the screen (Table 3).

The DSS helps in preparation of report instantaneously and reduces the time taken for computation of results, avoids the errors in calculation and preparation of reports. As this DSS provide a recommendation of fertilizer doses and suitability of water for tobacco crop, it will be highly helpful to the farmers to optimize their resources. This DSS is applicable for other crops also with slight modifications, if the same methods are adopted for estimation of soil and water quality parameters.

REFERENCES

Anonymous (2010). Annual Report, Tobacco Board, Ministry of commerce and Industry, Guntur, Andhra Pradesh. pp. 143-154.
Balena F (2005). Programming Microsoft Visual Basic NET. Microsoft Press, USA.
Bray RH, Kurtz LT (1945). Determination of total, organic and available forms of phosphorus in soils. Soil Sci. 59:39-45.
Davis DL, Nielsen MT (2007). Tobacco Production, Chemistry and Technology. Black well publications, London, P. 461.
Gaddis T, Lrvine K, Dention B (2003). Starting out with VB.Net Programming, Dream Tech Press, New Delhi, 2nd ed.
Ghosh AV, Bajaj JC, Hassan R, Singh D (1983). Laboratory manual on soil and water testing methods. Division of soil science and agricultural chemistry. IARI, New Delhi.
Hanway JJ, Heidal H (1952). Soil analysis methods used in Iowa State College, soil testing laboratory. Iowa Agron. 57:1-31.
Krishnamurthy V, Nagarajan K (2001). A manual on soil testing and irrigation water analysis for tobacco. Central Tobacco Research Institute, Rajahmundry.
Krishnamurthy V, Deosingh K (2002). Flue-cured tobacco soils of India: their fertility and management. Central Tobacco Research Institute, Rajahmundry, pp. 31-33.
Krishnamurthy V, Anuradha M (2011). Nitrogen nutrition of flue-cured tobacco. Tobacco Res. 37(1):1-17.
Manos BA, Ciani Th, Bournaris I, Vassiliadou J, Papathanasiou (2004). A Taxonomy Survey of Decision Support Systems in Agriculture. Agric. Econ. Rev. 5(2):80-92.
Nelson SL, Kelly J (2002). Office XP: The Complete Reference. Tata Mcgraw-Hill publishing Ltd, New Delhi.
Olsen SR, Cole CV, Watanabe FS, Dean LA (1954). Estimation of available phosphorus in soils by extraction with sodium bicarbonate. Circ. U.S. Department Agriculture. P. 939.
Pannell DJ (1996). Lessons from a decade of whole-farm modeling in Western Australia. Rev. Agric. Econ. 18:373-383.
Ravisankar H, Anuradham M, Chandrasekhararao C, Ravisankar H, Siva Raju K, Krishnamurthy V, Raju CA (2010). Expert system for identification and management of abiotic stresses in tobacco. Indian J. Agric. Sci. 80:151-154.
Ravisankar H, Anuradham M, Chandrasekhararao C, Nageswara RK, Krishnamurthy V (2009). Expert System for the diagnosis of nutrient deficiencies in flue-cured tobacco. Indian J. Agric. Sci. 79:45-49.
Stockle CO, Martin S, Campbell GS (1994). Crop Syst, a cropping systems model: water / nitrogen budgets and crop yield. Agric. Sys. 46:335-359.
Stone ND, Buick RD, Roach JW, Scheckler RK, Rupani R (1992). The planning problem in agriculture: Farm-level crop rotation planning as an example. AI Appl. 6(1):59-75.
Subbaiah BV, Asija GL (1956). A rapid procedure for the determination of available nitrogen in soils. Curr. Sci. 25:259-260.
Walkley AJ, Black IA (1934). Estimation of soil organic carbon by the chromic acid titration method. Soil Sci. 37:29-38.

20

Energy savings, cogeneration of electricity and water use in rural areas: Case study of Algarrobo irrigation community

Alberto Jesús Perea Moreno[1] and Federico Manzano Agugliaro [2]

[1]Department of Applied Physics, University of Cordoba, Campus Rabanales, 14071, Cordoba, Spain.
[2]Grupo Tragsa, Avda. Imperio Argentina no. 19, 29004, Málaga, Spain.

This study analyzed the effects of irrigation modernization on energy savings, cogeneration of electricity and water conservation, using the Algarrobo irrigation community (Spain) as a case study. Traditional surface irrigation systems and modern sprinkler systems occupy 73 and 27% of the irrigated area, respectively. Virtually all the irrigated area is devoted to field crops. Nowadays, farmers are investing on irrigation modernization by switching from surface to sprinkler irrigation because of the lack of labour and the reduction of net incomes as a consequence of reduction in European subsidies, among other factors. In this study, measures for energy savings, cogeneration of electricity and water management were proposed in order to improve the economic productivity of the community.

Key words: Irrigation community, hydraulic turbine, energy savings, renewable energy.

INTRODUCTION

The use of systems capable of generating clean and sustainable energy is having in recent years a huge growth in order to reduce the problems of climate change and resource depletion facing our planet. These systems are taking on more importance following the decision of the European Community, together with other countries to accept obligations to reduce emissions that cause climate change, as outlined in the Kyoto Protocol.

The use of the available potential energy of water between an upper and a lower height, had application for centuries: the water mills made possible the use of force provided by nature to perform different tasks.

Since the late eighteenth century, the use of water resources has been the most common electricity production. However, there has not been until recent years when added value has been ascribed for other types of energy production, the environmental benefits of its low impact from the emission of pollutants to the atmosphere, as opposed to coal combustion or oil.

On the other hand, some Irrigation Communities shows that, certain pumping systems are oversized, perhaps because they were built at a time when duties of water were much higher than the current ones, or because they are sized to supply 100% of the irrigable area and annually for various reasons, only irrigate a less percentage than the maximum irrigable. In any case, although 100% of the irrigable area was irrigated, they are always sized to the time of peak demand, usually one or two months a year, running well the rest of the year below its optimum operating point, may be even lower if the irrigated area is much less than the irrigable.

The design of the distribution network is a very important factor in providing efficiently irrigation water from the energy point of view. The topography determines the network design. It is very common that, there are significant slopes in the area irrigated by an irrigation community. In designing the distribution network, it is important to define various irrigation sectors so that each of them supplies the hydrants with uniform height. Thus, each pump unit consumes the energy demanded by the sector that supplies water and an efficient use of energy is obtained.

The efficiency concept has traditionally been used to design irrigation systems and to schedule irrigation. However, several authors have pointed out (mainly since the 1990s) that this concept is not appropriate for assessing the hydrological impact of irrigation in a basin (Willardson et al., 1994; Seckler, 1996; Perry, 1999; Seckler et al., 2003; Jensen, 2007; Perry, 2007). Efficiency does not take into account issues such as water reuse, the distinction between total water use and water consumption, the influence of location of use within the basin, and water quality. These issues are particularly important for water management in a context of water scarcity. The above mentioned authors, as well as others (Huffaker, 2008; Ward and Pulido-Velázquez, 2008), reported examples of misunderstandings in water management practices and water conservation programs due to an inadequate use of the efficiency concept.

This study applies the water accounting and water productivity concepts to the assessment of irrigation modernization in terms of water conservation, energy savings, and generation of electricity with hydraulic energy. The analysis has been applied to the case study of the irrigation community of Algarrobo, representative of large irrigation communities in interior Spain and in similar semi-arid areas. The objective of this work is to contribute to the optimization of water use in irrigation communities. The application of water accounting concepts to irrigation modernization constitutes another objective of this study.

MATERIALS AND METHODS

The irrigation community of Algarrobo, under study is located in the Axarquía (Málaga), and cover the towns of Algarrobo, Arenas, Sayalonga, and Vélez-Málaga (Spain). The irrigable area of the community comprises about 757 divided into four sectors.

Algarrobo has a subtropical climate. It has one of the warmest winters in Europe, with average temperatures of 17°C (62.6°F) during the day and 7 to 8°C (45 to 46°F) at night in the period from December to February. The summer's season lasts about 8 months, from April to November, although also in remaining 4 months temperatures sometimes reach around 20°C (68.0°F). The current crops of the area are distributed as follows:

(i) Tropical crops: avocado, mango, lychee, palms .- 100 ha,
(ii) Greenhouse: tomato, pepper, melon, etc. - 100 ha,
(iii) Outdoor-Vegetables: potato, strawberry, onion, beans, peppers, etc.- 350 ha,
(iv) Other woody crops: 207 ha.

The hydraulic irrigation scheme presented by this community is this: take water from Algarrobo River, leaving an ecological flow, at an elevation of 185 m. and carried through a natural pressure driving to a tank in Sector 2, which has a storage capacity of about 17000 m^3 and whose elevation of entrance is located at 156 m. In this tank also comes through pumping, the flow produced by the Waste Water Purifying Station of Algarrobo and water from the La Viñuela reservoir. From this tank, Sectors 1 and 2 will be directly irrigated, that cover 370 ha, and water is pumped to a tank located in Sector 4 (21000 m^3 capacity and 296 m elevation) to supply 387 has of Sectors 3 and 4. The last one has a higher pumping energy cost for the community and is required to raise a flow of 368 l/s for 16 h to meet the water needs of the Sectors 3 and 4.

This study seeks an improvement in the water use of Algarrobo River, energy savings in pumping the community and the generation of electricity from clean energy sources taking advantage of the steep topography of the area. We analyzed the average daily flows of the river over the past 25 years, and the ecological flow required. The ecological quality of rivers must be maintained by maintaining a minimum flow. Rivers must not dry-up or have their physical regimes significantly altered in order to conserve the hydrological and ecological functions of their drainage networks. This question must be borne in mind when planning and managing the water resources, especially in semi-arid zones. Ecological discharges, which take place as a result of the aquifer discharges in a natural regime, can be artificially maintained by reservoir management. The determination and mapping of ecological flows for semi-arid areas of EEA is, therefore, considered to be of paramount importance (INAG, 1995) (Figure 1).

Furthermore, we studied the availability of land for the inclusion of a storage ponds to allow better use of water resources of the Algarrobo River and evaluated from technical, economic and financial point the possibility to install a hydraulic turbine station that generates electricity from the difference in height between the ponds above mentioned and the current diversion dam that the irrigation community has.

RESULTS AND DISCUSSION

The results obtained from the analysis were as follow; The location of a new diversion dam located at elevation of 350 m. in Algarrobo River would allow gravity filling of two storage ponds (140000 m^3 and 50000 m^3) located at an elevation of 340 m. that have previously been calculated according to the analysis of the water needs of the sectors and the average daily flows of the Algarrobo River. From these ponds, it would be allowed to take water by gravity to the tank of Sector 4 so that the crop water requirements would be fulfill in Sectors 3 and 4 of the irrigation area, avoiding the pumping from sector 2 to sector 4 for a large part of the year, because increasing the elevation of the diversion dam would allow to fill the sector 4 pond using gravity.

On the other hand, there is the central location of Pelton turbines that allow cogeneration of electricity taking advantage of the steep topography of the land (150 m of drop), so that the water needed for irrigation of Sectors 1 and 2 is diverted by this station and then injected into the current diversion dam following the hydraulic scheme used nowadays by the irrigation community.

The nominal characteristics of the turbine are:

(i) Net head (m.): 155.00,
(ii) Nominal flow (m^3/seg.): 0.45,

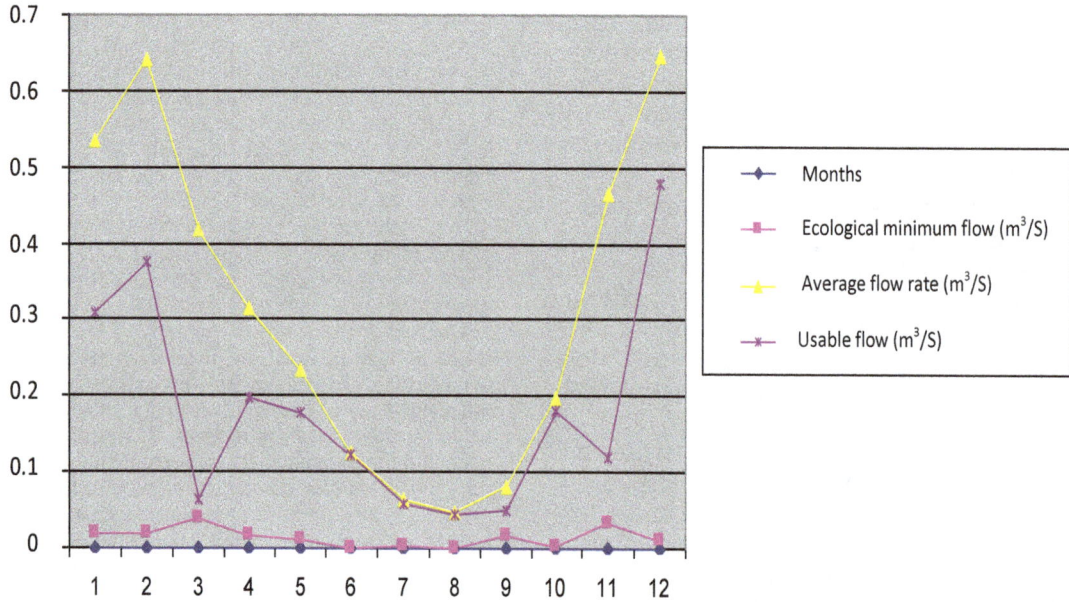

Figure 1. Average flow rates of Algarrobo River.

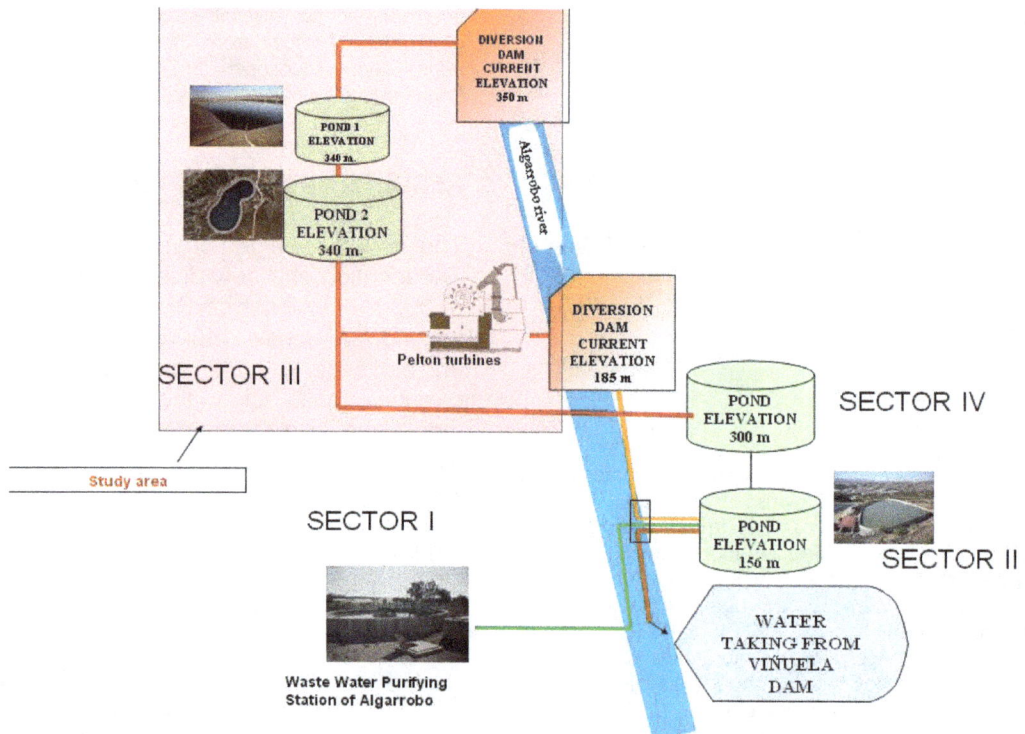

Figure 2. General diagram of the study.

(iii) Power (kw.): 604,00,
(iv) Spindle speed (rpm.): 750.00,
(v) Runaway speed (rpm.): 1450.00,
(vi) Operating hours (h): 2800.00,
(vii) Electricity production (kw/h.año): 1691200.00.

The aim is also to provide the excess water to the La Viñuela reservoir (main water resource of the region of Axarquia) by using the water intake of the community, shown in Figure 2. These improvements in the irrigation community mean incomes and savings for the community

Table 1. Savings and incomes of the alternative.

Concept	Volume (m^3)	Electricity production (kwh/year)
Energy saving pumping	2454447.29	
Water provided by La Viñuela reservoir	4000000.00	
Electrical cogeneration		1691200.00

which bears the cost of necessary works to carry out the purpose of this study, leading to improve the management of the main water resource (Algarrobo River), an energy saving for most of the year and the cogeneration of electricity from clean sources (Table 1).

REFERENCES

Huffaker R (2008). Conservation potential of agricultural water conservation subsidies. Water Resour. Res. 44:8.

INAG (1995). Comments to First draft report submitted by CEDEX/ITGE to ETC/IW EEA "Overview report on the key water resources issues in semi-arid/water scarcity of the EEA area, July 1995", October 1995.

Jensen ME (2007). Beyond irrigation efficiency. Irrig. Sci. 25(3):233-245.

Perry C (2007). Efficient irrigation; Inefficient communication; Flawed recommendations. Irrig. Drain. 56(4):367-378.

Perry CJ (1999). The IWMI water resources paradigm - definitions and implications. Agric. Wat. Manage. 40(1):45-50.

Seckler D (1996). The new era of water resources management: from "dry" to "wet" water savings. Research Report 1, International Water Management Institute. Colombo, Sri Lanka, 18 p.

Seckler D, Molden D, Sakthivadivel R (2003). The concept of efficiency in water-resources management and policy. In: Kijne W, Barkers R, Molden D (eds) Water productivity in Agriculture: limits and opportunities for improvement: CAB International, Wallingford, United Kingdom. pp. 37-51.

Ward FA, Pulido-Velázquez M (2008). Water conservation in irrigation can increase water use. In: Proceedings of the National Academy of Sciences of the United States of America. 105(47):18215-18220.

Willardson LS, Allen RG, Frederiksen HD (1994). Elimination of irrigation efficiencies. In: 13th Tech. Conf. USCID, USCID (Ed.), Denver (Colorado), USA, 17 p.

21

Impact of water deficit on growth attributes and yields of banana cultivars and hybrids

K. Krishna Surendar[1], V. Rajendran[1], D. Durga Devi[2], P. Jeyakumar[2], I. Ravi[3] and K. Velayudham[4]

[1]Vanavarayar Institute of Agriculture, Pollachi, India.
[2]Department of Crop PhysiologyTamil Nadu Agricultural University Coimbatore, India.
[3]National Research Centre for Banana (ICAR), Thiruchirapalli, India.
[4]Director of CSCMS, TNAU, Coimbatore-641 003, India.

Water deficit is one of the most important factors to limit banana productivity in the world, especially in dry and semidry areas where large fluctuation in the amount and distribution of the rain these areas faces. Some cultivars and hybrids have a set of physiological adaptations that allow them to tolerate water deficit and the degree of morphological and physiological adaptations may vary considerably among species. This study examined the relationship between the yield reduction by leaf area (LA), leaf area index (LAI) and specific leaf weight (SLW). Our results showed significant reduction in all growth attributes at all the stages due to water deficit. Association between all these growth attributes character and yield components was observed particularly at 5th and 7th MAP stage. It is clear that all these parameters could explain some of the mechanisms which indicate tolerance to drought and help in understanding the physiological responses that enable plants to adapt to water deficit and maintain growth and productivity during stress period and indicate important of these traits in future breeding programs for screening and selection of tolerant cultivars and hybrids of banana.

Key words: Banana, water deficit, leaf area, leaf area index, specific leaf weight and yield.

INTRODUCTION

Soil water deficit limits plant growth and field crops production more than any other environmental stresses (Zhu, 2002; Almeselmani et al., 2011). Its remains an ever-growing problem that severely limits banana production worldwide and causes important horticultural and agricultural losses particularly in arid and semiarid areas (Kallarackal et al., 1990). It induces many morphological and physiological responses on plants; so that banana plants are able to develop tolerance mechanisms which will provide to be adapted to limited environmental conditions (Turner, 1998). Banana plants respond and adopt to these stresses to survive under stress condition at the molecular and cellular levels as well as at the physiological and biochemical levels. Physiological responses to soil water deficit are the feature that is most likely to determine the response of the crop to irrigation. The banana plants are sensitivity to soil moisture stress is reflected in changes in reduced growth through reduced stomatal conductance and leaf size leads to reduction in photosynthetic pigments (Kallarackal et al., 1990) with increased leaf senescence (Turner, 1998). Leaf area is an important component that is closely related to the physiological processes controlling dry matter production and yield. Leaf area has been shown to influence the radiant energy interception, an important photosynthetic parameter in crop plants, showing positive relationship with net photosynthetic activity. Plants may respond to water deficit in different ways such as reducing leaf area, hence the transpiring surface (Meyer and Boyer, 1972). Leaf Area as one of the growth parameters also indicates the size of photosynthesizing apparatus. Leaf Area is a fundamental determinant of the total photosynthesis by the plant. Leaf Area showed a positive relationship with net photosynthetic activity. In banana, higher amount of LA on a shoot coincide with the emergence of the bunches (inflorescence) from the top of the pseudostem. After this, no new leaves are produced on that shoot because the bunch is terminal as the older leaves senescence (Turner, 1998). Turner (1998) found that water stress resulted in reduced LA leading to decreasing Leaf Area Index in banana. SLW is useful in understanding the means of the assimilates in leaf expansion. The SLA is a measure of LA per unit dry weight and it varies with cultivar, leaf position, growth stage and the environmental condition by Veerawirdh (1974). The SLW refers to photosynthetic efficiency and in turn higher total dry matter accumulation. It is the leaf dry weight per unit leaf area produced. Kramer (1983) found that water stress not only reduced LA but often increased leaf thickness, thereby increasing the weight per unit area, that is, in increased SLW. Thicker leaves aids in leaf water conservation because of the lower surface or lower volume ratio (Lopez, 1997). Drought has rarely been addressed in the past, but is gaining importance in the face of depleting natural resources (Iyyakkutty Ravi et al., 2013). The results of successful cultivation, especially of the water loving Cavendish clones, in drought prone areas with protected irrigation have provided the required momentum to perform research on drought in bananas (Iyyakutty Ravi et al., 2013). In subtropical and semi- arid banana cultivation zones have very limited rainy days and also had uneven distribution of rainfall, new crop management practices in terms of varieties selected, soil improvement (in terms of physical properties and nutrient enrichment), water management, etc. are being adopted (Iyyakutty Ravi et al., 2013). With this above background, the experiment aimed at evaluating the effects of the progressive water deficit, as well as to investigating the growth attributes behavior in twelve banana cultivars and hybrids submitted to water restriction during the different growth stages.

MATERIALS AND METHODS

The experiment was carried out at National Research Centre for banana, Thiruchirapalli, during 2011 to 2012. The experiment consists of two treatments as considered as main plot and twelve cultivars and hybrids as taken as sub plots were laid out in split plot design with three replications. The main plots are, M_1 (control) with the soil pressure maintained from -0.69 to -6.00 bar, M_2 (water deficit) with the soil pressure maintained from -0.69 to -14.00 bar.

Soil pressure of -14.00 bar was reached at 30 days and measured by using soil moisture release curve and measured the soil moisture by using the pressure plate membrane apparatus (Table 1 and Figure 1). The sub plots are: S_1: Karpuravalli (ABB); S_2: Karpuravalli x Pisang Jajee; S_3: Saba (ABB); S_4: Sanna Chenkathali (AA); S_5: Poovan (AAB), S_6: Ney poovan (AB), S_7: Anaikomban (AA), S_8: Matti x Cultivar Rose, S_9: Matti (AA); S_{10}: Pisang Jajee x Matti; S_{11}: Matti x Anaikomban, and S_{12}: Anaikomban x Pisang Jajee. The growth attributes of Leaf area, leaf area index and specific leaf weight were measured during 3rd, 5th, 7th, 9th month after planting and at harvest stages of the crop. The procedure for measuring leaf area, leaf area index and specific leaf weight are given as follows.

Leaf area (LA)

The leaf area was calculated by multiplying leaf length and breadth with the Constant factor 0.83 and number of green leaves and expressed in m^2 (Hewitt, 1955).

Leaf area = l x b x n x 'K'

Where, l = length of the leaf; b = breadth of the leaf; n = number of leaves, and 'K' = constant factor (0.83).

Leaf area index (LAI)

The leaf area index (LAI) of functional leaves was calculated by employing the formula of Williams (1946):

$$LAI = \frac{\text{Leaf area per plant}}{\text{Ground area occupied by the plant}}$$

Specific leaf weight (SLW)

The specific leaf weight (SLW) was calculated by using the formula of Pearce et al. (1968) and expressed as mg cm^{-2}:

$$SLW = \frac{\text{Leaf dry weight per plant (g)}}{\text{Leaf area per plant } (m^2)}$$

RESULTS

Leaf area (LA)

The leaf area was affected by water deficit in all the cultivars and hybrids as well as the interaction of M at S and S at M were significant (Table 2). Among the twelve cultivars and hybrids, Karpuravalli, Karpuravalli x Pisang Jajee, Saba, and Sannachenkathali had significant differences in leaf area under the irrigation at 50% available soil moisture level. The highest leaf area was observed in Karpuravalli with very lesser reduction was noticed under the water deficit (Figures 2 and 3). The lowest leaf area was observed in Matti, Pisang Jajee x Matti, Matti x Anaikomban and Anaikomban x Pisang Jajee cultivars and hybrids under the water deficit, respectively. There was a high and positive correlation

Table 1. General observations on germplasm performance under water deficit conditions (Anon, 2006, 2007; Iyyakutty Ravi et al., 2013, Uma and Sathiamoorthy, 2002; Uma et al., 2002).

Genomic group	Sub group / status	Genotypes (verities / types)	Reaction to water deficit
AA	Wild	*M. acuminata* ssp Burmannica	Highly susceptible
		M. acuminata ssp burmannicoides	Highly susceptible
		M. acuminata ssp malaccensis	Highly susceptible
		M. acuminata ssp zebrina	Highly susceptible
BB	Wild	Athiakol,	Susceptible
		Elavazhai, Attikol	Less tolerant
		Bhimkol,	Moderately tolerant
		M.balbisiana type Andaman	Tolerant
AAA	Ney Poovan	Ney Poovan and Nattu Poovan	Tolerant
	Unique	Thellachakkarakeli	Moderately tolerant
	Cavendish	Grand Naine, Robusta, Dwarf Cavendish, Williams	Susceptible
AAB	Mysore	Poovan	Moderately tolerant
ABB	Pisang Awak	Karpuravalli and Udhayam	Tolerant
	Monthan	Pidi Monthan and Ash Monthan	Moderately Tolerant

between leaf area and yield water deficit conditions.

Leaf area index (LAI)

The result on LAI had similar effect were showed in all the growth stages and also all the cultivars and hybrids by water deficit. The interaction effects of M at S and S at M were significant differed at all the cultivars and hybrids (Table 3). Water deficit decreased LAI in banana cultivars and hybrids. Among the twelve cultivars and hybrids, Karpuravalli, Karpuravalli x Pisang Jajee, Saba, and Sannachenkathali had significant differences in LAI under the main plot treatments. The highest LAI were observed in Karpuravalli due to the water deficit. The lowest LAI was observed in Matti, Pisang Jajee x Matti, Matti x Anaikomban and Anaikomban x Pisang Jajee cultivars and hybrids under the water deficit, respectively.

Specific leaf weight (SLW)

The data on SLW was affected under water deficit as well as the interaction of M at S and S at M were significant at all stages of growth (Table 4). Water deficit reduced SLW in all the twelve banana cultivars and hybrids. Among the twelve cultivars and hybrids, Karpuravalli, Karpuravalli x Pisang Jajee, Saba, and Sannachenkathali had significant differences in SLW under the main plot treatments. The highest SLW was observed in Karpuravalli under the water deficit than the other cultivars and hybrids. The lowest SLW content was

observed in Matti, Pisang Jajee x Matti, Matti x Anaikomban and Anaikomban x Pisang Jajee cultivars and hybrids under the water deficit, respectively. There was a high and positive correlation between SLW and yield water deficit conditions (Figure 4).

DISCUSSION

Leaf area is a fundamental determinant of the total photosynthesis of a plant. Leaf area always shows a positive relationship with net photosynthetic activity, because leaf enlargement is attributed to increase in number and width of grana and also high degree of stacking of grana (Flore et al., 1985). Leaf area development is based on the length and width of leaf, in general, was very sensitive to water deficit in banana as reported by Turner (1981). The leaf length of banana reduced during water stress situation, which is associated with reduced organ development. Gardner et al. (1981) opined that water stress decreases the leaf area due to reduced cell division and cell enlargement which could be caused by accumulation of unexpanded cells during the cycle. According to the results obtained in the present study, the cultivars of Karpuravalli, Karpuravalli x Pisang jajee, Saba and Sannachenkathali showed a lesser reduction in leaf area in the range of 8 to 12% due to water deficit over control. A 20 to 26% reduction in leaf area was registered by the cultivars of Poovan, Ney Poovan, Anaikomban and Anaikomban x Pisang jajee, whereas cultivars of Matti, Matti x Anaikomban, Matti x cultivar rose and Pisang jajee x Matti had higher

Table 2. Calculated pressure from stress treatment and soil moisture content from regression equation.

Soil moisture content (%)	Pressure (bar)	ASM (%)
33.46	-0.69	100.00
31.32	-2.46	93.60
30.19	-3.39	90.23
29.18	-4.22	87.21
28.14	-5.08	84.10
27.09	-5.94	80.96
26.12	-6.74	78.06
25.29	-7.43	75.58
24.91	-7.74	74.45
24.32	-8.22	72.68
23.78	-8.67	71.07
23.40	-8.98	69.93
23.11	-9.22	69.07
22.86	-9.43	68.32
21.28	-10.73	63.60
20.83	-11.10	62.25
19.51	-12.19	58.31
19.30	-12.36	57.68
18.63	-12.91	55.68
18.11	-13.34	54.12
17.81	-13.59	53.23
17.52	-13.83	52.36
17.10	-14.01	50.11
16.72	-14.47	49.01
16.00	-15.08	47.82

reduction in leaf area of about 38 to 48% over control. These results were confirmed by the findings of Levy et al. (1978) observing that leaf area increases with an increase in water supply because plants are able to photosynthesize more efficiently. This is because that an increased accumulation of photosynthates accelerates the pace of growth which in turn is reflected by vigorous plant growth. In banana, soil water regimes had a direct relationship on leaf width. There was an increase in leaf width with an increase in soil water regimes. This is because water is important for biochemical and physiological processes that lead to organ growth and development (Turner, 1972). A reduction in leaf area leading to reduced biomass accumulation and decreased growth and also leaf elongation of Kiwi fruit induced by water stress was a result of preferential partitioning of photosynthate to the roots and also shoots and thus affected leaf area development.

Leaf area index (LAI) is one of the principle factors influencing canopy net photosynthesis of the crop plants (Hansen, 1982). The capacity of a canopy of leaves in a plantation to intercept light and fix carbon is measured by the LAI. Turner et al. (2007) reported that the optimum

LAI for banana is 2 to 5. In banana plantation with LAI of 4.5 about 90% of the ground will be shaded at noon on a sunny day. This implies that about 90% of incoming radiation is being intercepted by the leaf canopy. Thus increasing LAI beyond this value is of little benefit to the plantation because most of the incoming solar radiation is already being intercepted (Turner et al., 2007). Drought stress induced changes in LAI, which duly reflected in biomass production (Kerby et al., 1990). Turner (1998) found that water stress resulted in reduced LA leading to decreased LAI in banana (Table 5). The lack of cell expansion due to water shortage would be determined by decreased LA rather than the number of leaves (Hsiao, 1973). In the present study also the effect of water deficit on LAI could be revealed. The cultivars like, Karpuravalli, Karpuravalli x Pisang jajee, Saba and Sannachenkathali showed a reduction of 8 to 12% in LAI, whereas the cultivars like Poovan, Ney Poovan, Anaikomban and Anaikomban x Pisang jajee recorded 8 to 12 and 19 to 25% reduction in LAI at 7th MAP over control. However, the other cultivars of Matti, Matti x Anaikomban, Matti x cultivar rose and Pisang jajee x Matti registered a higher reduction percent of about 38 to 43 over control. As per the report of De Silva et al. (1979), reduction in LAI was observed due to acceleration of senescence under drought. According to Hoffman and Turner (1993), leaf growth rate was more sensitive to water stress.

Specific leaf weight (SLW), a measure of thickness of leaf, has been reported to have a strong positive correlation with leaf photosynthesis in several crops as reported by Bowes et al. (1972). In many crop species, thicker leaves would have more number of mesophyll cells with high density of chlorophyll and, therefore, have a greater photosynthetic capacity than thinner leaves (Craufurd et al., 1999). Specific Leaf Weight is highly correlated with the development of reproductive organ namely flower and ultimately yield. As observed in the present study, Karpuravalli, Karpuravalli x Pisang jajee, Saba and Sannachenkathali recorded higher SLW with lesser reduction per cent of about 8 to 9 due to water deficit over control. The mechanism of maintaining higher SLW could be related to its thick leaves with more photosynthetic proteins per unit area of the leaf (Wells and Nugent, 1980). The higher reduction in SLW (24 to 26%) under stressed conditions in the cultivars of Matti, Matti x Anaikomban, Matti x cultivar rose and Pisang jajee x Matti could also be related to lesser number of mesophyll cells leads to lower photosynthetic efficiency (Gardner et al., 1985).

Conclusion

Plants respond to drought stress through alteration in physiological and biochemical processes. Our results showed that the growth attributes of leaf area, leaf area index and specific leaf weight decreased under the water deficit condition. The banana cultivars and hybrids of

Table 3. Effect of water stress on leaf area (m^2 $plant^{-1}$) at different growth stages of banana cultivars and hybrids.

Treatment	3rd MAP	5th MAP	7th MAP	9th MAP	Harvest	Mean
Main plot						
M_1	2.7	4.5	6.3	5.7	5.1	4.87
M_2	2.1	3.9	5.1	4.5	3.9	3.89
Mean	2.41	4.20	5.66	5.14	4.48	4.38
SEd	0.023	0.037	0.054	0.053	0.039	
CD (P= 0.05)	0.101	0.159	0.234	0.229	0.168	
Sub plot						
S_1	5.1	8.8	13.7	12.0	9.3	9.77
S_2	4.3	7.2	11.5	8.7	8.6	8.09
S_3	3.8	7.1	7.9	7.9	7.1	6.76
S_4	2.4	4.2	6.1	6.1	5.0	4.77
S_5	2.6	4.6	6.5	5.8	5.2	4.92
S_6	2.4	3.4	4.4	4.3	3.9	3.69
S_7	2.0	3.1	3.9	3.7	3.4	3.22
S_8	1.5	3.0	3.8	3.7	3.3	3.08
S_9	1.5	2.8	2.8	2.7	1.9	2.34
S_{10}	1.3	2.6	2.8	2.7	2.4	2.34
S_{11}	1.1	2.2	2.3	2.0	1.9	1.88
S_{12}	0.9	1.4	2.2	2.1	1.9	1.69
Mean	2.41	4.20	5.66	5.14	4.48	4.38
SEd	0.056	0.088	0.126	0.120	0.097	
CD (P= 0.05)	0.114	0.179	0.254	0.243	0.197	
Interaction effect						
M_1S_1	5.7	9.4	14.3	12.6	9.9	10.35
M_1S_2	4.9	7.8	12.1	9.3	9.2	8.67
M_1S_3	4.4	7.7	8.5	8.4	7.7	7.34
M_1S_4	3.0	4.8	6.5	6.5	5.4	5.23
M_1S_5	2.8	4.8	7.2	6.5	5.8	5.43
M_1S_6	2.7	3.7	5.1	5.0	4.5	4.20
M_1S_7	2.2	3.4	4.6	4.3	4.1	3.73
M_1S_8	1.8	3.2	4.5	4.4	4.0	3.59
M_1S_9	1.7	3.0	3.4	3.3	2.5	2.75
M_1S_{10}	1.5	2.8	3.3	3.3	3.0	2.75
M_1S_{11}	1.3	2.3	2.8	2.6	2.4	2.29
M_1S_{12}	1.0	1.6	2.7	2.7	2.5	2.10
M_2S_1	4.5	8.2	13.2	11.4	8.7	9.19
M_2S_2	3.8	6.7	10.9	8.1	8.0	7.50
M_2S_3	3.2	6.6	7.4	7.3	6.5	6.17
M_2S_4	1.8	3.7	5.7	5.7	4.6	4.30
M_2S_5	2.3	4.3	5.8	5.1	4.5	4.41
M_2S_6	2.1	3.2	3.8	3.7	3.2	3.18
M_2S_7	1.7	2.8	3.2	3.0	2.8	2.71
M_2S_8	1.2	2.7	3.2	3.1	2.6	2.57
M_2S_9	1.4	2.6	2.2	2.1	1.4	1.93
M_2S_{10}	1.1	2.4	2.2	2.1	1.8	1.93
M_2S_{11}	0.9	2.0	1.7	1.5	1.3	1.47
M_2S_{12}	0.7	1.3	1.6	1.5	1.3	1.28
Mean	2.41	4.20	5.66	5.14	4.48	4.38
SEd						
M at S	0.080	0.126	0.179	0.171	0.138	

Table 3. Contd.

S at M	0.080	0.125	0.178	0.170	0.138
CD (P= 0.05)					
M at S	0.177	0.278	0.399	0.384	0.303
S at M	0.161	0.253	0.359	0.344	0.278

Table 4. Effect of water stress on leaf area index (LAI) at different growth stages of banana cultivars and hybrids.

Treatment	3rd MAP	5th MAP	7th MAP	9th MAP	Harvest	Mean
Main plot						
M_1	0.69	1.13	1.57	1.43	1.27	1.22
M_2	0.52	0.97	1.27	1.14	0.97	0.97
Mean	0.60	1.05	1.42	1.28	1.12	1.09
SEd	0.006	0.009	0.013	0.013	0.010	
CD (P= 0.05)	0.027	0.042	0.057	0.056	0.046	
Sub plot						
S_1	1.28	2.20	3.44	2.99	2.32	2.44
S_2	1.08	1.81	2.88	2.18	2.15	2.02
S_3	0.94	1.78	1.99	1.96	1.77	1.69
S_4	0.60	1.06	1.53	1.52	1.25	1.19
S_5	0.64	1.14	1.63	1.45	1.29	1.23
S_6	0.60	0.86	1.11	1.08	0.97	0.92
S_7	0.49	0.78	0.97	0.92	0.86	0.80
S_8	0.38	0.74	0.96	0.93	0.83	0.77
S_9	0.38	0.70	0.70	0.67	0.49	0.59
S_{10}	0.33	0.65	0.69	0.67	0.60	0.59
S_{11}	0.28	0.54	0.57	0.51	0.46	0.47
S_{12}	0.22	0.35	0.54	0.53	0.47	0.42
Mean	0.60	1.05	1.42	1.28	1.12	1.09
SEd	0.014	0.022	0.031	0.029	0.024	
CD (P= 0.05)	0.028	0.045	0.064	0.060	0.049	
Interaction effect						
M_1S_1	1.42	2.34	3.58	3.14	2.46	2.59
M_1S_2	1.23	1.95	3.03	2.33	2.30	2.17
M_1S_3	1.09	1.93	2.13	2.11	1.92	1.83
M_1S_4	0.75	1.21	1.63	1.62	1.35	1.31
M_1S_5	0.71	1.21	1.80	1.62	1.46	1.36
M_1S_6	0.66	0.93	1.28	1.25	1.14	1.05
M_1S_7	0.56	0.84	1.14	1.09	1.03	0.93
M_1S_8	0.45	0.81	1.13	1.10	1.00	0.90
M_1S_9	0.42	0.74	0.84	0.81	0.63	0.69
M_1S_{10}	0.37	0.69	0.83	0.81	0.74	0.69
M_1S_{11}	0.32	0.58	0.71	0.66	0.61	0.57
M_1S_{12}	0.26	0.39	0.69	0.68	0.61	0.53
M_2S_1	1.13	2.05	3.29	2.84	2.17	2.30
M_2S_2	0.94	1.67	2.73	2.04	2.01	1.88
M_2S_3	0.80	1.64	1.84	1.82	1.62	1.54
M_2S_4	0.46	0.92	1.43	1.42	1.15	1.08
M_2S_5	0.58	1.08	1.46	1.28	1.12	1.10
M_2S_6	0.53	0.79	0.94	0.92	0.80	0.80
M_2S_7	0.43	0.71	0.81	0.75	0.69	0.68

Table 4. Contd.

Treatment						
M_2S_8	0.31	0.68	0.79	0.77	0.66	0.64
M_2S_9	0.34	0.66	0.55	0.52	0.34	0.48
M_2S_{10}	0.29	0.61	0.54	0.52	0.45	0.48
M_2S_{11}	0.24	0.50	0.42	0.37	0.32	0.37
M_2S_{12}	0.18	0.31	0.40	0.39	0.32	0.32
Mean	0.60	1.05	1.42	1.28	1.12	1.09
SEd						
M at S	0.020	0.031	0.045	0.042	0.035	
S at M	0.020	0.031	0.045	0.042	0.034	
CD (P= 0.05)						
M at S	0.045	0.071	0.100	0.094	0.078	
S at M	0.040	0.064	0.091	0.084	0.070	

Table 5. Effect of water stress on Specific Leaf Weight (SLW: mg / cm^2) at different growth stages of banana cultivars and hybrids.

Treatment	3rd MAP	5th MAP	7th MAP	9th MAP	Harvest	Mean
Main plot						
M_1	0.65	0.69	0.77	0.76	0.73	0.72
M_2	0.53	0.57	0.65	0.64	0.61	0.60
Mean	0.59	0.63	0.71	0.70	0.67	0.66
SEd	0.008	0.005	0.007	0.007	0.007	
CD (P= 0.05)	0.034	0.023	0.031	0.031	0.030	
Sub plot						
S_1	0.68	0.72	0.80	0.79	0.76	0.75
S_2	0.67	0.71	0.79	0.78	0.75	0.74
S_3	0.66	0.70	0.78	0.77	0.74	0.73
S_4	0.65	0.69	0.77	0.76	0.73	0.72
S_5	0.62	0.66	0.74	0.73	0.70	0.69
S_6	0.60	0.64	0.72	0.71	0.68	0.67
S_7	0.59	0.63	0.71	0.70	0.67	0.66
S_8	0.56	0.60	0.68	0.67	0.64	0.63
S_9	0.52	0.56	0.64	0.63	0.60	0.59
S_{10}	0.51	0.55	0.63	0.62	0.59	0.58
S_{11}	0.51	0.55	0.63	0.62	0.59	0.58
S_{12}	0.50	0.54	0.62	0.61	0.58	0.57
Mean	0.59	0.63	0.71	0.70	0.67	0.66
SEd	0.007	0.008	0.009	0.009	0.008	
CD (P= 0.05)	0.015	0.016	0.018	0.018	0.018	
Interaction effect						
M_1S_1	0.71	0.75	0.83	0.82	0.79	0.78
M_1S_2	0.70	0.74	0.82	0.81	0.78	0.77
M_1S_3	0.69	0.73	0.81	0.80	0.77	0.76
M_1S_4	0.68	0.72	0.80	0.79	0.76	0.75
M_1S_5	0.68	0.72	0.80	0.79	0.76	0.75
M_1S_6	0.66	0.70	0.78	0.77	0.74	0.73
M_1S_7	0.65	0.69	0.77	0.76	0.73	0.72
M_1S_8	0.62	0.66	0.74	0.73	0.70	0.69
M_1S_9	0.60	0.64	0.72	0.71	0.68	0.67
M_1S_{10}	0.59	0.63	0.71	0.70	0.67	0.66

Table 5. Contd.

M_1S_{11}	0.59	0.63	0.71	0.70	0.67	0.66
M_1S_{12}	0.58	0.62	0.70	0.69	0.66	0.65
M_2S_1	0.64	0.68	0.76	0.75	0.72	0.71
M_2S_2	0.63	0.67	0.75	0.74	0.71	0.70
M_2S_3	0.62	0.66	0.74	0.73	0.70	0.69
M_2S_4	0.61	0.65	0.73	0.72	0.69	0.68
M_2S_5	0.56	0.60	0.68	0.67	0.64	0.63
M_2S_6	0.54	0.58	0.66	0.65	0.62	0.61
M_2S_7	0.53	0.57	0.65	0.64	0.61	0.60
M_2S_8	0.50	0.54	0.62	0.61	0.58	0.57
M_2S_9	0.44	0.48	0.56	0.55	0.52	0.51
M_2S_{10}	0.43	0.47	0.55	0.54	0.51	0.50
M_2S_{11}	0.43	0.47	0.55	0.54	0.51	0.50
M_2S_{12}	0.42	0.46	0.54	0.53	0.50	0.49
Mean	0.59	0.63	0.71	0.70	0.67	0.66
SEd						
M at S	0.013	0.012	0.014	0.014	0.014	
S at M	0.011	0.011	0.012	0.013	0.012	
CD (P= 0.05)						
M at S	0.038	0.030	0.037	0.038	0.039	
S at M	0.022	0.023	0.026	0.026	0.025	

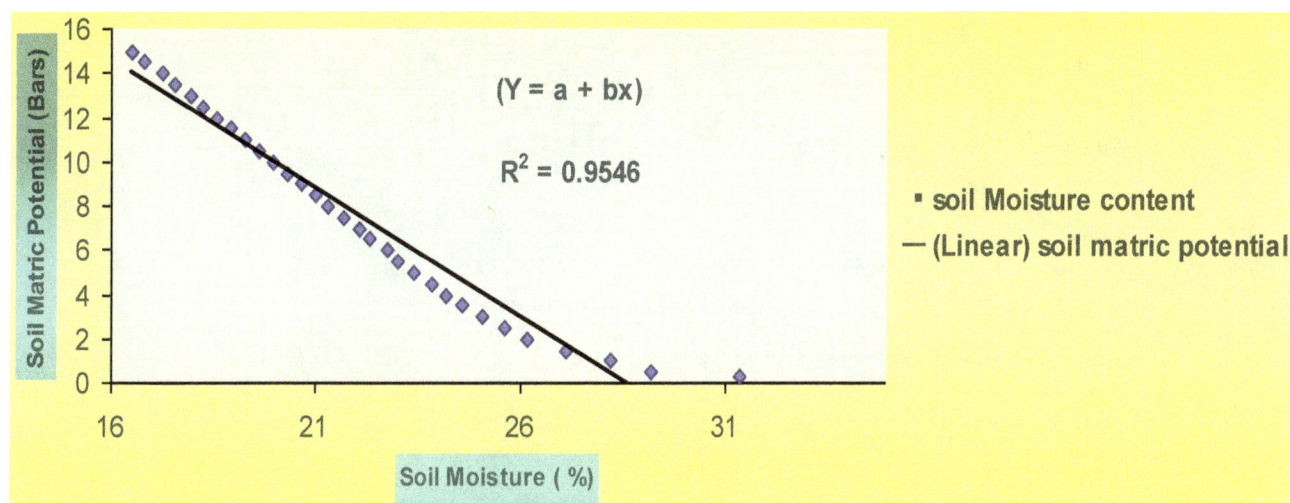

Figure 1. Pressure plate apparatus soil moisture release curve. Regression equation to find out pressure from soil moisture: [Y = a + bx]; where Y = Pressure (bar); X = soil moisture content (%); 'a' = 28.26158 and 'b' = - 0.8239.

Karpuravalli, Karpuravalli x Pisang jajee, Saba and Sannachenkathali with lesser reduction in leaf area, leaf area index and specific leaf weight and also smaller bunch yield reduction when the plants endured water deficit. The findings of this research also showed that the leaf area, leaf area index and specific leaf weight can be used as a drought tolerance index to selection tolerant genotypes under water deficit conditions in banana cultivars and hybrids.

ACKNOWLEDGEMENT

The research have been supported and facilitated by National Research Centre for Banana (ICAR), Trichy, Tamil Nadu, India. The authors extend their sincere thanks to Dr. M. M. Mustaffa (Director) NRC for banana, Dr. D. Durga Devi (Professor) TNAU and Dr. I. Ravi (Sr. Scientist) NRC for banana for given proper guidance during research.

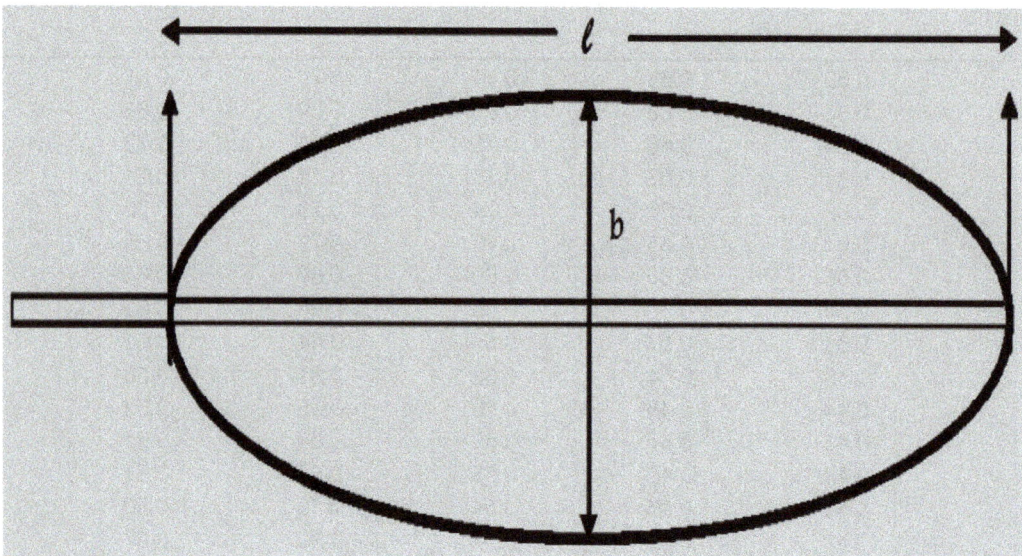

Figure 2. Diagram representing the banana leaf for measuring lamina length (l) and width (b). Source: Iyyakutty Ravi et al. (2013).

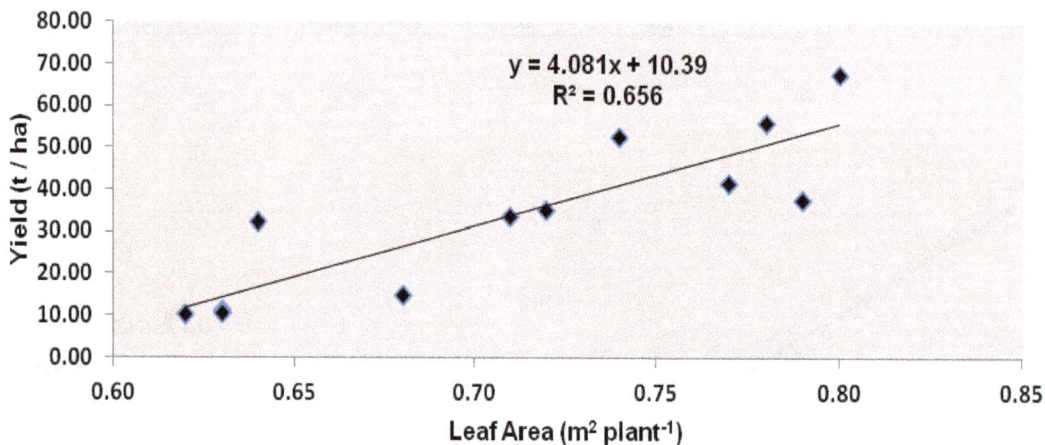

Figure 3. Correlation of leaf area (m^2 plant^{-1}) (LA) with yield.

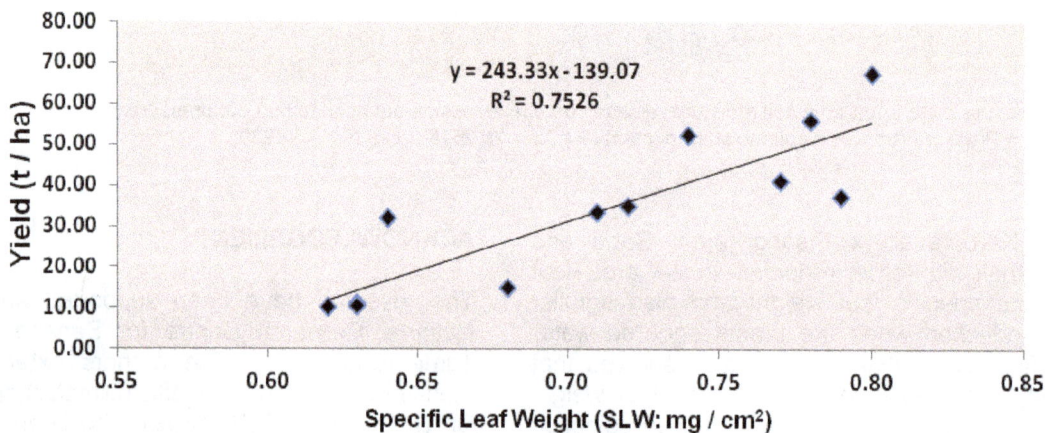

Figure 4. Correlation of specificleaf weight (mg/cm^2) (SLW) with yield.

Abbreviations: LA, Leaf area; **LAI,** leaf area index; **SLW,** specific leaf weight; ****,** highly significant; ***,** significant.

REFERENCES

Almeselmani M, Abdullah F, Hareri F, Naaesan M, Ammar MA, Kanbar OZ, Saud A (2011). Effect of drought on different physiological characters and yield component in different Syrian durum wheat varieties. J. Agric. Sci. 3:127-133.

Anon (2006). Annua lReport. Trichy, India: National Research Centrefor Banana (ICAR).

Anon. (2007). Annual Report. Trichy, India: National Research Centre for Banana (ICAR).

Bowes GW, Orgen L, Hageman RH (1972). Light saturated photosynthesis rate, RuBP carboxylase activity and specific leaf weight in soybeans grown under different light intensities. Crop Sci. 12:77-79.

Craufurd PC, Wheeler TR, Ellis RH, Summer field RJ, Williams JH (1999). Effect of temperature and water deficit on water use efficiency, carbon isotope discrimination and specific leaf area in peanut. Crop Sci. 39:136-142.

De Silva DLR, Hetherington AM, Mansfield TA (1979). Synergism between calcium ions and abscisic acid in preventing stomata1closure. New Phytol. 100: 473-482.

Flore JA., Lakso AN, Moon JW (1985). The effect of water stress and vapor pressure gradient on stomatal conductance, water use efficiency, and photosynthesis of fruit crops. Acta Hort. 171:207-218.

Gardner BR, Blad BL. Garrity DP, Wattes DG (1981). Relationships between crop temperature, grain yield, evapotranspiration and phonological development in two hybrids of moisture stressed sorghum. Irrig. Sci. 2:213-224.

Gardner EP, Pearce RB, Mitchell. KL (1985). Physiology of crop plates. Iowa state Univ. Press, Iowa.

Hansen AD, Hitz WD (1982). Metabolic responses of mesophytes to plant water deficit. Ann. Rev. Plant Physiol. 33:163-203.

Hewitt CW (1955). Leaf analysis as a guid to the nutrition of bananas. Emp. J. Exp. Agric. 23:11-16.

Hoffman HP, Turner DW (1993). Soil water deficits reduce the elongation rate of emerging banana leaves but the night/ day elongation ratio remains unchanged. Scient. Hortic. 54:1-12.

Hsiao TC (1973). Plant responses to water deficit. Ann. Rev. Plant Physiol. 24:519:570.

Iyyakkutty Ravi, Subbaraya U, Muthu M, Vaganan M Mustaffa M (2013). Phenotyping bananasfor droughtresistance. Frontiers Physiol. 4(9):1-15.

Kallarackal J, Milburn JA, Baker DA (1990). Water relations of the banana. III effects of controlled water stress on water potential, transpiration, photosynthesis and leaf growth. Austr. J. Plant Physiol. (17):79-90.

Kerby TA, Cassman RG, Keeley M (1990). Genotypes and plant densities for narrow row cotton system. II. Leaf area and dry matter partitioning. Crop Sci. 30:649-653.

Levy Y, Bielorai H, Shaheret R (1978). Longterm effects of different irrigation regimes on grapefruit tree development and yield. J. Am. Soc. Hortic. Sci. 117:325-417.

Lopez AS (1997). Impact of use of micro nutrients on productivity of soybean. Documentos-Empraba Soja, 180:367-398.

Meyer RF, Boyer JS (1972). Sensitivity of cell division and cell elongation to low water potentials in Soya bean hypocotyls. Planta 10:77-87.

Pearce RB, Brown RH, Blaster RE (1968). Photosynthesis of alfalfa leaves as influenced by age and environment. Crop Sci. 8:677-680.

Santos RF, Carlesso R (1998) Water deficit and morphologic and physiologic behavior of the plants. Rev. Bras. Eng. Agric. Ambient. 2:287-294.

Turner NC (1981). Techniques and experimental approaches for the measurement of plant water status. Plant Soil 58:339-366.

Turner DW (1998). The impact of environmental factors on the development and productivity of bananas and plantains. In Proceedings of the 13th ACORBAT meeting, Guayaquil, Ecuador, (Ed. L. H. Arizaga). Ecuador, CONABAN. pp. 635-663.

Turner DW (1972). Banana plant growth. 1. Groeth. 1. Gross morphology. Aust. J. Expt. Agric. Anim. Husband. 12:216-224.

Turner DW, Fortescue JA, Thomas DS (2007). Environmental physiology of the bananas (Musa spp.). Brazilian J. Plant. Physiol. 19:463-484.

Uma S, Sathiamoorthy S (2002). Namesand Synonyms of Bananas and PlantainsinIndia. Trichy, India: National Research Centre for Banana, ICAR.

Uma S, Sathiamoorthy S, Singh HP, Dayarani M (2002). Crop improvementin *Musa*-Evaluation of germplasm formale and female fertility. Indian J. Plant. Genet. Resour. 15:137-139.

Veerawirdh J (1974). Growth analysis of various soybean varieties as affected by different spacing. M.Sc. (Ag.) Thesis Tamil Nadu Agricultural University, Coimbatore.

Kramer PJ (1983). Water relations of plants. Academic press, New York, London, P. 489.

Wells JA, Nugent PE (1980). Effect of high soil water on quality of muskmelon. Hort. Sci. 15:258-259.

Williams RF (1946). The physiology of plant growth with special referance to the concept of net assimilation rate. Ann. Bot. 10:41-71.

Zhu JK (2002). Salt and drought stress signal transduction in plants. Ann. Rev. Plant Biol. 53:247-273.

22

Effects of vinasse irrigation on effluent ionic concentration in Brazilian Oxisols

Mellissa Ananias Soler da Silva[1,2], **Huberto José Kliemann**[2], **Alfredo Borges De-Campos**[2], **Beáta Emöke Madari**[1], **Jácomo Divino Borges**[2] and **Janine Mesquita Gonçalves**[2]

[1]Embrapa Rice and Beans, Rodovia GO-462, km 12 Zona Rural P. O. Box 179, CEP 75375-000 Santo Antônio de Goiás, GO, Brazil.
[2]Federal University of Goias, Campus Samambaia - Rodovia, km 0 – PO-box 131, CEP 74690-900, Goiânia, GO, Brazil.

The irrigation with vinasse can improve soil fertility. However, this use should take into account the characteristics of each soil because the vinasse has unbalanced amounts of mineral and organic elements which might lead to leach the ions, especially nitrate and potassium. The purpose of this study was to evaluate the impacts of vinasse irrigation on effluent ionic concentrations in Brazilian Ferralsols from two areas in Central Brazil: a sugarcane field, and a natural undisturbed savannah area. Soil samples from the two sites were placed into PVC columns with 120 cm height × 25 cm diameter with sugarcane. Undiluted vinasse was applied once on the surface of the soil columns at doses equivalent to 0, 300, 600 and 1200 m^3 ha^{-1}. After 0, 60, 90 and 120 days of irrigation, samples of the effluent were collected and the concentrations of dissolved organic matter (DOM), Cl, Ca, Mg, Na, K, total Fe, NH_4^+, NO_3^-, SO_4^{2-} as well as pH were determined. The ions concentration data were modeled in the chemical equilibrium model Visual Minteq v. 3.0. Results revealed vinasse's dose, days after irrigation and land use had a relevant effect on most nutrients effluent concentration. Contrasting pH values were observed for both soils and in the savannah soil was observed a decrease in pH at high vinasse doses. This paper thus revealed leaching of the DOM was strongly time dependent. High vinasse doses may lead to increase nutrient leaching and soil dispersion regardless the land use and time after irrigation.

Key words: Ion speciation, sugarcane, visual minteq, leaching, ethanol.

INTRODUCTION

The total sugarcane world production is nearly 1.5 billion tons and is mostly located in tropical developing countries of Latin America, Africa, and Southeast Asia. Brazil is the world leader in the production of sugarcane with almost 9 million ha tilled area, processing approximately 681 million tons in the 2010/2011 harvest (IBGE, 2011). Primarily used for sugar production, in the last three decades the cane plant has become a new paradigm of clean and renewable energy. It decisively contributes to the sustainability of the planet against global warming as currently the most efficient raw-material for ethanol production and bioelectricity are sugarcane juice and biomass, respectively (UNICA, 2008).

As ethanol production increases, the vinasse production increases as well. The vinasse is a nutrient-rich byproduct originated from sugarcane manufacture for producing ethanol. For each liter of ethanol, ten to eighteen liters of vinasse are produced (Freire and Cortez, 2000) with variable composition. The irrigation with vinasse can improve soil fertility (Silva, 2009).

However, this use should take into account the characteristics of each soil because the vinasse has unbalanced amounts of mineral and organic elements which might lead to leach the ions, especially nitrate and potassium.

Interest in treated or composted organic wastes on Brazilian agriculture is based on high carbon levels from organic compounds (organic carbon) and nutrient on it, on cation exchange capacity enhancement (CEC) and acidity soil neutralization (Abreu Júnior et al., 2005). Rise carbon and soil nutrients levels can improve its physical and chemical properties, increasing plant yield and improving agricultural products quality, decreasing production costs. However, these wastes can present environmental pollution potential, that is, its soil or water addition can insert inorganic elements or toxic organic compounds or pathogens within the food chain (Abreu et al., 2005).

In addition to the water that percolates through the soil system, the application of a material which has large amount of potassium can affect the quantity of ions leaching into the soil profile, the concentration of solutes and the distribution of pore sizes, pH, cation exchange capacity, the reactions of dissolution/precipitation, and the ionic exchanges between nutrients in the liquid phase with those in the solid phase in depth during leaching (Ernani et al., 2003). The soil chemical balance is affected by several combinations of the above mechanisms after vinasse application.

The land use may influence the behavior of the ions in the soil after vinasse irrigation. In cultivated soils with many years of vinasse fertilization, apparently the ion exchange complex is saturated with chemical components from this byproduct and, in this case, intense leaching would be expected with consecutive irrigations. The objective of this study was to evaluate if increased vinasse rates can contribute to groundwater contamination through ion displacement from soil sites on sugarcane cultivated compared with savannah soil cropped with sugarcane a columns experiment.

MATERIALS AND METHODS

Study area and sampling procedure

The study used disturbed soil samples of a dystrophic Ferralsol (Nachtergaele, 2005) from two distinct sites with the same topographic condition, a cultivated field and a natural savannah area which was taken as control, located inside the Cerrado Biome. Both areas are located in the municipality of Goianesia, State of Goias (15° 10' 00" S and 49° 15' 00" W). In the municipality the dominant climate is tropical wet-dry with an annual precipitation of 1,500 mm/yr and a rainy season from October to March (Figure 1).

Two trench 120 cm depth were opened, one in the savannah and another within commercial sugarcane area, for samples collection. The samples were collected by 20 cm layers tick up to 120 cm depth in the soil profile and separately stored in bags. Then, these samples were transported to Federal University of Goias/Agronomy Campus EA/UFG and stored in the lab. The soil was collected in

Figure 1. Location of experimental sampling site (Abreu, 2006).

both areas, in 2005.

A chemical soil analysis is shown on Table 1. In the cultivated area, which has been cultivated since 1984 with sugarcane and was limed in 1998, there was irrigation with vinasse instead water use, which corresponded to 700 m^3 ha^{-1} $year^{-1}$, in aspersion systems form. Soil from the savannah area, near the sugarcane cultivated, located at same soil type and topographic conditions was sampled.

The field samples were placed, without sieve, into PVC columns with 120 cm high by 25 cm in diameter which was internally coated with raffia bags to reduce preferential flow in the columns wall. A sugarcane thole of the variety 72,454 RB with sprouts of approximately 20 cm was planted in each column. The columns were equipped with a drain at the bottom that was attached to a 600 mL plastic bottle to sample the effluent.

One year before sampling the columns effluent, additional fertilization was performed in the soil columns with monoammonium phosphate to attend the plant needs. The fertilization, which was the same for all treatments except the control, was calculated as recommended by Raij et al. (1997) to sugarcane plants. The soil columns were incubated and cultivated for a year before vinasse irrigation just to allow the sedimentation and aggregation of soil particles improved by sugarcane root system cultivated inside these columns.

Undiluted vinasse was applied at the surface of the soil columns at different doses (0, 300, 600 and 1200 m^3 ha^{-1}) and the effluent was collected after 0, 60, 90 and 120 days of irrigation. The procedure was the same for the soils from the cultivated and control areas. The experiment was performed in triplicate for each soil and vinasse dose.

The amount of water used for irrigation during the experiment corresponded to the annual rainfall in the region of Goianesia municipality, turned into daily rainfall, which resulted in approximately 3.20 L $column^{-1}$ applied once a week during the evaluation period. The effluent samples were collected one day after irrigation, just in those predetermined four times, and the excess volumes during interval between sampling were discarded.

Table 1. Original soil analysis. Goianésia, GO, Brazil.

Treatment	Clay	Silt	Sand	pH CaCl$_2$	P(Mehl)	K	S available	Ca^{2+}	Mg^{2+}	H+Al	Al^{3+}	CEC*	SOM
	----------%----------				------------mg dm^{-3}----------			-----------------cmol$_c$ dm^{-3}-------------					%
					0 - 25 cm								
Cultivated	48.0	14.0	38.0	5.3	5.2	471.3	36.3	2.0	1.0	2.5	0.0	6.6	1.1
Savannah	52.0	17.0	31.0	4.3	1.0	45.5	6.9	1.1	0.5	6.1	0.5	7.8	2.6
					25 - 50 cm								
Cultivated	48.0	14.0	38.0	4.9	1.1	319.5	38.9	0.9	0.6	2.7	0.0	5.0	0.6
Savannah	52.0	17.0	31.0	4.2	0.4	19.5	5.9	0.3	0.2	5.2	0.6	5.7	1.4
					50 - 75 cm								
Cultivated	48.0	14.0	38.0	5.0	0.3	201.0	44.0	0.9	0.6	2.5	0.0	4.5	1.2
Savannah	52.0	17.0	31.0	4.3	1.3	10.2	4.9	0.2	0.1	3.8	0.4	4.2	1.0
					75 - 100 cm								
Cultivated	48.0	14.0	38.0	5.1	0.3	148.3	35.1	0.8	0.5	2.4	0.0	4.0	0.6
Savannah	52.0	17.0	31.0	4.3	0.1	12.8	6.9	0.2	0.1	3.5	0.2	3.9	0.8

CEC* Cation exchange capacity; SOM: Soil Organic Matter. (Source: Oliveira, 2006).

Table 2. Vinasse composition. Goianésia, GO, Brazil.

pH		4.02
CE	mS cm^{-1}	12,100.00
Fe		27.00
Cloretos		2,400.00
SST		37,450.00
NT		405.00
NH$_3$		107.00
SO4^{2-}		943.00
Na		18.50
Cr		0.04
Cd Total		0.06
Pb	mg L^{-1}	0.27
Ni		0.15
Cu		0.28
Cr total		0.04
Ca		823.00
Mg		295.00
K		3,920.00
Mn		0.35
Zn		0.23
DBO	mg L^{-1} O$_2$	11,133.33
DQO		31,000.00

Analytical methods

The effluent samples were immediately transferred to the Soil and Foliar Analysis Laboratory of the Federal University of Goias (LASF/UFG), and stored in the fridge under 5°C, until further analysis.

The measured variables on undiluted vinasse were determined according to Greenberg et al. (1992) (Table 2). The following measured variables were determined for the effluent t samples: dissolved organic matter (DOM) by Walkley-Black wet combustion without heating (Nelson and Sommers, 1996), Fe, Ca and Mg were determined by atomic spectroscopy (Wright and Stuczynski, 1996), Na and K were determined by flame emission spectrometry (Wright and Stuczynski, 1996), NO_3^- and NH_4^+ were determined by steam-distillation method using MgO and Devarda's alloy (Mulvaney, 1996), S-SO_4^{2-} was determined by turbidimetry using BaCl$_2$ (Faithfull, 2002) and pH was measured using electrode method (Thomas, 1996). The determination of chlorine (Cl$^-$) was made using a selective electrode performed according to Abreu et al. (2001).

Statistical analysis and ion speciation data modeling

The nutrients data were submitted to analysis of variance by the F test (Table 3). The relationships between vinasse doses and time after irrigation within each land use and its influence on the soil solution chemical parameters were obtained through multiple regression tests by general linear model (GLM), using the SAS software (2000).The Visual Minteq v. 3.0 chemical speciation model (Gustafsson, 2007) was used for ion speciation of soil effluent. The modeling took all solution phase measured variables, the Davies equation to calculate the activity coefficients, the Gaussian model of complex for the speciation of the organic matter, and the ionic strength was calculated by the program.

RESULTS AND DISCUSSION

The analysis of variance (Table 3) indicated there were no significant differences just for the element NO_3^- in any of the sources of variation assessed. Significant differences in pH and nutrients concentration were observed according to the vinasse doses, DAI, and

Table 3. Summary of analysis of variance for the measurable variables on soil solution.

Variation sources	pH CaCl$_2$	DOM	Cl$^-$	NH$_4^+$	NO$_3^-$	K$^+$	Na$^+$	Fe^{3+}	SO$_4^{2-}$	Ca^{2+}	Mg^{2+}
		-- mg dm^{-3}--									
						P > F					
T	*	*	*	*	ns	*	*	*	*	*	*
S	*	ns	ns	*	ns	*	*	*	*	*	*
D	*	ns	*	ns	ns	*	*	*	*	*	*
S × D	*	ns	ns	ns	ns	*	*	*	*	*	*
T × S	*	*	*	*	ns	*	*	*	ns	ns	*
T × D	ns	ns	*	ns	ns	*	*	ns	*	*	*
R^2	0.72	0.57	0.89	0.38	0.25	0.85	0.90	0.72	0.75	0.92	0.90
CV	5.41	94.33	47.53	9.88	21.18	97.84	31.54	136.29	160.55	61.92	88.92

T: days after application (0, 60, 90, 120 days); S: soil source (savannah; cultivated: crop with 21 years of application); D: doses (0, 300, 600, 1200 m^3 ha^{-1}); *: probability < 0.05; F: test F. DOM: Dissolved organic matter.

land use.

The surface application of vinasse on dystrophic Ferralsols leads to significant changes in the soil solution chemistry and in the percolate, in most cases regardless of the land use and time after vinasse application. On the other hand, the effects of land use and days after irrigation were noticed when the nutrients analysis were done individually. In this case, the responses of the savannah soil were distinct from that of the cultivated soil for most nutrients as discussed below.

Vinasse rates and time showed significant effects over pH on cultivated soil and savannah soil (control), and an interaction effect between these two variables over pH in cultivated areas has been observed (Table 4). However, on cultivated soil with passing of time, an increase on solution pH was observed. This apparently contrasting pH effects observed for the cultivated and savannah soils can be related to soil microorganism's action, which leads to increased pH (Rossetto, 1987; Silva and Ribeiro, 1998). Thus, although the vinasse has acidic character, after a given period of soil reaction along with biological reactions the pH can increase due to microorganism's action. This pH increase was observed for both land uses however at different vinasse doses. As a consequence, the changes in pH may strongly influence the solubility and retention of the ions (Camargo, 1991).

The absence of significant differences in the DOM concentrations for the vinasse doses (Table 3) in both soils may be due to stable binding between clay minerals and organic particles (Bartoli et al., 1992) which would prevent the leaching of organic compounds through the soil column regardless of the amount of applied vinasse. Different DOM retention mechanisms may have influenced the observed changes in DOM through time of incubation. However, it is unclear why the DOM concentration decreased in the savannah soil after 90 days of incubation for all vinasse doses whereas in the cultivated soil there was continuum increase in the DOM concentration (Table 4).

The sugarcane crops demand a high amount of potassium, an element present in several compounds, mostly proteins, and that is part of the chlorophyll, organic acids, and vegetal hormones (Santiago and Rossetto, 2005). Thus, the plant gives a fast response to K application (Rossetto et al., 2004). Rossetto et al. (2004) suggest a critical level of 82.11 mg dm^{-3} of potassium in the soil. In this study, the concentration of K in the cultivated soil was 471.12 mg dm^{-3} before the application of vinasse treatments, whereas it was 44.85 mg dm^{-3} in the savannah soil at 0 to 25 cm depth (Table 1). Thus, the K concentration in the cultivated soil was above the critical level whereas in the savannah soil, initially, it was below that level. This indicates an excess of K in the cultivated soil since the beginning of the experiment and that may have facilitated K leaching. The observed free form of K in the effluent (Figure 2d) may be due to the large concentration of counter ions provided by the vinasse, combined with the pre-existing amount in the soil occupying its charges. It is important to mention that the ion K$^+$, found in high concentration in the fresh vinasse (Table 2: 3920 mg dm^{-3}), is the element required in great amount by this sugarcane plants and is considered the major environmental problem concerning the byproduct in the ethanol industry (CETESB, 2006). Potassium leaching increased as vinasse application doses and time after application increased, thus suggesting saturation of exchange sites with free K passing through soil profile (Table 4). This result disagrees with Lyra et al. (2003) and Madejón et al. (2001) who worked with a Spodosol, Entisol and Inceptisol, respectively, and found no increase in K leaching after vinasse application. It indicates that the type of tropical soil may influence K leaching after vinasse application.

Although, the vinasse had significant Mg and Ca concentrations (Table 2: 295 mg dm^{-3} and 824 mg dm^{-3}, respectively) the concentration values observed in the effluent of the cultivated soil were higher than that of

Table 4. Regression equations for days after irrigation (T) and vinasse rates (V) for measurable variables on columns effluent on cultivated soil and control (savannah soil).

		Equation	R^2	F	P
pH	Control	$6.36-0.01T^*+0.001V^*+0.00007T^{2*}-0.000001V^{2*}+0.000002TV$	0.61	13.05	0.0001
	Cultivated	$6.49+0.009T^*-0.002V^*-0.0001T^{2*}+0.000001V^{2*}+0.000006TV^*$	0.59	11.30	0.0001
Cl^-	Control	$5.74-0.14T^*-0.0007V+0.0008T^{2*}+0.000002V^{2*}-0.000004TV$	0.90	69.03	0.0001
	Cultivated	$3.09-0.08T^*+0.0002V+0.0005T^{2*}+0.000001V^2+0.000005TV$	0.67	14.05	0.0001
DOM	Control	$0.004+0.0004T^*-0.00001V-0.000003T^{2*}+0.000000008V^2-0.00000002TV$	0.51	8.81	0.0001
	Cultivated	$-0.0001-0.00004T+0.000002V+0.000001T^{2*}-0.000000002V^2+0.00000004TV$	0.53	7.78	0.0001
Mg^{2+}	Control	$2.11+0.11T-0.04V-0.001T^2+0.00006V^{2*}+0.0001TV$	0.56	10.26	0.0001
	Cultivated	$46.07-1.04T-0.23V^*+0.005T^2+0.0002V^{2*}+0.002TV^*$	0.73	19.58	0.0001
K^+	Control	$-2.92+0.32T^*-0.02V+0.002T^{2*}+0.00006V^{2*}-0.0001TV$	0.79	30.01	0.0001
	Cultivated	$-15.38+0.47T+0.05V-0.001T^2+0.00009V^{2*}-0.001TV^*$	0.77	17.30	0.0001
Na^+	Control	$4.39-0.007T-0.002V-0.000004T^2+0.000005V^{2*}+0.00006TV^*$	0.72	21.62	0.0001
	Cultivated	$6.55-0.06T-0.01V+0.0003T^2+0.00001V^2+0.0002TV^*$	0.58	9.28	0.0001
Ca^{2+}	Control	$-1.61+0.55T-0.06V-0.005T^2+0.0001V^{2*}-0.00006TV$	0.64	14.16	0.0001
	Cultivated	$39.03-1.08T-0.12V+0.006T^2+0.0002V^{2*}+0.002TV^*$	0.88	49.45	0.0001
Fe^{3+}	Control	$-12.82+45.04T^*-4.33V^*-0.41T^{2*}+0.008V^{2*}-0.002TV$	0.76	24.57	0.0001
	Cultivated	$561.25-6.31T-1.59V^*+0.008T^2+0.001V^{2*}+0.02TV^*$	0.61	10.46	0.0001
SO_4^{2-}	Control	$-1.32-0.08T+0.01V+0.001T^{2*}+0.000009V^2-0.0002TV^*$	0.59	11.30	0.0001
	Cultivated	$-4.25+0.05T+0.008V+0.0004T^2+0.00003V^{2*}-0.0002TV$	0.65	10.43	0.0001
NH_4^+	Control	$23.06-0.04T+0.0005V+0.0001T^2-0.0000005V^2+0.000002TV$	0.30	0.84	0.5500
	Cultivated	$21.04-0.07T+0.003V+0.0003T^2-0.000001V^2+0.000002TV$	0.67	3.67	0.0400
NO_3^-	Control	$1.48-0.005T-0.00006V+0.00001T^2-0.0000001V^2+0.000003TV$	0.29	0.82	0.5600
	Cultivated	$1.23+0.001T-0.0003V-0.000004T^2+0.0000003V^2-0.000001TV$	0.29	0.74	0.6100

the vinasse (Table 4). We hypothesized that high Mg and Ca concentrations in the effluent may be due to ion displacement from soil exchange sites by the K ions after the application of high doses of vinasse. This result agrees with Gloaguen et al. (2007) who worked with ions leaching from sewage effluent application on a Brazilian dystrophic Ferralsol and found a similar increase in Mg and Ca in the soil effluent. Magnesium is an essential macroelement for the proper development and productivity of the sugarcane crop, and its critical level is of 48.62 mg dm^{-3} in the soil (Benedini et al., 2008). The suitable levels of Mg for plant needs, also influences the sugarcane response to the additions of lime and potassium (Rossetto et al., 2004; Silveira et al., 1980). In this study, the savannah soil was the only soil in need for magnesium. In the cultivated soil, as expected, the concentration of magnesium in the soil was around 121 mg dm^{-3}. After the fertigation with vinasse, there was an increase in Mg concentrations in the effluent, as the vinasse doses increased. This supports the hypothesis that increase in Mg leaching as free ion was due to the saturation of the ionic exchange complex (475.89, 1416.37, 2404.37 and 19,895.34 mg dm^{-3} of Mg in the cultivated soil at doses of 0, 300, 600, and 1200 m^3 ha^{-1} of vinasse, respectively). The initial concentration of calcium in the soil did not show problems in terms of

deficiency, since the critical level for the sugarcane crop is between 160.32 and 200.40 mg dm^{-3} (Azeredo et al., 1981; Benedini et al., 2008) and that was high for both soils (Table 1). However, the vinasse provided a large amount of this element to the effluent of the columns and there were concentrations well beyond those necessary for the development of the plant. Thus the excess of Ca was leached mainly in the free form (Figure 2d). Also, the high vinasse doses led to the formation of complex ion-organic carbon, such as the observed association Ca-DOM and that may have influenced the fate of DOM molecules.

In agreement with Gloaguen et al. (2007), the increased Na concentration in the leaching with the increase in the vinasse dose may be due to the opening of exchange complex caused by high monovalent ions concentration, Na and K in the vinasse (Table 4). The excess Na was leached after the saturation of soil exchange sites. Through the process of saturation of the soil exchange sites by Na, it is likely that other ions were replaced and released to the soil solution.

According to Klar (1987), the amount of sodium in irrigation water that would not harm the soil in terms of soil sodification, should be less than 69 mg dm^{-3}. This element at high concentrations, along with boron, bicarbonate, and chlorine, causes physiological disorders

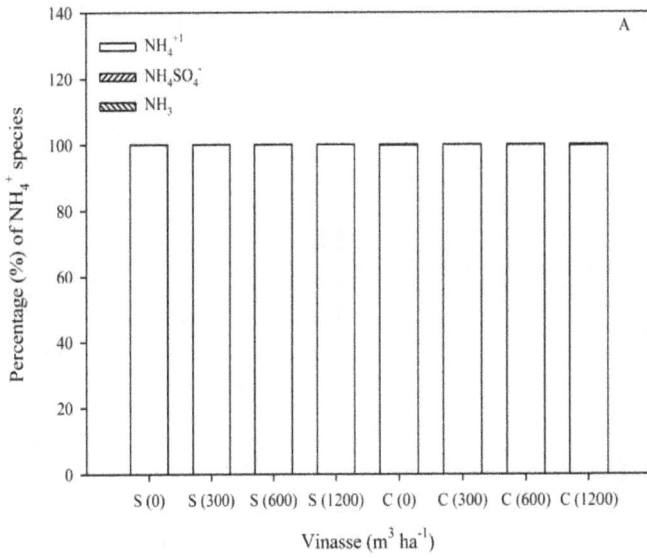

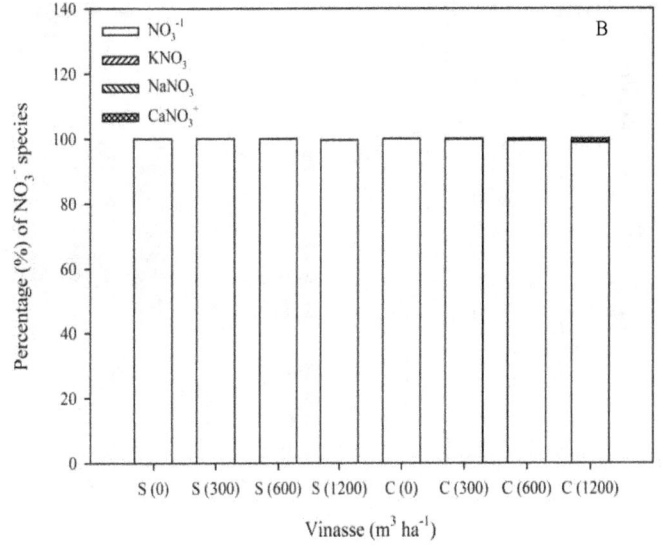

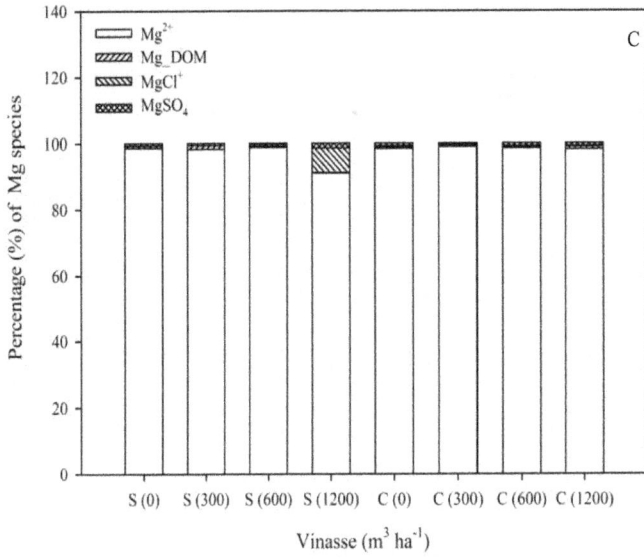

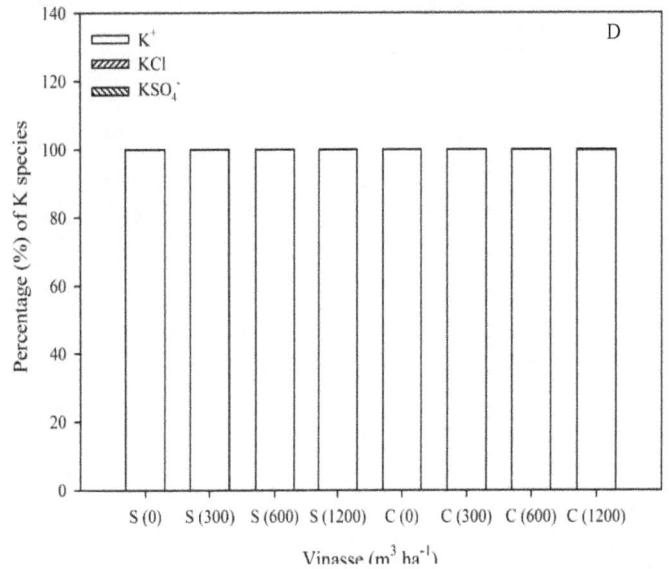

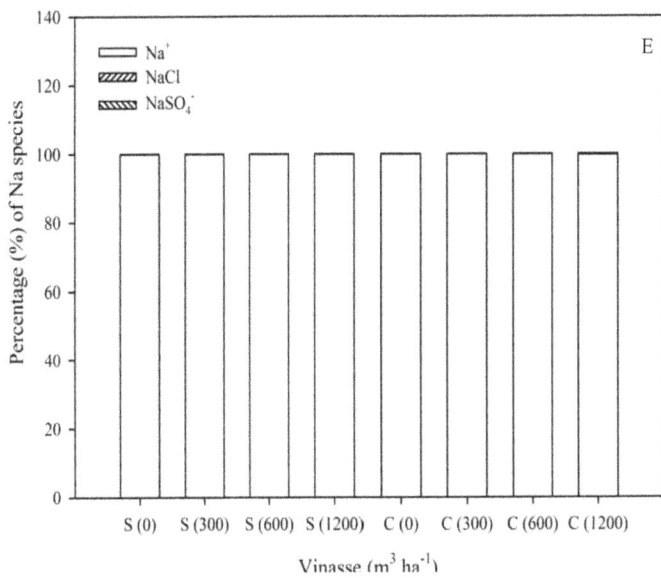

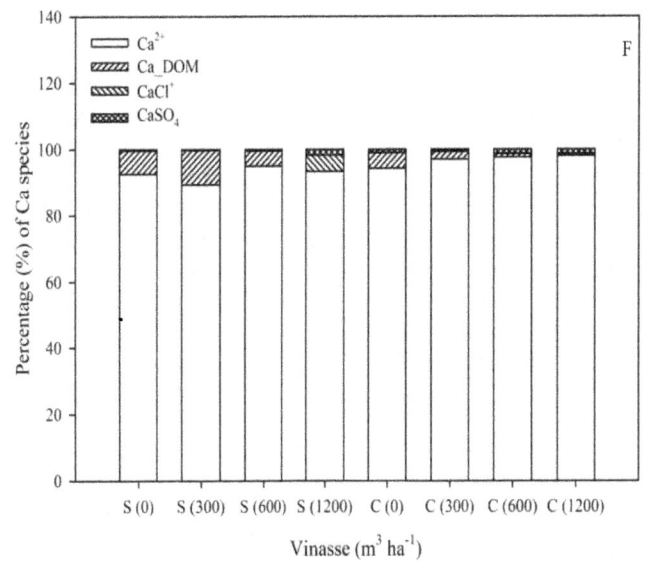

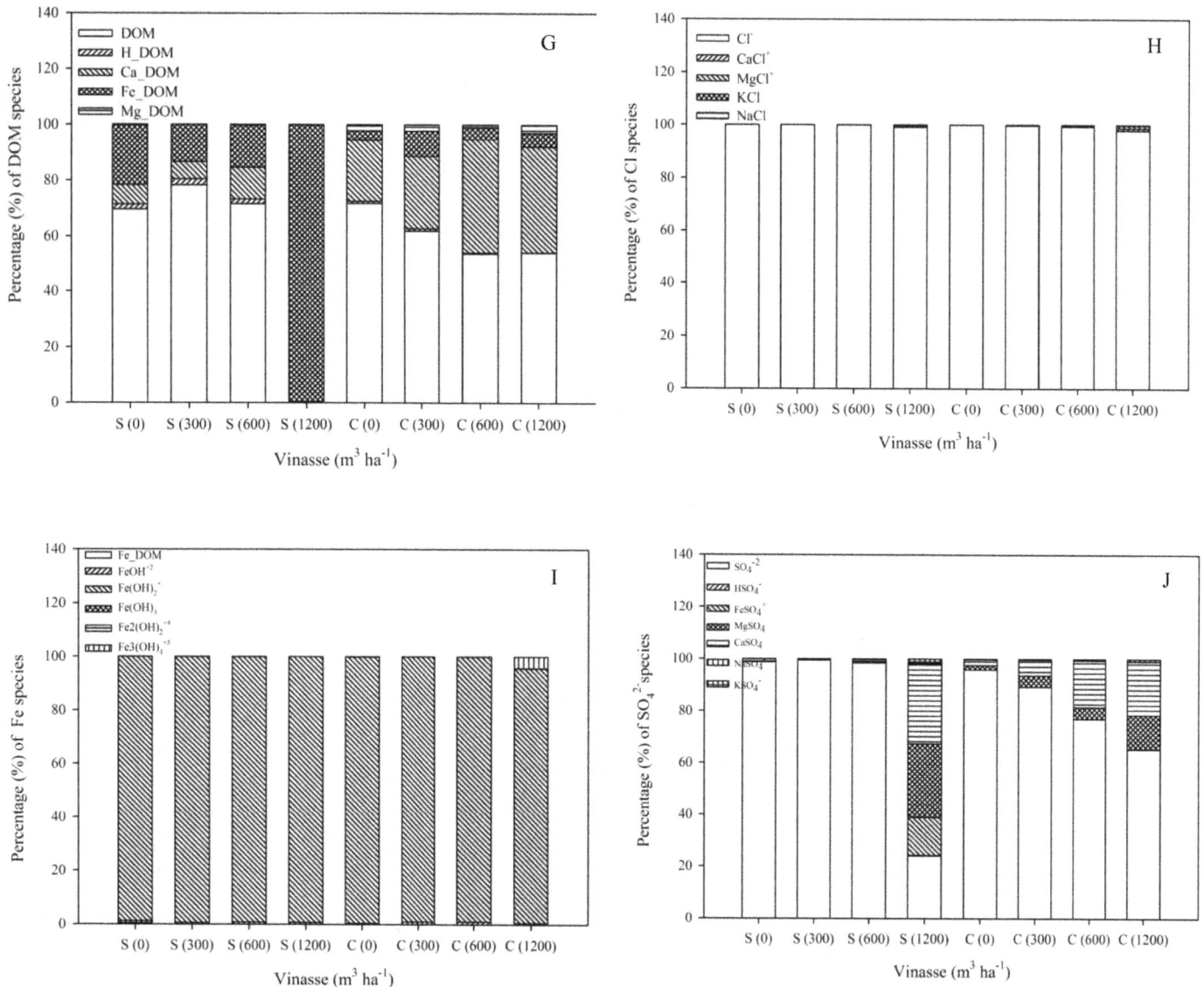

Figure 2. Concentration of ions and organic matter in the soil solution under different uses and vinasse rates ((S): Savannah soil; (C): Cultivated soil). Data from Minteq speciation analysis.

to the plants (Batista et al., 2002). Due to the frequent input of sodium from the use of the vinasse, the concentration of this element in the effluent showed a considerable enhancement as the vinasse doses increased. This behavior indicates that there may be a tendency to soil sodification after vinasse application. In this study, a concentration up to 20.50 mg dm^{-3} of Na in the effluent of columns exceeding almost 2 mg dm^{-3} that quantity on vinasse (Table 2) was observed, suggesting sodification due to the passage of Na through the 1.20 m of soil profile.

Chlorine concentrations on soil solution were extremely dependent on time (Table 4) and its quantity was higher in the beginning decreasing with time. Vinasse rates had presented a quadratic effect just on control. This behavior

of the chlorine co-ion may be due to the greater counter ion availability in the soil when higher doses of vinasse were used which enabled the formation of molecules thus reducing its concentration in the ionic form and influencing its retention in the soil charges (Figure 2b).

An interesting finding of this study was the high concentration of Fe on solution at the 1200 m^3 ha^{-1} vinasse dose, especially on control, where there was observed a quadratic effect of time and vinasse rates on Fe concentration (Table 4). On cultivated soil, an interaction effect between days after irrigation and vinasse rates occurred (Table 4). It was found that the Fe concentration in the vinasse was 27 mg dm^{-3} whereas in the effluent from the savannah soil the Fe concentration was about 10,000 mg dm^{-3}. Effluent solutions of orange color and large amounts of Fe deposited (crusts) on the surface of the sampling bottles were observed for the treatments with vinasse doses of 1200 m^3 ha^{-1}, regardless of the land use, and this can be explained by iron solubility.

We hypothesize that high vinasse doses could have disrupted soil structure thus solubilizing iron from Fe-oxide minerals in both types of use from this Ferralsol. As a consequence, iron solubility may have caused soil dispersion and that could promote soil structural problems, as reported by Amézketa (1999).

The high SO_4^{2-} leaching found at 0 day for cultivated soil was likely because of the replacement of the pre-existing soil solution by fresh effluent, in agreement with Gloaguen et al. (2007). During the first 60 days after irrigation, significant increase in sulfate leaching was observed when the 1200 m^3 ha^{-1} vinasse dose was applied probably because of the replacement on mineral particles net charge and sulfates ions free mobility (negative charge ion) in this soil type, which often presents negative charge (Table 4). On cultivated soil with 600 m^3 ha^{-1} and on savannah soil with 1200 m^3 ha^{-1} was observed an enhancing on calcium sulfate forms (Figure 2f) which can be explained by high free Ca^{2+} concentration on effluent from soils treated with high vinasse rates.

Generally, the ions Mg^{2+}, K^+, Na^+, Cl^-, NO_3^- and NH_4^+ were found in the free ionic form in the effluent, regardless of the amount of applied vinasse and land use (Figure 2). The other measured nutrients were bound to other ions in the solution phase. The other ions measured showed a decrease in their percentages of the free form as the vinasse doses increased resulting in a bond increment between other ions of soil solution phase.

Conclusions

The continuous use of vinasse in the soil may saturate the soil exchange capacity favoring the leaching of ions. The use of vinasse doses at 1200 m^3 ha^{-1} leads to significant increases in the levels of calcium and magnesium in the soil thus favoring the formation of complexes with sulfate and chlorine co-ions and with the dissolved organic matter. The leaching of these ions is facilitated by the formation of ionic pairs. Potassium leaching increased as vinasse application doses increased thus suggesting saturation of exchange sites with free K passing through soil profile. Large amounts of vinasse applied to the soil may favor soil dispersion, as evidenced in this study by the substantial increase in iron concentration in the effluent of the columns.

ACKNOWLEDGEMENTS

The authors thank go to CNPq and CAPES for the scholarship, to Goias State Federal University, to Jalles Machado Company to Embrapa Rice and Beans and to FAPEG for funding.

REFERENCES

Abreu Júnior CH, Boaretto AE, Muraoka T, Kiehl JC (2005). Uso agrícola de resíduos orgânicos potencialmente poluentes: Propriedades químicas do solo e produção vegetal. In: Solo SBdCd (Editor), Tópicos em Ciência do Solo. SBCS, Viçosa, pp. 391-470.

Abreu MF, Santos PH, Quaggio JA (2001). Determinação de cloreto em extratos de solo e planta. Instituto Agronômico de Campinas, pp. 2

Abreu RL (2006). Ficheiro: Goias Municip. Goianesia. In: http://pt.wikipedia.org/wiki/Ficheiro:Goias_Municip_Goianesia.svg (Hrsg.), JPEG. Wikipedia, Wikipedia.

Amézketa E (1999). Soil Aggregate Stability: A review. J. Sustain. Agric. 14:83-151.

Azeredo DF, Glória NA, Manhães MS (1981). Efeitos da calagem na cana-planta e nas características químicas de dois solos do Estado do Rio de Janeiro. In: STAB (Hrsg.), Congresso Nacional da Sociedade dos Técnicos Açucareiros e Alcooleiros do Brasil. Stab, Rio de Janeiro, pp. 71-88.

Bartoli F, Philippy R, Portal JM, Gerard B (1992). Poorly ordered hydrous Fe oxides, colloidal dispersion and soil aggregation. I. Effect of humic macromolecules on surface and colloidal properties of Fe (III) polycations. J. Soil Sci. 43:47-58.

Batista MJ, Novaes F, Santos DG, Suguino HH (2002). Drenagem como instrumento de dessalinização e prevenção de salinização de solos. Codevasf, Brasília P. 216.

Benedini MS, Faroni CE, Penatti CP (2008). Alternativas para redução de custos da adubação. Revista Coplana 50:22-24.

Camargo OA (1991). Reações e interações de micronutrientes no solo. In: Ferreira ME, Cruz MCP (Editors), Micronutrientes na agricultura. POTAFÓS/CNPq, Piracicaba. pp. 243-272.

CETESB (2006). Vinhaça – critérios e procedimentos para aplicação no solo agrícola. In: Sucroalcooleiro CAdS (Hrsg.). Companhia Ambiental do Estado de São Paulo. P. 12.

Ernani PR, Mantovani A, Scheidt FR, Nesi C (2003). Mobilidade de nutrientes em solos ácidos decorrentes da aplicação de cloreto de potássio e calcário, XXIX Congresso Brasileiro de Ciência do Solo. Sociedade Brasileira de Ciência do Solo, Ribeirão Preto, SP. pp. 46-47.

Faithfull NT (2002). Methods in agricultural chemical analysis: a pratical handbook. CABI Publishing, Oxon, UK. P. 206.

Freire WJ, Cortez LAB (2000). Vinhaça de cana-de-açúcar. Agropecuária, Guaíba. P. 203.

Gloaguen TV, Forti MC, Lucas Y, Montes CR, Gonçalves RAB, Herpin U, Melfi AJ (2007) Soil solution chemistry of a Brazilian Oxisol irrigated with treated sewage effluent. Agric. Water Manag. 88:119-131.

Greenberg AE, Clesceri LS, Eaton AD (1992). Standard methods for the examination of water and wastewater. American Public Health Association, Washington, D.C, P. 400.

Gustafsson JP (2007). Visual Minteq. USEPA, Sweden.

IBGE (2011). Pesquisa mensal de previsão e acompanhamento das safras agrícolas no ano civil. Levantamento Sistemático da Produção Agrícola 24:1-82.

SAS Institute Inc. S (2000). SAS/STAT software SAS Institute Inc., Cary, NC.

Klar AE (1987). Relações Água-solo-planta. Curso de Engenharia da irrigação-Parte A, Módulo III. ABEAS, Brasília, 106 pp.

Lyra MRCC, Rolim MM, Silva JAA (2003). Toposseqüência de solos fertirrigados com vinhaça: Contribuição para a qualidade das águas do lençol freático. Revista Brasileira de Engenharia Agrícola e Ambiental. 7:525-532.

Madejón E, Lopez R, Murillo JM, Cabrera F (2001). Agricultural use of three (sugar-beet) vinasse composts: Effect on crops and chemical properties of a Cambisol soil in the Guadalquivir river valley (SW Spain). Agric. Ecosyst. Environ. 84:55-65.

Mulvaney RL (1996). Nitrogen: Inorganic forms. In: Sparks DL (Editor), Methods of soil analysis: Chemical methods (part 3). SSSA, Madison,

Wisconsin. pp. 1123-1184.

Nachtergaele FO (2005). CLASSIFICATION SYSTEMS | FAO. In: Editor-in-Chief: Daniel H (Editor), Encyclopedia of Soils in the Environment. Elsevier, Oxford pp. 216-222.

Nelson DW, Sommers LE (1996). Total carbon, organic carbon and organic matter. In: Sparks DL et al. (Editors), Methods of Soil Analysis: Chemical Methods (part 3). SSSA, Madison, Wisconsin. pp. 961-1010.

Raij Bv, Cantarella H, Quaggio JA, Furlani AMC (1997). Recomendações de adubação e calagem para o Estado de São Paulo. Boletim técnico, IAC, 100. Fundação IAC, Campinas, P. 285.

Rossetto AJ (1987). Utilização agronômica dos subprodutos e resíduos da indústria açucareira e alcooleira. In: PARANHOS SB (Editor), Cana-de-açúcar: cultivo e utilização. Fundação Cargill, Campinas, pp. 435-504.

Rossetto R, Spironello A, Cantarella H, Quaggio JA (2004). Calagem para cana-de-açúcar e sua interação com a adubação potássica. Bragantia 63:105-119.

Santiago AD, Rossetto R (2005). Correção e adubação da cana-de-açúcar. In: Embrapa Adld (Hrsg.). Embrapa, Brasília.

Silva AJN, Ribeiro MR (1998). Caracterização de um Latossolo Amarelo sob cultivo contínuo de cana-de-açúcar no Estado de Alagoas: Propriedades químicas. Revista Brasileira de Ciência do Solo 22:291-299.

Silva MAS (2009). Fertirrigação com vinhaça: impactos ambientais. Coleção Expressão Acadêmica, 1. Editora UFG, Goiânia, P. 109.

Silveira JF, Siqueira JO, Guedes GAA (1980). Interação fósforo x potássio x calcário em cana-de-açúcar (cana planta). Brasil Açucareiro 95:18-21.

Thomas GW (1996). Soil pH and Soil acidity. In: Sparks DL (Editor), Methods of soil analysis: Chemical methods (part 3). SSSA, Madison, Wisconsin. pp. 475-490.

UNICA (2008). Qual a dimensão da cana-de-açúcar no Brasil e no mundo? UNICA, São Paulo.

Wright RJ, Stuczynski T (1996). Atomic absorption and flame emission spectrometry. In: Sparks DL (Editor), Methods of soil analysis: Chemical methods (Part 3). SSSA, Madison, Wisconsin. pp. 65-90.

23

Effect of soil matric potential on wolfberry (*Lycium barbarbum* L.) yield, evapotranspiration and water use efficiency under drip irrigation

Junshu Jia, Yaohu Kang and Shuqin Wan

Key Laboratory of Water Cycle and Related Land Surface Processes, Institute of Geographic Sciences and Natural Resources Research, CAS, Beijing, China.

The experiment was undertaken in order to investigate the effect of soil matric potential (SMP) on wolfberry yield, evapotranspiration (ET), water use efficiency (WUE) and the implications for soil water management under drip irrigation in the Yinchuan Plain arid region of Northwest China. The experiment consisted of five treatments, which maintained SMP at a depth of 0.2 m immediately under a drip emitter releasing water at -10 (S1), -20 (S2), -30 (S3), -40 (S4) and -50 kPa (S5), respectively, after wolfberry had been planted. The results showed that during the growing season, the target SMP value of the five different treatments was maintained within the experimental design range and the irrigation volume and ET declined as the target SMP value decreased. The temporal and spatial soil water content (SWC) changes observed in the soil profile suggested that the S1 and S2 treatments could meet the crop water absorption demand because the soil moisture was over 15 % and soil water was in excess of crop needs. However, the soil water content for S3, S4 and S5 was less than 15 %, which produced visible water stress in the crop. Statistical analysis for the change values in the electrical conductivity of the saturated soil-paste extract (ECe) showed the significant differences between the different SMP treatments in the middle layer (40 – 80 cm) and the plots S2 (-20 kPa) had the better effect of the soil salinity leached. The highest yield was achieved in the S2 plots and the lowest fresh to dry weight ratio was found in the S3 plots, which also produced the highest quality seed. WUE was in the order S3 > S5 > S2 > S4 > S1. Based on a comprehensive analysis that included irrigation volume, ET value, crop yield and WUE, it is recommended that the SMP threshold should be controlled in the range from -20 to -30 kPa, which was the most favorable drip irrigation schedule for increasing yields and irrigation efficiency. The study should provide information that could be used in other regions suffering from drought and water scarcity.

Key words: Deep percolation, irrigation volume, soil salinity, soil water content (SWC), vacuum tensiometers.

INTRODUCTION

The wolfberry (*Lycium barbarbum* L.) grown in the Ningxia Hui Autonomous Region is famous around the world. Ningxia wolfberry production represents 60% of Chinese wolfberry production and this has developed

rapidly over recent years, e.g. in 2010, the Forestry Bureau of the Ningxia Hui Autonomous Region reported that the land used for wolfberry plantations was over 4.67 × 10^4 hm^2 and production value was more than 2 billion

Yuan. However, in local areas, for some aspect poor irrigation management has led to water resources being wasted and poor practices, such as flood irrigation.

Drip irrigation is able to apply water at a low discharge rate and at a high frequency over a long time period. It can maintain high soil water content in the root zone (Keller and Bliesner, 1990) and can be used to overcome problems associated with water stress or over irrigation. In the past few years, sensor technology, that permits continuous on-farm monitoring of soil water status, has become increasingly accessible to commercial producers (Zotarelli et al., 2009). Tensiometers, which measure soil matric potential (SMP), can be used to more accurately balance specific crop water requirements. A number of studies on scheduling irrigation, using tensiometers to measure SMP, have been reported (Phene et al., 1989; Clark et al., 1996; Shock et al., 2000; Wilson et al., 2001). The SMP results presented by Kang (2004a) in a report on an applied method for drip irrigation scheduling are used as the SMP target value in China. They suggested that the SMP at 0.2 m depth immediately under the drip line was a good indicator of the soil water condition in the crop root zone layer.

In the north China plain, research has shown that when SMP was kept at -35 kPa, radishes absorbed more N, P, and K, accumulated more nutrients and had an increased dry matter weight (Feng et al., 2004). With potatoes, the highest yield and WUE values were achieved when the soil matric potential threshold was around -25 kPa and with an irrigation frequency of once a day (Kang et al., 2004b). Moreover, using tensiometers to measure SMP is a feasible and simple method that improves crop growth by controlling soil water-salts in arid and semi-arid areas (Kang et al., 2007). When there was not enough fresh water for irrigation in semi-humid areas, the soil matric potentials at 0.2 m depth immediately under drip emitters were kept higher than -20 kPa during the growing season. Saline water with salinity ranging from 2.2 to 4.9 dS m^{-1} could be applied to irrigate field grown tomatoes about 30 days after transplanting if appropriate management strategies were adopted (Wan et al., 2007a). Cucumber, a moderately salt-sensitive plant, could be irrigated with saline water (EC = 2.2 - 4.9 dS m^{-1}) if the SMP was maintained between -25 kPa and -35 kPa after the seeding stage (Wan et al., 2007b). In the northwest arid area, a 3-year experiment was carried out in a heavily saline field in Jinshawan, Ningxia Autonomous Region. When the SMP was maintained at over -10 kPa, soil salinity was well leached and waxy corn and oil sunflower planted in the experimental plots grew well. Tomato, cucumber, sorghum and some other plants also grew well, achieving normal or near normal yields (Jiao et al., 2007; Kang et al., 2007).

The objectives of this study were to maintain soil water levels, based on SMP at 0.2 m depth immediately under the drip emitter, and to evaluate the effect of SMP on wolfberry plant yield and WUE. The aim was to suggest a SMP threshold for wolfberry irrigation scheduling in the Yinchuan Plain located in the arid region of Northwest China.

MATERIALS AND METHODS

Experimental site and soil

The field experiment was conducted at the Zhongning L. barbarbum L. Drip-irrigation and Effective Cultivation Technique integration Experimental Area (latitude 37° 25' 30" N, longitude 105° 46' 12" E), Institute of Geographic Sciences and Natural Resources Research, Chinese Academy of Sciences. The study site was located at the Ningxia Yellow River Irrigation Area in arid Northwest China at an altitude of 1290 m. Average annual precipitation is 184 mm, of which 72% is received during June to September, and average annual evaporation is 2100 mm. Average annual temperature is 7.4°C with the coldest temperature being around -30 to -38°C and the warmest being 38°C. Throughout the year, the mean wind speed is 2.1 m s^{-1}. The groundwater table is approximately 20 - 40 m below the surface with a mineral content of 3 - 13 g L^{-1}. Data on the soil physical-chemical properties and nutrient composition within the different soil horizons are presented in Tables 1 and 2.

Experimental design

The experiment consisted of five treatments, which maintained the SMP at 0.2 m depth immediately under drip emitter at -10 kPa (S1), -20 kPa (S2), -30 kPa (S3), -40 kPa (S4) and -50 kPa (S5), respectively, after the wolfberry plants had been planted. Each treatment was replicated three times in a randomized complete block design. Each treatment plot had dimensions of 12 m × 5 m. The plant row spacing was 3 m and the interplant spacing was 1 m in each row. Each treatment was equipped with one pipe branch unit of the drip irrigation system. Each pipe branch unit had one valve, one flow meter, one pressure gauge and one fertilizer tank to control operating pressure and measure fertilizer-irrigation quantity. The drip tapes, with 0.5 m dripper spacing and a dripper discharge of 1.38 L h^{-1}, were placed 10 cm away from the plants.

Agronomic practices

Wolfberry (L. barbarbum L.), which is characterized as being drought resistant, salt tolerant, and tolerant of poor nutrient levels, is a typical halophyte and also a medicinal herb (Bai, 1999). In March 2009, the experimental land was leveled and the wolfberry plants were transplanted in April. The drip irrigation system was installed in May 2009. Meanwhile, all treatment plots were irrigated with the same quantity of water at the same frequency in order to ensure initial seeding survival.

Irrigation

In 2010, irrigation treatments were carried out from 15 May to 22 September. Irrigation was applied only when the SMP at 0.2 m depth immediately under drip emitter (measured with a vacuum tensiometer) reached the respective target value for that treatment plot. Irrigation volume at each application was about 5 mm for all the treatments. Based on experience, irrigation for recovery and winter water was added to allow salts to leach away and preserve soil moisture in March and November respectively. These application volumes were around four times that of the irrigation treatments (about 20 mm).

Table 1. Physical and chemical properties of the experimental soil.

Soil layers (cm)	ECe (dS m^{-1})	PH	Bulk density (g cm^{-3})	K_s (cm h^{-1})	Soil texture
0-20	1.77	8.01	1.45	1.385	Silty clay
20-40	1.40	8.07	1.45	1.053	Silty clay
40-60	1.21	8.09	1.34	2.968	Silty clay
60-80	2.08	8.04	1.39	3.059	Silty clay
80-100	3.06	8.00	1.35	4.838	Silty clay

Table 2. Nutrient content of the experimental soil.

Soil layers cm	Organic carbon g kg^{-1}	Total nitrogen mg kg^{-1}	Total phosphorus mg kg^{-1}	Nitrate nitrogen mg kg^{-1}	Available phosphorus mg kg^{-1}	Available potassium mg kg^{-1}
0-20	5.7	627.0	528.4	52.2	7.0	147.1
20-40	6.7	629.2	508.3	47.6	3.4	91.8
40-60	7.0	561.1	566.9	14.6	1.5	30.6
60-80	5.7	403.4	528.1	1.4	1.5	22.7
80-100	4.7	421.7	553.5	1.3	1.9	34.6

Fertilizer

After the experiment had started, fertilizer solution was put into the tanks at each irrigation event until the last irrigation. During the growing period, a top dressing of urea (46% N) and potassium dihydrogen phosphate (34% K; 51% P) was applied with the irrigation applications, amounting to 600 kg ha^{-1} and 360 kg ha^{-1}, respectively.

Observation

The weather data

A weather station was located within 200 m of the experimental field and used to measure rainfall, temperature, relative humidity, solar radiation and wind speed.

Soil matric potential (SMP)

One tensiometer with a vacuum gauge was installed at 0.2 m depth immediately under the emitters in each treatment to determine irrigation timing. The tensiometers were observed three times daily at 8:00, 14:00 and 18:00 h. When the data reached the target SMP, the drip-fertilization was started.

Soil water content (SWC %)

In 2010, soil samples were collected three times from each plot. They were taken before the treatments started (March 30), in the middle of experimental period (June 20) and after the experimental period had ceased (September 22). Using a 0.05 m diameter soil auger, soil cores were extracted at eight different depths (0 - 0.1, 0.1 - 0.2, 0.2 - 0.3, 0.3 - 0.4, 0.4 - 0.6, 0.6 - 0.8, 0.8 - 1.00, 1.00 - 1.20 m) and at six different horizontal distances from the emitter (0, 0.1, 0.2, 0.3, 0.4 and 0.5 m). The samples were immediately weighed and were oven-dried at 105° for 8 h in order to obtain the gravimetric soil water content θ (%).

Soil salinity (ECe)

In 2010, the collected soil samples at before the treatments started (March 30) and after the treatments (September 22) were air-dried and sieved through a 1-mm sieve. A plot soil samples were mixed by the weighted mean method in order to get three mixed samples along 0 - 0.5 m horizontal distance from the emitter with three difference depths 0 - 0.4, 0.4 - 0.8 and 0.8 - 1.2 m. Each mixed sample of the extrat electrical conductivity of the saturated soil (ECe) was measured by a conductivity meter (DDS - 11A, Shanghai Rex Instruments, China).

Yield

Plots were harvested on 4, 18, 28 July, 6, 20 August and 12 September in 2010. Four plants, each with a listing number, were selected from each plot. The weight of the fruits and the weight of 100 fresh seeds were recorded per plot for each harvest date. Then the picked fruits were air-dried over time and the dry weight recorded. The ratio of fresh to dry weight was calculated as the weight of the fresh seeds divided by dry weight.

Calculations

In the present study, crop evapotranspiration (ET) during the growing period was estimated using the water balance relationship from soil moisture and irrigation data as

$$ET = I + P \pm \Delta S - R - D \tag{1}$$

where I is irrigation (mm), P is precipitation (mm), R is surface

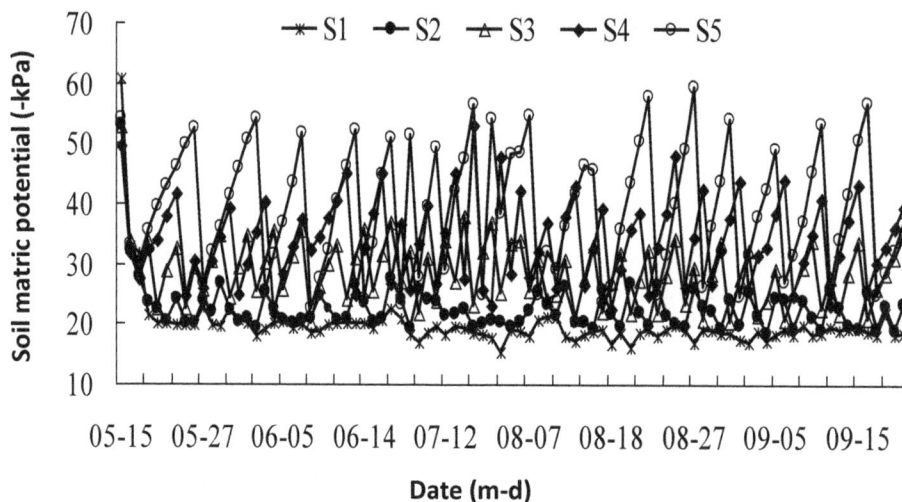

Figure 1. Soil matric potentials at 0.2 m immediately under drip emitter for different treatments.

runoff (mm), $\triangle S$ is the water storage change of the soil profile over the growing season (mm), and D is the drainage or deep percolation below the crop root zone (mm).

To estimate $\triangle S$, the soil water content in the soil profile (down to 90 cm) just before sprouting and after harvesting was determined by gravimetric measurement. Surface runoff (R) was negligible due to the controlled water application on each plot. Deep percolation (D) was estimated according to Darcy's equation (Azevedo et al., 2003; Kang et al., 2004b) as:

$$D = -K\left(\bar{\theta}\right)\left(\frac{\Psi_{m2} - \Psi_{m1}}{Z_2 - Z_1} + 1\right) \qquad (2)$$

where D is deep percolation at 80 cm depth between t_1 and t_2; Ψ_{m2} and Ψ_{m1} are matric potentials at 90 and 70 cm, respectively; Z_1 and Z_2 are soil depths under the crop root zone ($Z_1 = 70$ cm, $Z_2 = 90$ cm); θ is the mean of the volumetric soil water content at Z_1 and Z_2 and $K(\theta)$ is the unsaturated hydraulic conductivity (mm h^{-1}) estimated according to Lei et al. (1988):

$$K\left(\theta\right) = K_s\left(\frac{\theta}{\theta_s}\right)^m \qquad (3)$$

where K_s is saturated hydraulic conductivity; θ_s is saturated volumetric water content; m is the regression coefficient and θ is the volumetric water content.

K_s was estimated by the flow of water measured using the invariable hydraulic head method and θ_s was calculated from the mass water content and soil bulk density. Both K_s and θ_s (3.058 cm h^{-1} and 0.433 cm^3 cm^{-3}, respectively) were measured using soil obtained from a depth of 80 cm in the field where the experiment was conducted. The constant, m, was fitted as 3.7295 for this soil. θ was derived from a soil water characteristic curve based on field measurements. A set of nine vacuum tensiometers were installed at nine different soil depths (0.6, 0.65, 0.7, 0.75, 0.8, 0.85, 0.9, 0.95, 1.0 m) in a relatively flat and homogenous position in the experimental field. After a large water event, SMP was recorded form the different depths of tensiometers at 8 h intervals until the lowest values of SMP were recorded. At the same time, soil

samples were collected from the zone around the bottom of each tensiometer installed to measure soil water content. The soil water retention curves were drawn using a series of soil water content measurements and the corresponding SMP. Therefore at 80 cm soil depth, θ is expressed as

$$\theta = 0.5589e^{-0.0035\Psi_m} \qquad (4)$$

where Ψ_m is soil matric potential measured in the field within a range of -1 to -50 kPa.

According to Huang et al. (2004) the water use efficiency (WUE, kg ha^{-1} mm^{-1}) was calculated as

$$WUE = \frac{Y}{ET} \qquad (5)$$

Where Y is fruit yield (kg ha^{-1}) (Kundu et al., 2008).

Statistical analysis

Statistical analysis of *L. barbarbum* L. fruit yields was undertaken using SPSS software (Statistics Package for Social Science). The ANOVA was performed at $P < 0.05$ level of significance to determine if significant yield differences existed among different SMP treatments.

RESULTS AND DISCUSSION

Soil matric potential (SMP)

Figure 1 shows the daily changes in soil matric potential at 0.2 m depth, immediately under a drip emitter, for the different soil matric potential treatments at 8:00 h each day. Soil matric potential varied considerably over the experimental period, while its fluctuation cycle and range increased as the SMP target level decreased. As the SMP target level decreased, the interval time between

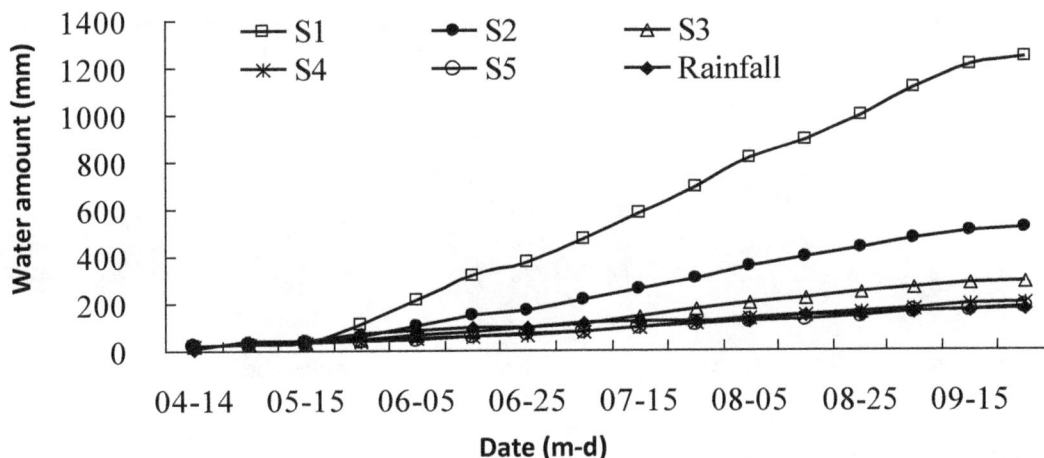

Figure 2. Accumulative amount of irrigation and rainfall during the growing season.

two irrigation events increased. The SMP target levels for the different treatments were controlled to within the experimental design range.

Irrigation and rainfall

Figure 2 shows the accumulative irrigation volume for the different SMP treatments and the accumulative rainfall during the growing season. The irrigation frequency differed between treatments. S1 received the most frequent irrigation and irrigation occurred almost twice a day, except for short periods when precipitation exceeded 5 mm. S5 had the lowest irrigation frequency and the average irrigation was once every five days. As the SMP control level increased, irrigation accumulative volume rose.

The volumes were 1263, 542, 319, 224 and 203mm (including irrigation for recovery and the winter water) for S1, S2, S3, S4 and S5, respectively. The largest irrigation volume (S1, 1263 mm) was 6.2 times greater than the lowest irrigation volume (S5, 203 mm). However, it should be noted that there was only S1 treatment of which the accumulative irrigation volume reached the average value of flooding-water at local plantation (He, 2004).

The cumulative precipitation depth for 2010 is shown in Figure 2. Total precipitation during the experiment was 219.5 mm, which was higher than the average annual precipitation. Overall precipitation distribution was more uniform from April to October except in July where precipitation was only 8.6 mm. There were 28 precipitation events, with five events of more than 20 mm, three events of 10 to 20 mm and five events of 5 to 10 mm. A major 29.2 mm rain event occurred at the beginning of growing season on 20 April when the wolfberry needed more water as it was at the new twig growth stage.

Soil water content (SWC)

Figure 3 shows the spatial distribution of soil moisture in a vertical transect for each SMP treatment before (30 March), during (20 June) and after (22 September) the experimental period. It can be seen that the soil water content transect was affected by the SMP target levels. As the SMP declined, the average soil moisture and the wetted perimeter decreased. Before the experimental period began (30 March), the vertical transect for soil water content was about 6 to 11% lower with a layered distribution. Soil moisture in the 30 to 60 cm horizon was nearly 11% higher than the soil moisture levels in the surface and deep soil horizons. Under drip irrigation, the results for the measurements taken on the 20 June and after the experimental period had ceased (22 September), showed that the vertical transect for average soil moisture was 17.5 and 17.1% for treatment S1 and 16.0 and 16.7% for treatment S2, respectively. The soil moisture content of the major root horizon (30 to 60 cm) retained about 17 to 18% of the water. In contrast, S3, S4 and S5 retained less than 15% of the soil water throughout the whole of the experimental period.

These results suggest that a soil water content of 15 to 18% in the root layer was most favorable for wolfberry growth (Science and Technology Department of National Forestry Bureau et al., 2008) in the Yinchuan Plain. The S1 and S2 treatments have been shown to meet crop water demand but treatments S3, S4 and S5 did not.

Soil salinity (ECe)

Figure 4 shows the changes in the extract electrical conductivity of the saturated soil (ECe) value at the difference soil depths for each SMP treatment before (30 March) and after (22 September) the experimental period. It can be seen that the soil electrical conductivity

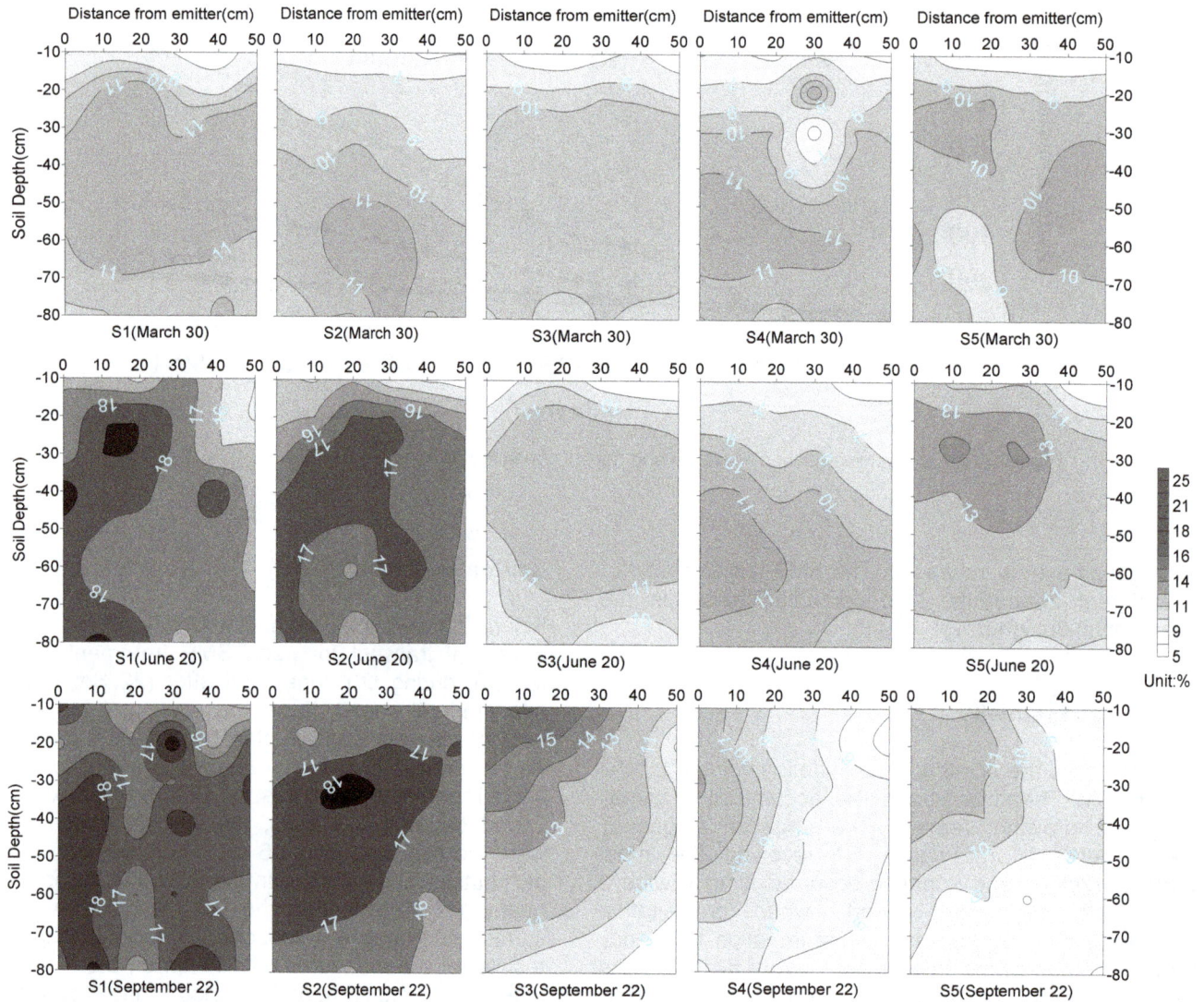

Figure 3. The vertical transect of spatial distribution of soil water content θ for different treatments.

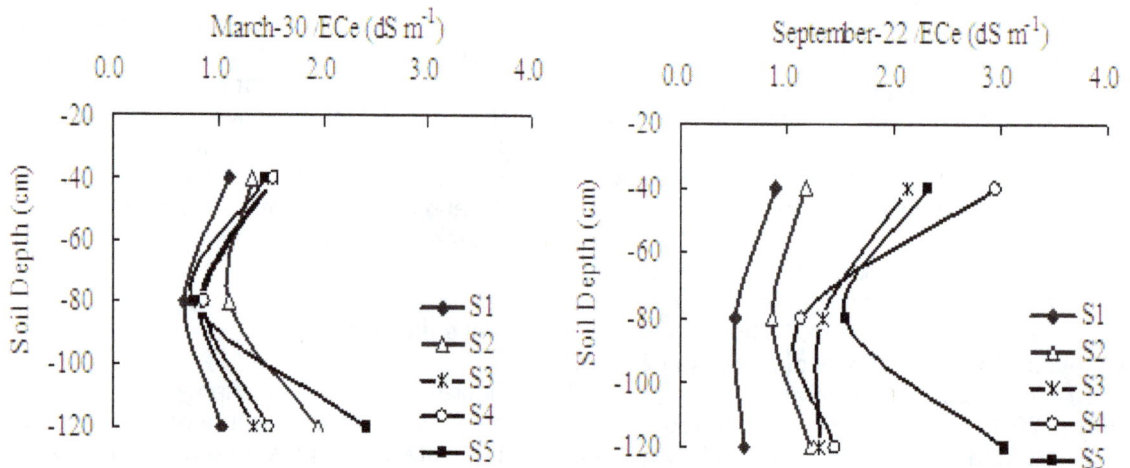

Figure 4. The changes of the soil salinity (ECe) for different treatments.

Table 3. Crop yield, evapotranspiration and agronomic properties.

Parameter	S1	S2	S3	S4	S5
I (mm)	1220	497.4	271.5	177.2	156.5
P (mm)	167	167	167	167	167
D (mm)	-96.65	-69.17	7.37	4.86	15.11
△S (mm)	25.33	32.88	7.54	-59.96	-12.38
ET (mm)	1265.0	562.3	438.3	409.0	351.0
Fruit yield (kg ha^{-1})	1278.3	1453.3	1257.4	912.6	993.6
Dry yield (kg ha^{-1})	307.7	361.5	328.2	229.4	247.0
Weight of per 100 fresh seeds (g)	69.31	66.96	63.65	66.76	61.52
Ratio fresh to dry weight	4.15	4.02	3.83	3.98	4.02
WUE for fruit (kg ha^{-1}.mm^{-1})	1.01	2.58	2.87	2.23	2.83

(ECe) was affected by the SMP target values. Before the experimental period (30 March), the average ECe was 0.93, 1.46, 1.22, 1.29 and 1.54 dS m^{-1} in the S1, S2, S3, S4 and S5 plots, respectively. The soil salinity in the middle horizon (40 to 80 cm) was lower than in the upper horizon (0 to 40 cm) and the deeper horizon (80 to 120 cm) on the vertical transect.

Under drip irrigation, after the experimental treatment period had ceased (22 September), the average ECe values increased as the target SMP decreased. Compared to the samples taken before the experimental treatments began (30 March), the soil transect of average ECe for S1 and S2 declined by 27.9 and 24.8%, respectively. This was in contrast to treatments S3, S4 and S5 where ECe increased 30.9, 43.8 and 50.0%, respectively.

When the target SMP was above -30 kPa, the soil salts in S1 and S2 were leached away due to the higher irrigation accumulative volume, especially in the middle horizon (40 - 80 cm) and the deeper horizon (80 - 120 cm) where the ECe reduction with the S2 treatment were 0.24 and 0.70 dS m^{-1}, respectively, significantly greater than occurred with the S1 treatment. Statistical analysis of the experimental treatments showed that the change in ECe values was significantly affected ($P < 0.05$) by the target SMP in the middle horizon (40 to 80 cm). The S2 treatment (-20 kPa) showed the greatest degree of saline leaching. When the target SMP was below -30 kPa, water to leach the salts out of the soil became limiting so in plots S3, S4 and S5, salt began to accumulate in the soil.

Wolfberry evapotranspiration (ET)

Table 3 shows changes in soil water storage (ΔS), drainage below crop root zone (D), and ET values for the five SMP treatments during the plant growing period (15 April to 22 September). Cumulative ET values declined as SMP decreased. This showed that wolfberry ET was clearly affected by the SMP threshold. The highest ET

value (S1 treatment) was 914 mm more than the lowest value (S5 treatment), while soil subjected to the S1 treatment held the highest volume of water and had the greatest depth percolation among the five treatments. When the target SMP was above -30 kPa, the soil in the S1 and S2 treatment plots was waterlogged, which reduced soil disturbance and leached fertilizer out of the root zone. When the target SMP was below -30 kPa, the wolfberry growing in the S4 and S5 plots suffered serious water stress and the crop ET declined significantly. When the target SMP was controlled at -30 kPa (S3), depth percolation and soil water storage improved.

In order to clarify the effects of the target SMP on crop ET, regression analysis was carried out (Figure 5). The regression model was expressed as follows:

$$ET = 0.9886\psi^2 + 79.128\psi + 1891.6$$
$$(R^2 = 0.9318) \qquad\qquad (6)$$

Wolfberry yield

Statistical analysis of the experimental treatments showed that both the fruit yield and the dry yield as significantly affected ($P < 0.05$) by the target SMP. The S2 plots (-20 kPa) produced the highest yield with a fruit yield of 1453.3 kg ha^{-1} and a dry yield of 361.5 kg ha^{-1}. The lowest fruit yield (912.6 kg ha^{-1}) and dry yield (229.4 kg ha^{-1}) were produced by the S4 plots. LSD comparisons between groups showed that treatment S2 had a significantly higher yield than treatments S4 and S5. S1 and S3 also had significantly higher yields than S4. The dry yield for S1, S2 and S3 were higher than the average yield at the local plantation. He (2004) reported the dry yield geted 300 to 450 kg ha^{-1} after the plant transplanted 2 to 3 years, and S4 and S5 dry yields were lower due to a soil water deficit.

Each treatment produced weight per 100 fresh seeds values that were higher than the 47.6 to 58.6 g achieved

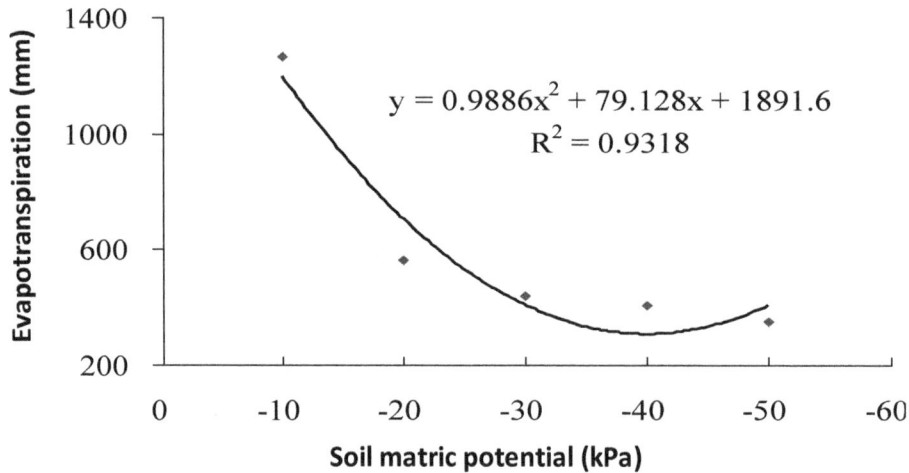

Figure 5. Relation of the crop ET to the target SMP for the five treatments.

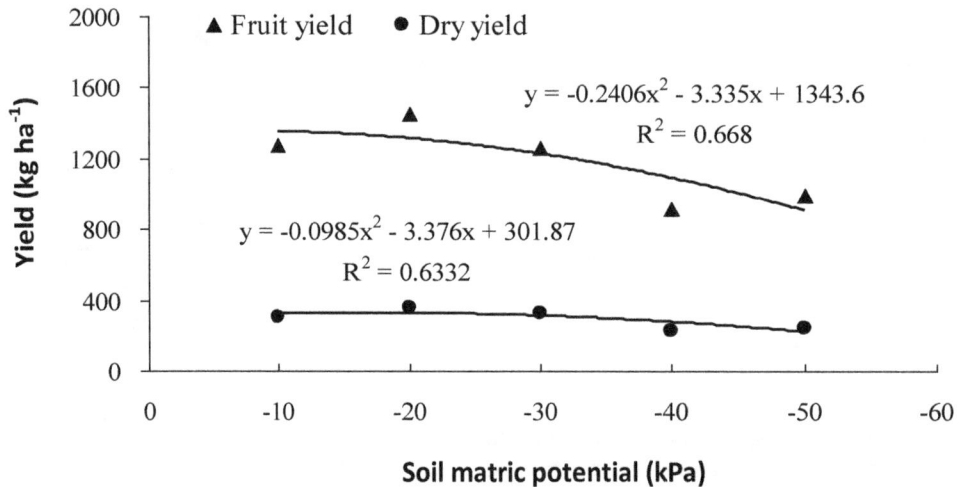

Figure 6. Relations of the crop yield to the target SMP for the five treatments.

at the local plantation and the ratio of fresh to dry weight was also lower than the 4.37 achieved by the local plantation (He, 2004). The greatest weights per 100 fresh seeds (69.31 g) and ratios of fresh to dry weigh (4.15) were achieved with the S1 treatment because there was an excess of soil water. The lowest ratio of fresh to dry weight (3.83) occurred with the S3 treatment because the wolfberry seeds contained more dry matter than the other treatments.

A regression model was constructed using correlation analysis between the target SMP for the five treatments and *L. barbarbum* L. yield (Figure 6).

$$Yield\,(fruit) = -0.2406\psi^2 - 3.335\psi + 1343.6$$
$$(R^2 = 0.668) \tag{7}$$

$$Yield\,(dry) = -0.0985\psi^2 - 3.376\psi + 301.87$$
$$(R^2 = 0.6332) \tag{8}$$

Water use efficiency (WUE)

Water use efficiency (WUE) is the relationship between yield and ET and was first calculated using tomato yields divided by ET (Kang et al., 2004a; Wang et al., 2007). This can be represented as an incremental gain in dry matter per unit water taken up and transpired by the plant (Draycott, 2006; Hassanli et al., 2010). The results from this study (Table 3) showed that the WUE for the five SMP treatments were ordered: S3 > S5 > S2 > S4 > S1. The WUE value (-30 kPa) was highest for S3 and lowest for S1 (-10 kPa), which indicated that the water was used most effectively by plants subjected to the S3 treatment. These results suggest that WUE could be a good criterion for evaluating the effectiveness of irrigation.

SUMMARY AND CONCLUSIONS

Wolfberry irrigation volume, ECe, ET, yield and WUE were affected by soil matric potential (SMP). Irrigation volume and ET decreased as the target SMP value declined. During the growing season, the accumulative water volume for irrigation was 1263, 542, 319, 224 and 203 mm in 2010 for treatments S1, S2, S3, S4 and S5, respectively, but there was only S1 treatment value in the accumulative irrigation volume data as the flooding-water at the local plantation.

Drip irrigation and a target SMP above -30 kPa (S1 and S2 treatments) should meet the crop water demand. However, when the soil moisture is over 16%, soil disturbance was reduced and fertilizer leached out of the root zone as the soil water was in excess. The soil water content for S3, S4 and S5 was less than 15% and produced visible water stress in the crop. When the target SMP for S1 and S2 was above -30 kPa, the soil salinity was leached due to the higher irrigation accumulative amount, but the soil salinity was accumulated in the plots of the target SMP for S3,S4 and S5 because the water of the soil salinity leached was shortage. Statistical analysis of the change ECe values showed the significant differences between the different SMP treatments in the middle layer 40 to 80 cm, the plots S2 (-20 kPa) had the better effect of the soil salinity leached. Cumulative ET values decreased as the target SMP declined, the highest ET (1265 mm) was with the S1 treatment and was 72.3% larger than the lowest value (S5 treatment). Statistical analysis of the experimental treatments showed that the target SMP level had a significant impact on the crop yield. The highest yield was achieved in the S2 plots. The lowest ratio of fresh to dry weight was found in the S3 plots, which also produced the highest quality seeds. The WUE for the five SMP treatments were ordered as S3 > S5 > S2 > S4 > S1. In general, the SMP at 0.2 m depth immediately under drip emitter can be used as an index for scheduling drip irrigation for wolfberry in the Yinchuan Plain, Chain. Yields and WUE were greatest when the SMP threshold was controlled in the range from -20 to -30 kPa.

ACKNOWLEDGEMENTS

This study was supported by the National Key Technology R&D Program of China (Grant No. 2009BAC55B07), the National Science Foundation for Young Scientists of China (Grant No. 51009126), and the Knowledge Innovation Program of the Chinese Academy of Sciences (Grant No. KSCX2-YW-N-080).

REFERENCES

Azevedo PV, Silva BB, Silva VPR (2003). Water requirements of irrigated mango orchards in northeast Brazil. Agric. Water Manage. 58:241-254.

Bai SN (1999). Study on Wolfberry (Lycium barbarbum L.) (in Chinese). Ningxia people ' s publishing house, Yinchuan.

Clark GA, Albregts EE, Stanley CD, Smajstrla AG, Zazueta FS (1996). Water requirements and crop coefficients of drip-irrigated strawberry plants. T ASABE. 39(3):905-913.

Draycott AP (2006). Sugar Beet. Oxford: Blackwell Publishing. P. 474.

Feng LP, Kang YH, Wang, GD, Wan SQ (2004). Effect of soil matric potential on radish nutrient uptake. Agricultural Research in the Arid Areas (in Chinese). 22(2):8-14.

Forestry Bureau of Ningxia Hui Autonomous Region 2010. Lycium barbarbum L. plantation over 4.67×104 hm² and creation output value of 2 billion Yuan by ecological effect. http://news.Cntv.cn/20101209/111002.shtml.

Hassanli AM, Ahmadirad S, Beecham S (2010). Evaluation of the influence of irrigation methods and water quality on sugar beet yield and water use effiviency. Agric. Water Manage. 97:357-362.

He L (2004). Lycium L. and Glycyrrhiza L. (in Chinese). Scientific Technical Documents Publishing House, Beijing, pp. 1-84.

Huang M, Gallichand J, Zhang L (2004). Water yield relationships and optimal water management for winter wheat in the Loess Plateau of China. Irrig. Sci. 23:47-54.

Jiao YP, Kang YH, Wan SQ, Liu W, Dong F (2007). Effect of soil matric potential on waxy corn growth and irrigation water use efficiency under mulch drip irrigation in saline soils of arid areas. Agricultural Research in the Arid Areas (in Chinese). 06:144-151.

Kang YH (2004a). Applied Method for Drip Irrigation Scheduling. Water Saving Irrigation (in Chinese). 03:11-12.

Kang YH, Wang FX, Liu HJ (2004b). Potato evapotranspiration and yield under different drip irrigation regimes. Irrigation Sci. 23:133-143.

Kang YH, Wan SQ, Jiao YP, Tan JL, Sun ZQ (2007). Saline soil salinity and water managemengt with tensiometer under drip irrigation. Proceeding of annual Symposium of Chinese Society of Agricultural Engineering (CSAE, Beijing, in Chinese), pp. 1-7.

Keller J, Bliesner RD (1990). Sprinkle and Trickle Irrigation, Van Nostrand Reinhold. New York.

Kundu M, Chakraborty PK, Mukherjee A, Sarkar S (2008). Influence of irrigation frequencies and phosphate fertilization on actual evapotranspiration rate, yield and water use pattern of rajmash (Phaseolus vulgaris L.). Agric. Water Manage. 95:383-390.

Lei ZD, Yang SX, Xie SZ (1988). Soil Water Dynamics (in Chinese). Tsinghua University Press, Beijing. P. 33.

Phene CJ, Allee CP, Pierro JD (1989). Soil matric potential sensor measurements in real-time irrigation scheduling. Agric. Water Manage. 16:173-185.

Science and Technology Department of National Forestry Bureau, Chinese Academy of Forestry Sciences (2008). Lycium L. of cultivation practical technology (in Chinese). Chinese forestry publishing house, Beijing.

Shock CC, Feibert EBC, Saunders LD (2000). Irrigation criteria for drip-irrigated onions. Hortsci. 35(1):63-66.

Wan SQ, Kang YH, Wang D, Liu SP, Feng LP (2007a). Effect of drip irrigation with saline water on tomato (Lycopersicon esculentum Mill.) yield and water use in semi-humid area. Agric. Water Manage. 90:63-74.

Wan SQ, Kang YH, Wang D, Liu SP, Feng LP (2007b). Effects of saline water on cucumber yields and irrigation water use efficiency under drip irrigation. Transactions of the CSAE (in Chinese). 23(3):30-35.

Wang FX, Kang YH, Liu SP, Hou XY (2007). Effects of soil matric potential on potato growth under drip irrigation in the North China Plain. Agric. Water Manage. 88:34-42.

Wilson CR, Pemberton BM, Ransom LM (2001). The effect of irrigation strategies during tuber initiation on marketable yield and development of common scab disease of potato in Russet Baurbank in Tasmania. Potato Res. 44:243-251.

Zotarelli L, Scholberg JM, Dukes MD, Munoz-Carpena R, Icerman J (2009). Tomato yield, biomass accumulation, root distribution and irrigation water use efficiency on a sandy soil, as affected by nitrogen rate and irrigation scheduling. Agric. Water Manage. 96:23-34.

24

Combined effect of tillage system, supplemental irrigation and genotype on bread wheat yield and water use in the dry Mediterranean region

Mohammed Karrou

ICARDA, P. O. Box 6299, Rabat – Instituts, Morocco.

Approaches that can help reduce future water shortages and increase food production in the Mediterranean region include the capture of more rainwater and better use of scarce irrigation water. The objective of this study is to evaluate the additive effect of combining zero tillage, supplemental irrigation and adapted varieties of bread wheat on yield and water use. For this purpose, a trial was conducted at the International Center for Agricultural Research in the Dry Areas' (ICARDA) research station in Tel Hadya, Aleppo, Syria in 2008/2009 and 2011/2012 under two tillage systems: zero tillage (ZT) vs. conventional tillage (CT); two water regimes: rainfed (RD) vs supplemental irrigation (SI); and five genotypes. Results showed that in 2008/2009, grain yield increased on average, from 4515 kg ha^{-1}, under CT in the rainfed regime, to 5929 kg ha^{-1} when ZT and SI were combined. ZT and SI increased respectively, the yield by 548 and 729 kg ha^{-1} as compared to CT under rainfed condition. The combination of ZT and SI increased water use by 110 mm as compared to CT under rainfed condition. In year 2011/2012, ZT tended to have no effect on yield and a negative influence on water use. Irrigation increased yields from 5055 to 5927 kg ha^{-1} under CT tillage and from 4996 to 5448 kg ha^{-1} under zero tillage. The genotypes Cham 6, Cham 8 and Shuha were the most responsive to SI and ZT in year 1. In year 2, Shuha and Cham 10 tended to be in general, more performing. The preliminary results showed that in dry year, the combination of the two technologies, zero tillage and supplemental irrigation, increased water use and yield. In wetter year, the effect of SI was still positive; but that of ZT was not significant on yield and negative on water use. Further studies are needed, for a longer period, to consolidate the obtained results using more diversified genetic material.

Key words: Wheat, zero tillage, irrigation, genotype, yield, evapotranspiration, productivity.

INTRODUCTION

Increasing scarcity of water due to more frequent droughts and growing demand for resources for industrial, tourism and domestic uses (due to population growth) will result in the future in less water for agricultural production. The crops that will suffer most from this situation are cereals and this is for two reasons: First, because when there is drought this crop are the least to benefit from irrigation water. Second, because these species are mainly grown in arid zones where the possibilities of irrigation are very low. Consequently, the reduction of cereals production will affect seriously the poor that live in dry areas, because this product is the main human nutritional food.

Despite the significant increase in wheat yields since the beginning of the green revolution, there are still management and environmental causes underlying water and environmental problems (FAO, 1996). Severe drought and climate change that the Mediterranean basin has been facing (Giorgi and Lionello, 2008) imposed the development of new strategies that will allow less water loss, a better use of scarce resource and the preservation of natural resources and hence the sustainable increase of land and water productivity. To reach this objective there is a need, not only for a better management of irrigation, but also for the use of improved cultural practices and cultivars that are adapted to drought and global warming conditions.

Plant breeders have always considered early flowering time as one of the most important traits to select adapted wheat varieties for water limited environments (Passioura, 2006). With climate change, this criterion will be more emphasized in breeding programs as the length of the growing period will shrink more and earlier cultivars are needed to escape terminal drought and heat stress. However, plants that flower early may achieve large harvest indices but do not produce enough biomass to set a large enough number of seeds to generate a good yield (Fischer, 1979). Plants that flower too late will have, usually, high seeds abortion because of heat stress and too little water left for post-flowering photosynthesis and remobilization of carbohydrates from the vegetative organs to the grains (Passioura, 2006).

Consequently, ensuring a balance between the source (vegetative biomass production) and sink (seeds production) under early flowering is an important strategy to increase and stabilize yields in rainfed areas. Both breeding and field management can play an important role in capturing more water and improving its efficient use. So, in addition to the selection of varieties that have the characteristics described above, other technologies such as early planting facilitated by zero tillage technique can help the plants take advantage from the early rains and escape terminal drought and heat. Moreover, zero tillage can also reduce rainwater losses by evaporation and increase transpiration early in the season and hence improve yield and water productivity of wheat (Mrabet, 2000a, b; Cantero-Martinez et al., 2007).

In this study, a no-till drill was used to plant directly the seeds without any previous cultivation, but a very small amount of residues was retained on the soil to simulate the farmer technique that consists of using all the straw as forage in summer. Therefore, the beneficial effects the authors seek from this zero tillage is the early planting to take advantage from early rains and the conservation of water because the soil is not disturbed. A very little effect of the residues is expected. In fact, in the southern part of the Mediterranean basin, the no-till is practiced with little or no straw left on the soil surface and this is the technique of conservation tillage or zero tillage that was tested. The application of supplemental irrigation is another option that can help to compensate the rainfall deficit late in spring. This technique can reduce the seeds abortion and increase soil moisture during the post-anthesis

period and seed size. Research conducted in dry areas (Oweis and Hachum, 2006; Karrou and Boutfirass, 2007) showed that supplemental irrigation is a technology that can improve significantly, water productivity and save the resources without reducing land productivity.

Oweis et al. (1999) demonstrated that water productivity under supplemental irrigation after heading was as high as 2.5 kg of wheat grain per cubic meter of water, compared to 500 g under rainfed conditions and 1 kg under full irrigation. To take advantage from more water conserved under zero tillage due to non soil disturbance (no-tillage) and from supplemental irrigation, it is important to use adapted varieties to these systems. Laaroussi (1991) and Boutfirass (1997) showed in semi arid rainfed areas of Morocco, genotypic differences in yield and water use efficiency under rainfed and SI conditions. Haul and Cholick (1989) and Ciha (1982) found that the ranking of wheat cultivars across tillage systems, including ZT, changed for grain yield.

Further to the positive effects of supplemental irrigation at critical stages, zero tillage and drought tolerant varieties on sustainability of wheat productivity in rainfed areas, information concerning the beneficial impacts of combining these two techniques and the use of more adapted improved varieties is still limited. The objective of this study is to evaluate the effects of this combination on durum wheat grain yield, water use and water productivity.

MATERIALS AND METHODS

The experiment was carried out during 2008/2009 and 2011/2012 seasons at ICARDA's main research station, Tel Hadya, Aleppo, in Northern Syria (36°01'N.36° 56'E; elevation 284 m asl). Mean annual rainfall in the area was 320 mm with considerable year to year variation ranging from 200 mm to over 500 mm. The soil at Tel Hadya station is generally deep, over 1 m and fine textured (Ryan et al., 1997) and is classified as fine clay (montmorillonitic, thermic Calcixerollic Xerochrept). The soil has good structure and is well drained, with a basic infiltration rate of about 11 mm/h. At field capacity and at the permanent wilting point, mean soil moisture content in the top 100 cm of the soil is about 48 and 24% by volume, respectively.

The factors studied were tillage system (Conventional tillage vs Zero tillage), water regime (Rainfed vs supplemental irrigation) and the genotype (5 genotypes of bread wheat). The experimental design used was a split-split plot with tillage as the main plot, water regime as the sub-plot and genotype as the split-split plot with 3 replications. In the zero-till plot, seeds were sown directly with a no-till drill without any previous soil cultivation. For the conventional tillage plot, the soil was plowed twice with an offset disk and this was followed by a roll-packing operation. In the case of zero tillage, a very small amount of residues were kept on the soil surface to simulate the conditions of the farmer where all the straw after harvest is used as forage. Rainfed treatment received only rainfall. However, supplemental irrigation plot received, in addition to rainfall, amounts of irrigation water of 35 mm at boot stage and 35 mm at kernel milky stage. The genotypes tested were Cham 6 (V1), Cham 8 (V2), Shuha (V3), Cham 10 (V4) and Raaid-3 (V5). The experiment was planted on November 17 in 2008 and November27 in 2011 and the seeding rate was 140 kg ha^{-1}.

Table 1. Mean monthly precipitation, September to May at Tel Hadya, northern Syria, during 2008/2009-2011/2012).

Year	Mean monthly rainfall (mm)									
	Sep	Oct	Nov	Dec	Jan	Feb	Mar	Apr	May	Total
2009/2010	21	15	16	36	27	72	32	20	3	241
2011/2012	1.1	27	93	65	117	4	27	4	5	343
Long-term mean (1980/2012)	4	20	44	56	63	54	46	30	16	333

Phosphorus (P) and nitrogen (N) were applied at planting as Superphosphate (45%) and urea (46%) at rates of 80 Kg P ha^{-1} and 60 Kg N ha^{-1} respectively for P and N. At stem elongation, 30 Kg N ha^{-1} were added as urea. The measurements taken were grain yield, actual evapotranspiration (ETa) and grain water productivity (WP) at harvest. Water productivity was calculated as the ratio of grain yield to actual evapotranspiration. Soil moisture was measured at planting and harvest using a neutron probe device. Measurements were taken at 0-15, 15-30, 30-45, 45-60, 60-75, 75-90, 90-105, 105-120, 120-135 and 135-150 cm. ETa was calculated using the water balance equation. All data were analyzed using SAS (1997) statistical software. The analysis of variance (ANOVA) was performed to examine the various treatment differences and interactions. LSD test at 5% and 1% probability levels was applied to compare the differences among the treatment means. Data on rainfall for the two seasons are presented in Table 1. The total amounts received were 241 mm in 2008/2009 and 343 mm in 2011/2012. In year 2, rainfall was very high and concentrated mostly between November (planting period) and January with 93 mm in November, 65 mm in December and 117 mm in January. In March, rainfall was 27 mm; but in February and April there was only around 4 mm per year. In year 1, rainfall was less during November (16 mm), December (36 mm) and January (27 mm). However, in February and April, the quantities were 72 mm and 20 mm, respectively.

RESULTS AND DISCUSSION

Grain yield per hectare obtained under different tillage systems, water regimes and genotypes is presented in Table 2. Due to the difference in rainfall pattern between years, the interaction year x tillage system for the different parameters measured was significant. Consequently, the data were discussed separately by year. In year 1, the analysis of variance showed that the effects of supplemental irrigation, the genotype, the interactions zero tillage x genotype and supplemental irrigation x genotype were statistically highly significant. However, the effects of ZT and the interactions ZT x SI and ZT x SI x G were not significant; although zero tillage increased yield by 600 kg ha^{-1} as compared to the conventional tillage. During this year, supplemental irrigation increased productivity from 4789 to 5586 kg ha^{-1}, on average; in year 2, only water regime and genotype effects were significant.

The yield was increased on average, from 5025 to 5687 kg ha^{-1} by the application of irrigation water. The positive effect of ZT observed in year 1 was also demonstrated in Iran by Hemmat and Eskandari (2004a, b, 2006); in Morocco by Mrabet (2000a, b, 2008) and in Tunisia by Vadon et al. (2006). The higher yield in zero

tillage treatment than in conventional tillage in the first year in this study was probably due to the uptake of more water (30 mm) under the former system. This result confirmed the finding of Passioura (2006) who demonstrated that capturing 30 mm of rainwater could be translated into an increased yield of about 1 t/ha. In year 2, there was no effect of tillage system on wheat and some scientists (Pala, 2000; Thomas et al., 2007) showed even negative response. This negative effect was attributed mainly to disease infestation (Thomas et al., 2003) which was not observed in our trials and to too high early biomass production (Kumudini et al., 2008) that increased more completion for water later and reduced yield. It seems that the positive effect of zero tillage shown by many scientists in the southern Mediterranean region was offset in our study by the high precipitation during most of the 2011/2012 growing season. The positive effect of supplemental irrigation during the two seasons confirmed the results of Oweis and Hachum (2006) and Karrou and Oweis (2012).

In year 1, all the genotypes responded positively, but differently, to ZT and SI, except Raaid-3 for which the difference was not statistically significant. The finding on tillage confirmed those of Hernandez et al. (2004) who showed that the grain yields of the genotypes of wheat were higher by 7% under zero tillage than under conventional tillage. The genotypes Cham 6 and Cham 8 tended to respond more to the conservation tillage and to supplemental irrigation. Moreover, for these 2 genotypes, the beneficial effect of combining these two techniques as compared to their individual influences was noticed. The genotypes Cham 10 and Raaid-3 were the least to be affected by this combination. These results confirmed the genotypic variation observed by Laaroussi (1991) and Boutfirass (1997) under supplemental irrigation; and Haul and Cholick (1989) and Ciha (1982) under conservation tillage. In year 2, Shuha and Cham 10, tended to yield more, in average, than the others.

Table 3 shows that in year 1, there was a significant effect of tillage system, SI, genotype and ZT x SI on the actual evapotranspiration (ETa). In average, zero tillage and supplemental irrigation had, respectively, higher ETa than conventional and rainfed treatments. Supplemental irrigation increased ETa from 342 to 403 mm under conventional tillage and from 372 to 452 mm under conservation tillage. These increase amounted to 61 and 80 mm, respectively. Zero tillage increased water use from 342 to 372 mm under rainfed conditions and

Table 2. Effect of tillage system (TS), water regime (WR) and genotype on grain yield (kg ha^{-1}) of bread wheat in Tel Hadya, Syria during the cropping season 2008/2009 and 2011/2012.

Parameter	Genotype	Conventional tillage		Mean	Zero tillage		Mean
		Rd [*]	SI [*]		Rd [*]	SI [*]	
Year 1	Cham 6	4307	5020	4664	4920	6270	5595
	Cham 8	4810	5710	5260	5570	6950	6260
	Shuha	4720	5800	5260	5730	6520	6125
	Cham 10	4650	5380	5015	5610	5980	5795
	Raaid 3	4090	4320	4005	3490	3920	3705
	Mean	4515	5244	4841	5063	5929	5496
	Statistical analysis	**CV** = 11.2					
		ANOVA:TS effect: not significant at α = 5%; WR effect: highly significant at α = 5% with LSD = 420; Genotype effect: highly significant with LSD = 375; TS x WR effect: not significant; TS x Genotype effect: highly; significant WR x Genotype effect: significant;Interaction TS x WRI x Genotype effect: not significant					
Year 2	Cham 6	5117	5894	5506	4697	5417	5057
	Cham 8	4708	6286	5497	5376	5355	5366
	Shuha	5484	6164	5824	5565	5421	5493
	Cham 10	5116	6146	5631	5099	5901	5500
	Raaid 3	4849	5144	4997	4241	5144	4693
	Mean	5055	5927	5091	4996	5448	5222
	Statistical analysis	**CV** = 9.8					
		ANOVA: TS effect: not significant at α = 5%; WR effect: significant at α = 5% with LSD = 270; Genotype effect: Highly significant at α = 1% with LSD = 430; TS x WR effect: Not significant; TS x Genotype effect: Not significant; WR x Genotype effect: Not significant; Interaction TS x WR x Genotype effect: Not significant					

[*] Rd and SI are Rainfed and Supplemental irrigation, respectively.

from 403 to 452 mm in irrigated plots. Our results confirmed those of Power et al. (1986) who reported greater water storage and grain yield under no till than under conventional tillage. Merrill et al. (1996) attributed the increase in soil water use to greater root growth. The application of zero tillage under rainfed conditions and supplemental irrigation in conventional tillage system increased, respectively, water use by 30 and 61 mm as compared to the conventional tillage under rainfed (without irrigation) situation. More importantly, results showed the more significant effect of combining ZT and SI on water use. Literally, evapotranspiration increase due to this combination was 110 mm, so, there was an additive effect of the improved technologies of tillage and supplemental irrigation.

In year 2 (Table 3), the effects of tillage, supplemental irrigation, genotype, tillage x SI and tillage x SI x genotype on ETa were significant. The results obtained were different than the ones of year 1 because ETa under conventional tillage was higher under zero tillage in year 2. They also contrast those of Lopez-Bellido et al. (2007a, b) who found that under Mediterranean conditions, no-till was not more efficient than conventional tillage in soil water accumulation and

productivity. However, the effect of supplemental irrigation remained positive. The application of irrigation increased, in average, ETa from 356 to 461 mm. Conventional tillage, as compared to zero tillage, increased water use from 336 to 377 mm under rainfed conditions and from 415 to 508 mm in irrigated plots; the difference was 41 and 93 mm respectively. Supplemental irrigation increased ETa from 377 to 508 mm under conventional tillage and from 336 to 415 mm under zero tillage and this corresponded to 131 and 79 mm, respectively.

The difference between year 1 and year 2 can be explained by the difference in rainfall distribution and amount. Rainfall in 2011/2012 was around 104 mm more and well distributed during the period of November-March. The soil under plowing was loose and allowed probably more water infiltration and storage in the soil and greater root growth, hence the increased amount of water used. Kirkegaard et al. (1994) found that direct drilling reduced the total root length in the soil profile at anthesis by 40% but there was no effect of stubble retention; and Martinez et al. (2008) demonstrated that the penetration resistance was higher under no-till as compared to conventional tillage.

Table 3. Effect of tillage system (TS), water regime (WR) and genotype on actual evapotranspiration (mm) of bread wheat in Tel Hadya, Syria during the cropping season 2008/2009 and 201120/12.

Parameter	Genotype	Conventional tillage		Mean	Zero tillage		Mean
		Rd[(*)]	SI[(*)]		Rd[(*)]	SI[(*)]	
Year 1	Cham 6	306	399	353	335	428	382
	Cham 8	374	398	386	356	407	382
	Shuha	330	407	369	358	479	419
	Cham 10	339	395	367	406	523	465
	Raaid 3	359	416	388	404	419	412
	Mean	342	403	373	372	452	412
		CV = 9.8					
	Statistical analysis	**ANOVA:** TS effect: not significant at α = 5%; WR effect: highly significant at α = 5% with LSD = 20; Genotype effect: significant with LSD = 32; TS x WR effect: significant; TS x Genotype effect: highly significant; WR x Genotype effect: not significant; Interaction TS x WR x Genotype effect: not significant					
Year 2	Cham 6	353	515	434	325	435	380
	Cham 8	350	498	424	341	414	378
	Shuha	421	511	466	328	417	373
	Cham 10	340	511	426	351	400	376
	Raaid 3	419	503	461	337	408	373
	Mean	377	508	442	336	415	376
		CV = 4.00					
	Statistical analysis	**ANOVA:** TS effect: highly significant at α = 1% with LSD = 8; WR effect: highly significant at α = 1% with LSD = 8; Genotype effect: with LSD = 3; TS x WR effect: highly significant at α = 1%; TS x Genotype effect: not significant; WR x Genotype effect: not significant; Interaction TS x WR x Genotype effect: highly significant at α = 1%					

[(*)] Rd and SI are Rainfed and Supplemental irrigation, respectively.

The genotypes responded differently to zero tillage in year 1 and to zero tillage and supplemental irrigation in year 2. In year 1, the genotype Cham 10 used the highest amount of water under ZT when compared to CT and then it was followed by the genotype Shuha. ETa of the genotypes Cham 6 and Raaid-3 were less affected by the variation of the tillage system; however, that of Cham 8 was not sensitive. In year 2, Cham 6, Cham 8 and Cham 10, responded more to the application of water under conventional tillage than the others. Under zero tillage, Cham 6 responded more and Cham 10 less than the other genotypes to supplemental irrigation. Zero tillage had a negative effect on ETa for all genotypes, but the degree of response differed from one genotype to another and with water regime. Under rainfed conditions, the effect of tillage system variation on ETa of Cham 6, Cham 8 and Cham 10 tended to be very small; while in the case of Shuha and Raaid, the reduction due to the application of zero tillage was statistically significant. Under SI, zero tillage was the most depressive for Cham 10 and less for Cham 6 and 8.

Water productivity was affected significantly only by the genotype in year 1 and by tillage system, water regime and the interaction tillage system x water regime x genotype in year 2 (Table 4). In year 1, although the interaction tillage system x genotype was not significant, the genotypes Cham 6, Cham 8 and Shuha tended to have the highest WP under ZT; while Cham 10 and Raaid-3 tended to use water more efficiently under the conventional tillage. The genotypes that used more water under zero tillage had their WP reduced. Positive effect of zero tillage was shown by Cantero-Martinez et al. (2007), Hemmat and Eskandari (2006) and Bouzza (1990). In year 2, rainfed and zero tillage treatments gave the highest water productivity and Raaid-3 was in average, the least efficient in water use. However, the genotypes responded differently to tillage system and water regime variation. Cham 6 and Cham 10, under conventional tillage, and Cham 6, Cham 8 and Shuha under conservation tillage, responded negatively to the application of irrigation water. For the other genotypes, the difference between the two water regimes was not significant.

Conclusion

From this experiment we can conclude that the

Table 4. Effect of tillage system (TS), water regime (WR) and genotype on water productivity (Kg/m3/ha) of bread wheat in Tel Hadya, Syria during the cropping season 2008/2009 and 2011/2012.

Parameter	Genotype	Conventional tillage		Mean	Zero tillage		Mean
		Rd[*]	SI[*]		Rd[*]	SI[*]	
Year 1	Cham 6	14.0	12.6	13.3	14.9	14.6	14.8
	Cham 8	12.9	14.3	13.6	15.9	17.1	16.5
	Shuha	14.3	14.4	14.4	16.2	13.9	15.1
	Cham 10	13.6	13.6	13.6	14.0	11.5	12.8
	Raaid 3	11.4	10.4	10.9	8.6	9.4	9.0
	Mean	13.2	13.0	13.2	13.6	13.1	13.6
	Statistical analysis	CV = 13.3 **ANOVA:** TS effect: not significant at α = 5%; SI effect: not significant at α = 5% Genotype effect: highly significant with LSD = 1.4; TS x WR effect: not significant; TS x Genotype effect: not significant; WR x Genotype effect: not significant; Interaction TS x WR x Genotype effect: not significant					
Year 2	Cham 6	14.5	11.5	13.0	14.4	12.5	13.5
	Cham 8	13.5	12.6	13.1	15.8	13.0	14.4
	Shuha	13.0	12.0	12.5	17.0	13.0	15.0
	Cham 10	15.0	12.0	13.5	14.5	14.8	14.7
	Raaid 3	11.6	10.2	10.9	12.6	12.7	12.7
	Mean	13.5	11.7	12.6	14.9	13.2	14.0
	Statistical analysis	CV = 10.1 **ANOVA:** ZT effect: significant at α = 5% with LSD = 0.7; SI effect: significant at α = 5% with LSD = 0.7; Genotype effect: highly significant with LSD = 1.1; ZT x SI effect: not significant; ZT x Genotype effect: not significant; SI x Genotype effect: not significant; Interaction ZT x SI x Genotype effect: significant at α = 5%					

[*] Rd and SI are Rainfed and Supplemental irrigation, respectively.

combination of zero tillage and supplemental irrigation late in the season can ensure an additive positive effect of the two technologies on total water uptake and land productivity in dry years. However, in wetter years, the beneficial effect of zero tillage disappears; but that of supplemental irrigation remains because there was rainfall deficit in spring during the formation of the grain. The genotype effect on yield and water use was in general significant; but the interaction with the other factors as tillage system and water regime, was not observed. However for water productivity, the genotypes responded differently to tillage system and water regime variation. Cham 6 and Cham 10 under CT, and Cham 6, Cham 8 and Shuha under conservation tillage, responded negatively to the application of irrigation water. For the other genotypes, the difference between the two water regimes was not significant. Further studies are needed, for a longer period, to consolidate the obtained results using more diversified genetic materials.

REFERENCES

Boutfirass M (1997). Economie et efficience d'utilisation de l'eau en agriculture par l'irrigation d'appoint des céréales. Mémoire d'ingénieur en chef. INRA-Settat, Maroc.

Bouzza A (1990). Water conservation in wheat rotations under several management and tillage systems in semi-arid areas. PhD Dissertation, University of Nebraska, Lincoln, USA.

Cantero-Martinez C, Angas P, Lampurtanes J (2007). Long-term yield and water use efficiency under various tillage systems in Mediterranean rainfed conditions. Ann. Appl. Biol. 150(3):293-305.

Ciha AJ (1982). Yield and yield components of four spring wheat cultivars grown under three tillage systems. Agron. J., 74:317-320.

Fischer RA (1979). Growth and water limitation to dryland wheat yield in Australia: A physiological framework. J. Aust. Inst. Agric. Sci. 45:83-94.

FAO (1996). Food for all. In 'World food summit; 13-17 November 1996, Rome, Italy. FAO Corporate document repository. (http://www.fao/docrep/x0262E/x0262E00.htm).

Haul EF, Cholick FA (1989). Cultivar x tillage interaction of hard red spring wheat cultivars. Agron. J. 81:789-792.

Giorgi F, Lionelio P (2008). Climate change projections for the Mediterranean region. Global and planetary change, 63(2-3):90-104.

Hemmat A, Eskandari I (2004a). Conservation tillage practices for winter wheat-fallow farming in the temperate continental climate of north-western Iran. Field Crops Res. 89:123-133.

Hemmat A, Eskandari I (2004b). Tillage system effects upon productivity of a dryland winter wheat-chickpea rotation in the northwest of Iran. Soil. Tillage Res. 78:69-81.

Hemmat A, Eskandari I (2006). Dryland winter wheat response to conservation tillage in a continuous cropping system in northwest Iran. Soil. Till. Res. 86:99-109.

Hernandez Martinez M, Medina Cazares T, Ramirez Ramierz A, Grageda Cabrera OA, Arreola Tostado IM, Vuelvas Cisneros MA, Aguilar Acuna JL (2004). Response of weeds and wheat genotypes to zero tillage vs conventional tillage. In "Memoria XVI congreso latinoamericano de Malezas y XXIV; Congreso Nacional de la Asociacion Mexican de la Ciencia de la Malaeza Manzanillo, Colima,

Mexico del 10 al 12 de Noviembre de 2003, pp. 231-237. Eds. Pedrero, G. M., Valenzuela, J. A. D., Diaz, G. M., Ruiz, R. A. O.

Karrou M, Boutfirass M (2007). La gestion intégrée de l'eau en agriculture pluviale. A book. Edition INRA, DIC Rabat, Morocco.

Karrou M, Oweis T (2012). Water and land productivities of wheat and food legumes with deficit supplemental irrigation in a Mediterranean environment. Agric. Water Manage. 107:94-103.

Kirkegaard JA., Angus JF, Gardner PA, Muller W (1994). Reduced growth and yield of wheat with conservation cropping. 1. Field studies in the first year of the cropping phase. Austr. J. Agric. Res. 45(3):511-528.

Kumudini S, Grabau L, Van Sanford D, Omielan J (2008). Analysis of yield formation processes under no-till and conservation tillage for soft red winter wheat in the South-Central Region. Agron. J. 100(4):1026-1032.

Laaroussi M (1991). Efficacité de l'irrigation de complément de variétés de blé relativement tolérantes au stress hydrique. Mémoire de troisième cycle. Département du génie rural, IAV Hassan II, Rabat, Maroc.

Lopez-Bellido RJ, Lopez-Bellido L, Benitez-Vega J, Lopez-Bellido FJ (2007a). Tillage system, preceding crop and nitrogen fertilizer in wheat crop. I. Soil water content. Agron. J., 99:59-65.

Lopez-Bellido RJ, Lopez-Bellido L, Benitez-Vega J, Lopez-Bellido FJ, (2007b). Tillage system, preceding crop and nitrogen fertilizer in wheat crop. II. Water utilization. Agron. J. 99:66-72.

Martinez E, Fuentes JP, Silva P, Valle S, Acevedo E (2008). Soil physical properties and wheat root growth under no-tillage and conventional tillage systems in a Mediterranean of Chile. Soil Till. Res. 99:232-244.

Merrill SD, Black AL, Bauer A (1996). Conservation tillage affects root growth of dryland spring wheat under drought. Soil Sci. Soc. Am. J. 60:575-583.

Mrabet R (2000a). Differential response of wheat to tillage management in a semiarid area of Morocco. Field Crops Res. 66:165-174.

Mrabet R (2000b). Long-term no-tillage influence on soil quality and wheat production in semiarid Morocco. In Proceedings of the 15th ISTRO Conference Tillage at the Threshold of the 21st Century: Looking Ahead, July 2-7, 2000, Fort Worth, Texas, USA.

Mrabet R (2008). No-tillage systems for sustainable dryland agriculture in Morocco. INRA Publication. Fanigraph Edition. P .153.

Oweis T, Hachum A, Kijne J (1999). Water Harvesting and Supplemental Irrigation for Improved Water Use Efficiency in the Dry Areas. SWIM P. 7. Colombo, Sri Lanks: International Water Management Institute.

Oweis T, Hachum A (2006). Water harvesting and supplemental irrigation for improved water productivity of dry farming systems in West Asia and North Africa. Agric. Water Manage. 80(1-3):57-73.

Pala M (2000). Challenges and opportunities for conservation tillage-direct drilling in CWANA region: ICARDA/NARS's experience. Options Méditerranéennes 69:161:165.

Passioura J (2006). Increasing crop productivity when water is scarce: from breeding to field management. Agric. Water Manage. 80:176-196.

Power JF, Wilhelm WW, Doran JW (1986). Crop residue effects on soil environment and dryland maize and soybean production. Soil Till. Res. 8:101-111.

Ryan J, Masri S, Garabet S, Diekmann J, Habib H (1997). Soils of ICARDA's agricultural experiment stations and sites: climate, classification, physical and chemical properties, and land use. ICARDA, Aleppo, Syria.

SAS (1997). Statistical Analysis System. SAS/STAT user's Guide. Release 6.03. Statistical Analysis System institute, Cary, NC.

Thomas GA, Thompson JP, Amos RN (2003). A long term fallow management experiment on a vertisol at Hermitage Research Station in southern Queensland, Australia. In: proceedings of 16th Conference of the international Soil Tillage Research Organization, Brisbane, Australia, pp. 1223-1228.

Thomas RJ, de Pauw E, Qadir M, Amri A, Pala M, Yahyaoui A, El Bouhssini M, Baum M, Ineguez L, Shideed K (2007). Increasing the resilience of dryland agro-ecosystems to climate change. J. Agric. SAT Res. 4(1):1-37.

Vadon B, Lamouchi L, Elmay S, Maghfour A, Mahnane S, Benaouda H, El Gharras O (2006). Organisations paysannes: un levier pour développer l'agriculture de conservation au Maghreb. Options Méditerranéennes 69:87-99.

25

Oil quality and aroma composition of 'Chemlali' olive trees (*Olea europaea* L.) under three irrigation regimes

Hechmi Chehab[1], Manel Issaoui[2], Guido Flamini[3], Beligh Mechri[2], Faouzi Attia[4], Cioni Pier Luigi[3], Dalenda Boujnah[1] and Mohamed Hammami[2]

[1]Institute of Olivier of Sousse, Tunisia.
[2]Laboratoire de Biochimie UR 'Nutrition Humaine et Altérations Métaboliques' Faculté de Médecine, Rue Avicenne, 5019 Monastir, Tunisie.
[3] Dipartimento di Chimica Bioorganica e Biofarmacia, via Bonanno 33, 56126 Pisa, Italy.
[4]Agronutrition Parc Activestre-3 avenue de l'Orchidée 31390 carbonne- France.

The present work focused on the chemical composition of monovarietal virgin olive oil from the cultivar *Chemlali* cultivated in the South of Tunisia: sub-arid zone under three different irrigation regimes: stressed, moderate and well irrigation treatment with the restitution of 50, 75 and 100% of crop evapotranspiration (ETc), respectively. Quality characteristics (acidity and peroxide value) and chemical data (antioxidant compound, fatty acids volatile compounds and oxidative stability) were studied in addition to the pomological characteristic of olive fruit. Results show that there were significant differences observed in oil composition according to the irrigation regime applied. Total phenols, bitterness intensity and LOX products content showed the highest values for low irrigation regime, whereas polyunsaturated fatty acid and oxidative susceptibility values had highest values for olive oil from well irrigated trees. Analytic characteristic of fruits showed the highest values of pulp/stone ratios from olive trees irrigated by the highest amount of water.

Key words: Virgin olive oil, phenols, volatile compounds, oxidative stability.

INTRODUCTION

World-wide production of olive oil during the last 20 years increased by almost 70% (from 1.7 to 2.8 million tons) (Zampounis, 2006). Olive oils makes up a small proportion (<3.5%) of the volume in the world vegetable oil market. However, in terms of product value, only olive oil has a 15% share of world trade (Luchetti, 2000). The price of olive oil can be two to five times higher than that of other vegetable oils depending on the country, category of the oil and year of production (Luchetti, 2000). Spain is the primary world producer of olive oil, followed by Italy, Greece, Tunisia and Turkey. World consumption generally follows a parallel path to the production rate. This enormous growth is due to several factors, which are the acknowledgements of olive oil's health-promoting potential as a part of the Mediterranean diet and the global promotional campaigns initiated by the International Olive Council (Dag et al., 2008).

In Tunisia 'Chemlali' is the main cultivar; 1.3 million ha are grown in South and Central, and accounts for 80% of Tunisia's oil production (Baccouri et al., 2008). The chemical and organoleptic characteristics of olive oil depend on several factors (Salvador et al., 2001).

According to Aparicio and Luna (2002), these factors are clustered into four main groups: environmental (soil, climate), cultivation (ripeness, harvesting), technological (fruit storage, extraction procedure), and agronomic factors (fertilisation, irrigation). Among these factors irrigation is a major determinant of olive oil quality (Gomez-Rico et al., 2007). High quality olive oil cannot be obtained from fruit that have suffered a severe water stress (Gomez-Rico et al., 2007). As water supplies decrease and better quality water supplies are reserved for more sensitive crops to drought conditions than olive tree (Palese et al., 2006). In fact, studies have shown that irrigation can increase olive production (Grattan et al., 2006; Samish et Spiegel, 1961; Moriana et al., 2003) thereby increasing total fruit yield and oil production per tree (Grattan et al., 2006). However, studies differ regarding their overall performance to applied water. Chemical and sensory characteristics, however, allow distinguishes clearly between virgin olive oils from irrigated and non-irrigated olive trees (Gomez-Rico et al., 2006; Patumi et al., 2002; Bedbabisa et al., 2010).

Thus, total content of polyphenols, which contribute to the oil bitter taste, is lower in the virgin olive oil harvested from irrigated zones (Gomez-Rico et al., 2006; Patumi et al., 2002). This is of great importance for varieties characterised by high values of astringent, throat-catching or bitter sensory descriptors. This paper presents the first investigation on the characterization of virgin olive oils from *Chemlali* cultivar implanted in the South of the country submitted to different irrigation regimes by fresh water using a localized irrigation in high density olive orchard. In this study, we describe the composition of Tunisian virgin olive oils through classical quality indexes, fatty acids composition and minor components, which are related to both oxidative stability and oil purity.

MATERIALS AND METHODS

Plant material and growing areas selected

The study was carried out on monovarietal virgin olive oils from the main Tunisian cultivar, *Chemlali*. The orchard is localized in the South, trees are planted at a density of 6mx6m (278 trees ha^{-1}) since 2002. The soil is sandy-loam with alkaline pH. The mean precipitation and temperature registered were 250 mm year^{-1} and 30°C, respectively. In the olive orchard, water was delivered three times each week (from Juin to September) using a localized irrigation system with four drip nozzles of 8 l/h each per tree (two per side), set in a line along the rows at a distance of 0.5 m from the trunk.

Olive *Chemlali* cultivars were tested in a factorial combination with three irrigation levels well irrigated (T3), moderate (T2) and stressed (T1) receiving a seasonal water irrigation amount equivalent to 100, 75 and 50% of crop evapotranspiration (ETc) calculated using the Penman–Monteith (Allen et al., 1998) with a single estimated crop coefficient (Kc = 0.6) and a coverage coefficient (Kr = 0.5).(D'Andria et al., 2004). ETc = Kc.Kr.ET0, when ET0 is the reference evapotranspiration calculated by Penman–Monteith equation (Allen et al., 1998).

Morphological study

Immediately before harvest during 2008, 100 fruits were randomly sampled from around the canopy of each tree to determine the maturation index (MI) according to a 0 to 7 scale (Boskou, 1996) and the fruit fresh weight. The variety characterization was realized according to the method adopted by the International Olive Oil Council (IOOC) called 'methodology for the characterization of olive tree varieties'. The mean average weight of 40 fresh fruits was measured for each sample. Then, olives were de-stoned and the flesh (pulp) was separated in order to measure the pulp to stone ratio. The moisture content was determined from 40 g of olive fruits which were dried in an oven at 80°C to constant weight. Oil content, expressed as a percentage of dry weight, was determined by extracting dry material with hexane at 68°C using a *Soxhlet* apparatus, following the procedure described in AOCS Regulation Official Method (A.O.C.S, 1995). The fruit ripeness index of was attributed by qualitative evaluation of the olive skin and flesh colours (Uceda, 1975). This system is routinely used by the olive oil industry to characterize the degree of ripeness of olives arriving at their facilities. This evaluation was performed in triplicate.

Reagents and standards

The (*p*-hydroxyphenyl) ethanol (*p*-HPEA) was obtained from Janssen Chemical Co. (Beerse, Belgium). Pure analytical standards of volatile compounds were purchased from *Fluka* and *Aldrich* (Milan, Italy).

Analytical methods

Quality parameters

Determinations of free acidity and peroxide value were carried out following analytical methods described in the EEC 2568/91 and EEC 1429/92 European Union Regulation (European Union Commission Regulation 1991).

Fatty acid methyl esters (FAMEs) analysis

The FAMEs were prepared as described by EU official method. The chromatographic separation was carried out using a Hewlett-Packard (HP 5890) chromatograph, a split/splitless injector, and a flame ionization detector (FID) linked to an HP Chemstation integrator. A fused silica capillary column HP-Innowax (30 m x 0.25 mm x 0.25 μm) was used with nitrogen as the carrier gas at a flow rate of 1ml min^{-1}; flame-ionization detection temp. 280°C; injector temperature 250°C and an oven temperature programmed from 180 to 250°C. Results were expressed as relative percent of total area (Issaoui et al., 2007).

Total phenols

These were determined colorimetrically as previously reported (Montedoro et al., 1992) and the results are expressed as 3, 4-DHPEA equivalents. Evaluation of the intensity of bitterness was carried with the procedure described by Beltranet al. (2007).

Rancimat assay

Rancimat assay oxidation stability was evaluated by the Rancimat apparatus (Mod. 743, Metrohm Ω, Switzerland) using an oil sample of 3 g warmed to 120°C and an air flow of 20 L h^{-1}. Stability was

Table 1. Influence of irrigation regimes on maturity index, pomological parameters of *Chemlali* olive fruit.

Parameter	*Chemlali* olive oil		
	T1	T2	T3
Maturity index	2.9[b]	2.6[b]	3.1[a]
Fesch fruit weight (g)	0.7[bc]	0.8[b]	1.4[a]
Stone weight (g)	0.20[b]	0.21[b]	0.25[a]
Pulp/stone ratio	2.5[c]	2.8[b]	4.6[a]
Fruit damage (%)	2[c]	3[b]	24[a]
*Oil content Soxhlet (%)	29.7[a]	24.2[b]	24.8[b]

T1, T2 and T3, irrigated treatments with 50, 75% and 100% of ETc, respectively. Mean values (n = 3). Values in each row with different superscript letters present significant differences (p<0.05) between the different irrigation strategies for each parameter. *Expressed as dry matter.

expressed as induction time (h).

Volatile compound analyses

Extraction: Solid phase micro extraction was used as a technique for headspace sampling of virgin olive oils. Sampling was performed with Supelco SPME with film thickness of 100m; 75m carboxen/polydimethylsiloxane (CAR/PDMS) were used to sample the headspace of 2 ml of virgin olive oil inserted into a 5 ml glass vial and allowed to equilibrate for 30 min after the equilibration time, the fiber was exposed to the headspace for 50 min at 25°C room temperature. Once sampling was finished, the fiber was withdrawn into the needle and transferred to the injection port of the GC and GC-MS system (Campeol et al., 2001).

Identification: GC analyses were accomplished with an HP-5890 series II instrument equipped with a HP-5 capillary column (30 m x 0.25 mm, 0.25 µm film thickness), working with the following temperature programme: 60°C for 10 min, ramp of 5°C min^{-1} to 220°C; injector and detector temperatures, 250°C; carrier gas, nitrogen (2 ml min^{-1}); detector FID; split ratio, 1:30; injection, 0.5 µl. the identification of the components was performed by comparison of their retention times with those of pure authentic samples and by means of their linear retention indices (LRI) relative to the series of *n*-hydrocarbons. The relative proportions of the constituents were obtained by FID peak area normalization.

GC-EIMS analyses were performed with a Varian CP 3800 gas-chromatograph equipped with a DB-5 Capillary column (30m x 0.25 mm; coating thickness= 0.25µm) and a Varian Saturn 2000 ion trap mass detector. Analytical conditions were as follows: injector and transfer line temperature at 250 and 240°C, respectively; oven temperature was programmed from 60 to 240°C at 3°C min^{-1}; carrier gas, helium at 1 ml min^{-1}; splitless injection. Identification of the constituents was based on comparison of the retention times with those of pure standards of volatile compounds comparing their linear retention indices relative to the series of *n*-hydrocarbons, and on computer matching against commercial (NIST 98 and Adams 1995) and homemade library mass spectra built from pure substances and components of known oils and MS literature data (Stenhagen et al., 1974; Adams, 1995). Moreover, the molecular weights of all the identification substances were confirmed by GC-CIMS, using MeOH as CI ionizing gas (Flamini et al., 2003).

Statistical analysis

All parameters analyzed were carried out in triplicate. Significant differences among varieties studied were determined by an analysis of variance which applied a Student test, using the Statistical Package for the Social Sciences (SPSS) programme, release 11.0 for Windows.

RESULTS AND DISCUSSION

Pomological parameters

Table 1 lists the olive fruit characteristics and composition, as affected by the different irrigation treatments studied and the fruits ripening index for the *Chemlali* cultivar. Infect, ripening index was affected by the irrigation level (Table 1). In the same ripening date, the average level was 2.9 for olives produced from trees submitted to 50% ETc. The average level of fruits obtained from trees submitted to a restitution of 75 and 100% ETc were 2.6 and 3.1, respectively. Individual fruit size increased proportionally with applied water (r=0.924). As a consequence, the pulp stone ratio increased two times at well irrigated treatment. The irrigation treatment apparently did affect the oil accumulation in the *Chemlali* fruit since statistically significant differences in the oil yield were observed in the present study only at a volume of irrigation more than 50% ETc. In fact, Lavee and Wodner (1991) and Motilva et al. (1999) did not observe a slight delay in oil accumulation in fruits from non-irrigated olive trees as a consequence of water stress at the end of the summer season.

Quality indices

The quality parameters evaluated in the olive oil samples from the *Chemlali* olive oil of different irrigation percentage are shown in Table 2. A trend of increasing free acidity (FFA) of virgin olive oil with increased irrigation levels was observed in the analyzed samples. Fatty acid content of 0.8% as oleic acid divides the categories 'virgin' and 'extra virgin' olive oil according to

Table 2. Influence of irrigation regimes on analytical parameters, fatty acid profiles and oxidative stability of *Chemlali* olive oil.

Parameter	Chemlali olive oil		
	T1	T2	T3
Acidity (%)	0.2[c]	0.4[b]	1.6[a]
Peroxide value (meqO$_2$ kg^{-1})	10.6[c]	13.5[b]	20.0[a]
Category	Extra Virgin	Extra Virgin	Virgin
C16 :0 (%)	18.1[a]	16.9[b]	12.0[c]
C16 :1	2.1[a]	2.1[a]	0.4[b]
C18 :0	2.4[b]	2.5[a]	2.7[a]
C18 :1	58.7[b]	60.9[a]	55.1[c]
C18 :2	16.2[b]	15.9[c]	28[a]
C18 :3	1.0[a]	1.1[a]	1.0[a]
SFA	20.5[a]	19.0[b]	14.7[c]
PUFA	17.1[b]	17.1[b]	29[a]
MUFA	60.8[b]	63.0[a]	55.5[c]
UFA/SFA	3.8[c]	4.2[b]	5.7[a]
MUFA/PUFA	3.5[b]	3.7[a]	1.9[c]
O/L	3.6[a]	3.8[a]	1.9[b]
IV	87.0[b]	82.8[c]	102.9[a]
OS	887.4[b]	898.2[b]	1415.7[a]
Oxidative stability (h)	10.5[a]	10.0[a]	4.8[b]
Oxidative stability (day kg^{-1})	145.9[a]	139.4[b]	67.3[c]
Oxidative stability (year kg^{-1})	0.4[a]	0.4[a]	0.2[b]

T1, T2 and T3, Irrigated treatments with 50, 75 and 100% of ETc, respectively. Mean values (n = 3). Values in each row with different superscript letters present significant differences (p<0.05) between the different irrigation strategies for each parameter. *Expressed as dry matter.

EU-legislated standards (European Union Commission Regulation 2003). Consequently, the oil from olives submitted to 50 and 75% ETc were categorized as 'extra virgin' olive oil, while the oil produced from olive tress submitted to 100% ETc fell into the 'virgin' category. These results suggest that free acidity increased during the short period (several hours at most) between the time the fruit was removed from the trees and the time the oil was extracted and collected. The positive correlation between irrigation level and FFA content could be due to increased sensitivity of fruit with higher water content, a thinner cuticle layer to mechanical injury (Table 1). Such relative sensitivity of olive fruit from high irrigated compared to rain-fed trees was reported previously in Spain for the *Cornicabra* cultivar (Gomez-Rico et al., 2006). Olive oil quality indices observed in the most irrigated olives was probably due to the relatively high fruit damage produced by an olive fly attack that affected mainly the well irrigated olive trees (Gomez-Rico et al., 2007). The same behaviour was observed with the peroxide value for *Chemlali* virgin olive oil submitted to the different irrigation levels (Table 2).

Fatty acid composition

Table 2 reports the level of fatty acids in the *Chemlali*

olive oils subjected to three irrigation regimes (50, 75 and 100% Etc, respectively). The oleic acid level exhibited a clear variation in relation to the irrigation level. Hence, virgin olive oil from trees submitted to 50% ETc had 58.7% of oleic acid. The level of this compound increased proportionally with water restitution of about 75% ETc until 60.9%. However, we noticed a significant decrease when 100% ETc was applied to olive trees (Table 2). The linoleic acid showed significant increase when amount of water exceeded 75% ETc. Moreover, the palmitic acid content had values varied from 18.1 to 12% which were higher in oils from tress submitted to low irrigation regimes than those reported in virgin olive oils from tress subjected to high irrigation volume (Table 2). There was a negative linear relationship between Kc and the palmitic acid rate (r = -0.944). Another important saturated acid is the stearic acid; its content (2.4%) was barely affected by irrigation regimes (Table 2). With regard to the effect produced in the fatty acid composition by the irrigation regime, we noticed that, the oleic linoleic acid ratio decrease of about the half when 100% of water was used. Similarly, the MUFA/PUFA and SFA/UFA ratios were practically unaffected by the irrigation regimes of 50% ETc and 75% ETc. However, statistically a significant difference was established between the later analyzed samples and virgin olive oils from high irrigation regimes. As a consequence, the MUFA/PUFA ratios

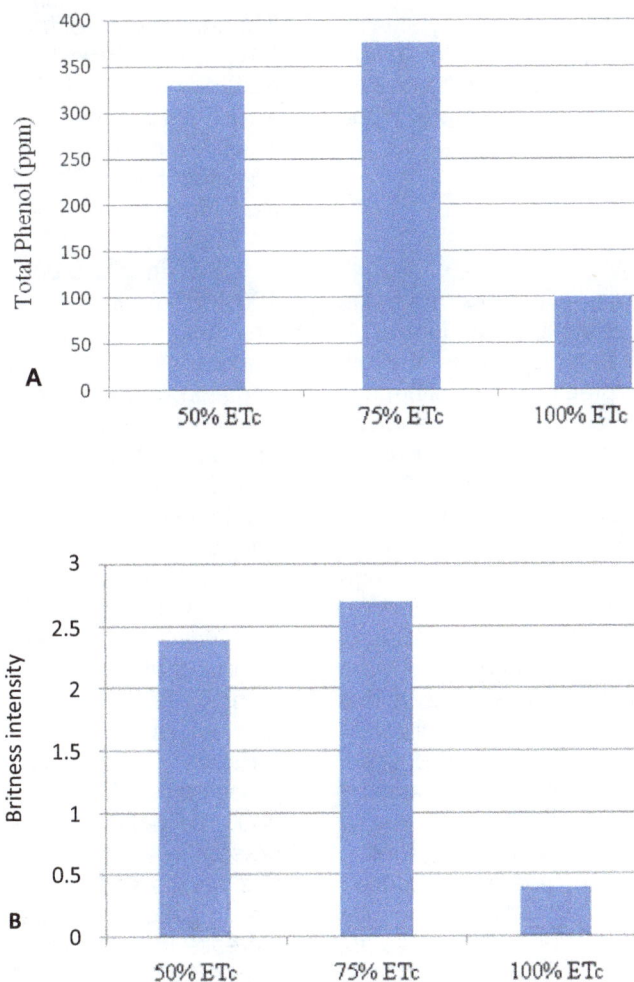

Figure 1. Influence of irrigation regimes on total phenols and bitterness intensity category.

obtained in virgin olive oil obtained from olive trees submitted to 100% ETc were about 2 fold less important than other samples. The irrigation regime superior to 75% ETc, affected significantly the oil fatty acid ratio influences the organoleptic characteristics of the oil because oil with a high content of saturated fatty acids could influence the organoleptic characteristics of the oil. This gives rise to the defect defined as a "fatty sensation" (Patumi et al., 2002; Solinas, 1990). The potential oxidative susceptibility is therefore much higher in the virgin olive oil produced from the well irrigated trees than other VOO (Table 2). Gomez-Rico et al. (2007) and Salas et al. (1997) found that irrigation induces an increase in palmitic and linoleic acids and a decrease in oleic and linolenic acid content in virgin olive oil. Other investigations observed contrast trends in *Arbequina* virgin olive oil obtained by a young super high density orchard, where different regulated deficit irrigation was applied. Moreover, Ingelese et al. (1996) and Patumi et al. (1999) found that the fatty acid composition of different

Italian varieties was affected mainly by cultivar and not by irrigation practices.

Phenols

Figure 1 reports the concentration of the phenolic compounds (mg kg^{-1}) found in virgin olive oils obtained from *Chemlali* olives subjected to the different irrigation managements studied. Our results showed that an irrigation regime higher than 75% ETc is very detrimental to the concentration of total phenols. Hence, we noticed a sharp decrease of about 3.5 times in *Chemlali* virgin olive oil obtained from olives well irrigated with 100% ETc. In fact, phenolic biosynthesis in plants is known to be highly sensitive to environmental conditions. Several investigations (Patumi et al., 1999; Tovar et al., 2002) have reported that different water status of olive trees under different irrigation strategies, implied changes in the activity of enzymes responsible for phenolic compounds synthesis in drupes, such as L-phenylalanine ammonia lyase (PAL), whose activity is grater under higher water stress conditions, leading to superior phenolic contents in the olive flesh and therefore in the virgin olive oils obtained.

Intensity of bitterness

The bitterness intensity of virgin olive oils from the different irrigation strategies are depicted in Figure 1. Bitter taste is one of the characteristic attributes of virgin olive oil. Its intensity varies greatly and influences consumer attraction and acceptance. Our results showed that an irrigation volumes superior to 75% ETc decrease the bitterness intensity of the oils from 2.8 to 0.4. The same volume of water transforms Chemlali virgin olive oils from bitter oils to non bitter oils. The decrease in bitterness due to the irrigation technique used, though very low in the oils studied, is indeed relevant from a marketing point of view. Oils with a high level of bitterness are less preferred by consumers and therefore the descent of this attribute could be desirable in rich phenol virgin olive oil such as the Chétoui olive oil (Issaoui et al., 2010). On the contrary, the same effect could negatively affect Chemlali virgin olive oil sensory profile, in fact due to its natural low phenolic content, a descent of these compounds would lead to a virgin olive oil which is too mild and flat, and moreover its reduced oxidative stability could reduce significantly the shelf-life of this product (Gomez-Rico et al., 2006; Patumi et al., 2002 ; Aparicio et Luna 2002).

Volatile compounds

From the biochemical point of view, the volatile compounds are considered as direct metabolites produced in plant organs by intracellular biogenic pathways (Aparicio and Luna, 2002). The increased activity and synthesis of enzymes can lead to an accumulation of metabolites that influence the biogenesis of volatiles. This explains why the biochemical pathways are responsible for the particular profiles of the monovarietal virgin olive oils although external parameters (climate, soil, harvesting and extraction conditions) modulate the final sensory profile (Aparicio and Luna, 2002). Table 3 shows the level of these individual volatile compounds in Chemlali virgin olive oils as influenced by different irrigation managements. Because of the importance of the green perception in the virgin olive oil flavour, the C6 compounds- also called "green volatiles" (Aparicio and Morales, 1998) were quantified in all the analyzed samples. Virgin olive oil obtained from olive trees submitted at 50% ETc had the highest level of C6 aldehydes (59.0%), flowed by samples of olive trees treated by 75% Etc (14.0%),

whereas the oil obtained from trees submitted to 100% ETc had the lowest level (11.6%). It is also noteworthy that the content of (E)-2-hexenal, which gives the typical "green note" to extra virgin olive oil, is by far the major C6 aldehyde compound in studied oils of trees submitted to irrigation treatments T1 and T2 (Table 3). Hence, we found that, the rate of (E)-2-hexenal decreased from 54.4 (in oils produced from olive trees submitted to 50% ETc) to 7.2% (in oils produced from olive trees submitted to 100% ETc) with increasing irrigation volumes (Table 3). The hexanal content (related to apple, green, and cut grass sensory notes) seems to be barely affected by the irrigation management. In terms of C6 alcohols, we identified only the 1-hexanol in virgin olive oil analyzed the value of this compounds ranged between traces to 11.2%. They have less sensory significance than aldehydes because of their higher odour threshold values and their sensory descriptions being associated with fruity, soft green and aromatic sensory notes (Luna et al., 2006). Our results showed that irrigation had a great influence on the level of this compound; hence we noticed that the level of this compound is the highest in sample T2 when irrigation treatment 75% of Etc is applied (Table 3).

Esters, compounds associated with sweet, fruity and green leaves notes (Aparicio and Luna, 2002; Luna et al., 2006) as hexyl acetate and (Z)-3-hexenyl acetate, were present in aroma of all studied virgin olive oils (Table 3). Our results showed that esters were very influenced by the irrigation regimes applied. Hence, we noticed an increase in both hexyl acetate and (Z)-3-hexenyl acetate from 1.6 and 2.3% to 21.2 and 4.3% when we increased irrigation volumes until 75% ETc. The hydrocarbons of olive oils have been studied by different authors as possible markers to distinguish virgin olive oil from different olive varieties or different geographic origins (Aparicio and Luna, 2002 ; Issaoui et al., 2010; Ben Temime et al., 2006; Issaoui et al., 2009). The hydrocarbons 1-nonene, 3-methyl -4-heptanone, 5-methyl-3-heptanone, 1-heptanol, 3-octanone, 3-octanol, myrcene, limonene, phenyl acetaldehyde, (E)-β-ocimene, (E)-2-octenal, phenylethylalcohol, borneol, E)-2-hexenyl butyrate, ethyl octanoate, decanal, (E, Z)-2,4-decadienal, valencene and (E, E)-α-farnesene have been detected in the aroma fraction of the tested oils. The contents of limonene did exceeded 10% in Chemlali virgin olive oils. Our results showed that their level was dependent to the irrigation regimes applied because their response to the irrigation was quite high. It seems that the high volume of water decrease the abundance of limonene in the virgin olive oils (Table 3). This component could play a very important role in the fragrance of this precious food as indicated by Vichi et al. (2003) and Baccouri et al. (2008). In contrast, the (E)-β-ocimene showed an inverse behaviour to the limonene. Hence, E)-β-ocimene increased with increasing irrigation volume (Table 3). We noticed also a positive correlation between the

Table 3. Influence of irrigation regimes on aromatic composition (%) of *Chemlali* olive oil.

Parameter	Chemlali olive oil		
	T1	T2	T3
Hexanal	4.6[b]	5.7[a]	4.4[b]
(E)-2-hexenal	54.4[a]	8.3[b]	7.2[b]
1-hexanol	tr	11.2[a]	7.2[b]
1-nonene	ND	7.2[a]	4.3[b]
3-methyl -4-heptanone	1.4	ND	ND
5-methyl-3-heptanone	0.9	ND	ND
1-heptanol	0.7[a]	tr	0.5[a]
3-octanone	0.4[a]	tr	0.9[a]
3-octanol	0.7	ND	ND
Myrcene	1.2[a]	tr	1.3[a]
(Z)-3-hexenyl acetate	2.3[b]	4.3[a]	4[a]
1-hexyl acetate	1.6[b]	21.2[a]	17.4[a]
Limonene	10.8[a]	2.6[b]	3.3[b]
Phenyl acetaldehyde	ND	ND	0.5
(E)-β-ocimene	1.4[c]	7.6[b]	11.5[a]
(E)-2-octenal	1.2	ND	1.5
Nonanal	0.5	ND	ND
Phenylethylalcohol	1.0[b]	2.2[a]	1.4[b]
Borneol	tr	5.0[a]	2.8[b]
(E)-2-hexenyl butyrate	ND	0.4	ND
Ethyl octanoate	ND	1.0	ND
Decanal	5.0[c]	10.9[b]	15.3[a]
(E, Z)-2,4-decadienal	ND	1.4	0.9
Valencene	5.8[a]	tr	1.5[b]
(E, E)-α-farnesene	3.7[b]	6.9[a]	7.5[a]
Total (%)	97.2[a]	95.9[a]	94.0[a]

T1, T2 and T3, Irrigated treatments with 50, 75 and 100% of ETc, respectively. Mean values (n = 3). Values in each row with different superscript letters present significant differences (p<0.05) between the different irrigation strategies for each parameter. *Expressed as dry matter. Tr, Traces; ND, not detected.

appearance of decanal and the irrigation volumes applied to the olive tress (r = 0.996) and also a positive correlation between decanal and the percentage of the fruit damage (r = 0.843). In fact, decanal is the result of autoxidation of oleic acid, for that reason it was more abundant in oils rich in free fatty acid (r = 0.889), peroxide value (r = 0.955) and oxidative susceptibility (r = 0.831).

The sum of the products of lipoxygenase oxidation (LOX) pathways, which generally are the major components of the olive oil volatile fraction, were higher in *Chemlali* virgin olive oils produced from tress submitted to a low irrigation volumes (50%), than in higher ones (75 and 100% ETc). This value decreased from 62.9 (at a restitution of 50% of crop evapotranspiration) and 50.7 (at a restitution of 75% of crop evapotranspiration) to 40.2% (at a restitution of 100% of crop evapotranspiration). It became clear that the changes in the volatiles of *Chemlali* virgin olive oils under the different irrigation practices applied in the orchards. Hence, the volatile compounds most affected by the water status of olive

trees were (E)-2-hexenal, hexanal, limonene, 1-hexanol, (Z)-3-hexenyl acetate, 1-hexyl acetate and (E)-β-ocimene. *Chemlali* virgin olive oils obtained from olive trees under stress due to deficit irrigation are characterized by high green and citrus like sensory notes than those from high irrigated trees.

Oxidative stability

The observed decrease of oxidative stability does affect markedly the *Chemlali* olive oil shelf life or quality when the amount of water added is more than 75% ETc (*Table 2*). In fact, as a consequence of antioxidant compounds decrease we showed also an oxidative stability (OSI) decrease. Hence, a positive correlation was established between total phenols and OSI (r = 0.97), also between oleic linoleic acid ratio and OSI (r = 0.985) and with (E)-2-hexenal (r = 0.584). In contrast, the potential oxidative susceptibility is therefore much higher in the *Chemlali* virgin olive oil produced from the well irrigated trees by 100% ETc (lower oxidative stability) than other VOO. In fact, as the irrigation volumes increased, oxidative susceptibility and iodine value increased (r = -0.998 and r = -0.961, respectively).

Conclusion

The analysis of *Chemlali* virgin olive oils has shown that there were quantitative differences among the chemical profiles of oils tested, in spite of issued from the same cultivar. Then, as the harvesting period and extraction conditions were similar for the samples analyzed, these results confirmed that irrigation practices is responsible for some of the differences observed in olive oil quality since, for the same raw material, different final products are obtained. Increased irrigation quantity increased the free acidity level of the oil and decreased the total phenol content and the aromatic compounds produced through LOX pathway of VOO. A restitution of 75% of crop evapotranspiration was sufficient to achieve good quality of *Chemlali* olive oil, while higher water volumes gave low quality. Moreover, additional studies are necessary to confirm normal composition values for *Chemlali* olive oils cultivated in other irrigation conditions for more than one year.

ACKNOWLEDGEMENTS

This research was supported by the Tunisian Ministry of Enseignement Superieur, Scientific Research and Technology (UR03/ES-08). Part of this work was carried out at the Dipartimento di Chimica Bioorganica e Biofarmacia, Universita` di Pisa, Italy. We wish to thank the personnel of the laboratory of "Human Nutrition and Metabolic Disorder" Faculty of Medicine of Monastir, Institute of Olivier of Sousse"

Abbreviations: FAMEs, Fatty acid methyl esters; **FID,** flame ionization detector; **PAL,** L-phenylalanine ammonia lyase; **LOX,** lipoxygenase oxidation.

REFERENCES

Adams RP (1995). Identification of Essential Oil Components by Gas ChromatographyMass Spectroscopy, Allured Publ. Corp, Carol Stream, IL.

Allen RG, Pereira LS, Raes D, Smith M (1998). Crop evapotranspiration. Guidelines for Computing Crop Water Requirements. FAO Irrigation and Drainage Rome P. 56.

AOCS (1995). American Oil Chemists Society: Official and Recommended Practices of the AOCS. 4th Edn. AOCS Press, Champaign, IL (USA) 1995, Official Method Aa pp. 4-38.

Aparicio R, Luna G (2002). Characterisation of monovarietal virgin olive oils. Eur. J. Lipid Sci. Technol. 104:614-627.

Aparicio R, Morales MT (1998). Characterization of olive ripeness by green aroma compounds of virgin olive oil. J. Agric. Food

Chem. 46:1116-1122.

Baccouri O, Bendini A, Cerretani L, Guerfel M, Baccouri B, Lercker G, Zarrouk M, Daoud Ben Miled D (2008). Comparative study on volatile compounds from Tunisian and Sicilian monovarietal virgin olive oils. Food Chem. 111:277-296.

Bedbabisa S, Ben Rouinab B, Boukhrisa M (2010). The effect of waste water irrigation on the extra virgin olive oil quality from the Tunisian cultivar Chemlali. Sci. Hortic. 125:556-561.

Beltran G, Ruona MT, Jimenez A, Uceda M, Aguilera MP (2007). Evaluation of virgin olive oil bitterness by total phenol content analysis. Eur. J. Lipid Sci. Technol. 108:193-197.

Ben Temime SE, Campeol E, Cioni PL, Daoud D, Zarouk M (2006). Volatile compounds from chatoui olive oil and variations induced by growing area. Food Chem. 99:315-325.

Boskou D (1996). Olive oil: Chemistry and technology. AOCS Press, Champaign, IL, USA. P. 85.

Campeol E, Flamini G, Chericoni S, Catalano S, Cremonini R (2001). Volatile compounds from three Cultivars of Olea europea from Italy. J. Agric. Food Chem. 49:5409-5411.

D'Andria R, Lavini A, Morelli G, Patumi M, Tiranziani S, Calandrelli D, Fragnito F (2004). Effects of water regimes on five pickling and double aptitude olive cultivars (Olea europaea L.). J. Hort. Sci. Biotechnol. pp. 79-18.

Dag A, Ben-Gal A, Yermiyahu U, Basheer L, Nir Y, Kerem Z (2008). The effect of irrigation level and harvest mechanization on virgin olive oil quality in a traditional rain-fed 'Souri' olive orchard converted to irrigation. J. Sci. Food. Agric. 88:1524-1528.

European Union Commission Regulation (1989/2003). Characteristics of olive and olive pomace oils and their analytical methods. Off. J. Eur. Commun. L295:57.

European Union Commission Regulation (2568/1991). Characteristics of olive and olive pomace oils and their analytical methods. Off. J. Eur. Commun. L248, 1991, P. 1.

Flamini G, Cioni PL, Morelli I (2003). Volatiles from leaves, fruits and virgin oil from Olea europaea Cv. Olivastra Seggianese from Italy. J. Agric. Food Chem. 51:1382-1386.

Gomez-Rico A, Salvador MD, Fregapane G (2007). Virgin olive oil and olive fruit minor constituents as affected by irrigation management based on SWP and TDF as compared to ETc in medium-density young olive orchards (Olea europaea L. cv. Cornicabra and Morisca). Food Res. Int. 42:1067-1076.

Gomez-Rico A, Salvador MD, Moriana A, Perez D, Olmedilla N, Ribas F, Fregapane G (2006). Influence of different irrigation strategies in a traditional Cornicabra cv. olive orchard on virgin olive oil composition and quality. Food Chem. 100:568-578.

Grattan SR, Berenguer MJ, Connell JH, Polito US, Vossen PM (2006). Olive oil production as influenced by different quantities of applied water. Agric. Water Manage. 85:133-140.

Issaoui M, Ben Hassine K, Flamini G, Brahmi F, Chehab H, Aouni Y, Hammai M (2009). Discrimination of some monovarietal olive oils according to their oxidative stability, volatiles compounds and sensory analysis. J. Food Lipids 16:164-186.

Issaoui M, Dabbou S, Echbili A, Rjiba I, Gazzah N, Trigui A, Hammami M (2007). Biochemical characterisation of some Tunisian Virgin Olive Oils obtained from different cultivars growing in Sfax National Collection. J. Food Agric. Environ. 5:17-21.

Issaoui M, Flamini G, Brahmi F, Dabbou SB, Hassine K, Taamali A, Zarrouk M, Hammami M (2010). Effect of the growing area conditions on differenciation between Chemlali and Chétoui olive oils. Food Chem. 119:220-225.

Lavee S, Wodner M (1991). Factors affecting the nature of oil accumulation in fruit of olive (Olea europaea L.) cultivar. J. Hort. Sci. 66:583-591.

Luchetti F (2000). Introduction: In Handbook of Olive Oil: Analysis and Properties (Eds, Harwood, J. Aparicio, R.) Aspen Publishers, Inc., Gaithersburg, MD, USA, pp. 1-16.

Luna G, Morales MT, Apaaricio R (2006). Characterization of 39 varietal virgin olive oils by their volatile compositions. Food Chem. 98:243-252.

Montedoro GF, Servili M, Baldioli E, Miniati E (1992). Simple and hydrolyzable phenolic compounds in virgin olive oil.1.1. Their extraction, separation and quantitative and semiquantitative. Afr. J.

Agric. Res. 40:1571-1576.

Moriana A, Orgaz F, Fereres E, Pastor M (2003). Yield responses of mature olive orchard to water deficits. J. Am. Soc. Hortic. Sci. 128:425-431.

Motilva MJ, Romero MP, Alegre S, Girona J (1999). Effect of regulated deficit irrigation in olive oil production and quality. Acta Horticulturae., 474:377-380.

Palese AM, Celano G, Masi S, Xiloyannis C (2006). Treated wastewater for irrigation of olive trees: effects on yield and oil quality. In: Olivebioteq 2006, November 5-10 Mazara del Vallo, Marsala (Italy), II, pp. 123-129.

Patumi M, d'Andria R, Fontanazza G, Morelli G, Giori P, Sorrentino G (1999). Yield and oil quality of intensively trained trees of three cultivars of olive under different irrigation regimes. J. Hortic. Sci. Biotechnol. 74:729-737.

Patumi M, d'Andria R, Marsilio G, Fontanazza G, Morelli G, Lanza B (2002). Olive and olive oil quality after intensive monocone olive growing (Olea europaea L., cv. Kalamata) in different irrigation regimes. Food Chem. 77:27-34.

Salas J, Pastor M, Castro J, Vega V (1997). Influencia del riego sobre la composicion y caracteristicas del aceite de oliva. Grasas Aceites 48:74-82.

Samish RM, Spiegel P (1961). The use of irrigation in growing olives for oil production. Israel J. Agric. Res. 11:87-95.

Salvador MD, Aranda F, Gomez-Alonso S, Fregapane G (2001). Cornicabra virgin olive oil: a study of five crops seasons. Composition, quality and oxidative stability. Food Chem. 74:267-274.

Solinas M (1990). Olive oil quality and its determining factors. In proceedings of problems on olive oil quality congress. Sassari, Italy. pp. 23-55.

Stenhagen E, Abrahamsson SMc, Lafferty FW (1974). Registry of mass spectral data. John Wiley and Sons, New York. P. 3136.

Tovar MJ, Romero MP, Alegre S, Girona J, Motilva MJ (2002). Composition and organoleptic characteristics of oil from Arbequina olive (Olea europaea L.) trees under deficit irrigation. J. Sci. Food Agric. 82:1755-1763.

Uceda M, Frias L (1975). Harvest dates. Evolution of the fruit of content, oil composition and oil quality. In Proceedings of II Seminario Oleicola International; International Olive-oil Council: Cordoba, Spain. P. 125.

Vichi S, Pizzale L, Conte LS, Buxaderas S, López-Tamames E (2003). Solidphase microextraction in the analysis of virgin olive oil volatiles fraction: Characterisation of virgin olive oils from two distinct geographical areas of northern Italy. J. Agric. Food Chem. 57:6572-6577.

Zampounis V (2006). In Olive oil: Chemistry and Technology, 2nd add. (Ed, Boskou, D.) Champaign: AOCS Press, pp. 21-40.

Permissions

List of Contributors

Khumbulani DHAVU
Bioresources Engineering, School of Engineering, University of KwaZulu-Natal, Private Bag X01, Scottsville, South Africa

Hiroshi YASUDA
Arid Land Research Center, Tottori University, 1390 Hamasaka, Tottori, 680-0001, Japan

Ömer Faruk KARACA
Department of Biosystems Engineering, Faculty of Engineering and Architecture, University of Bozok, 66200 Yozgat, Turkey

Kenan UÇAN
Department of Biosystems Engineering, Faculty of Agriculture, University of Kahramanmaras Sutcu Imam, 46100 Kahramanmaras, Turkey

Tadesse Getacher
College of Business and Economics Mekelle University, Ethiopia

Amenay Mesfin
Research Economist, International Water Management Institute (IWMI), Ethiopia

Gebrehaweria Gebre-Egziabher
Research Economist, International Water Management Institute (IWMI), Ethiopia

D. Khalkho
Indira Gandhi Krishi Vishwavidyalaya, Raipur (Chhattisgarh) – 492012, India

N. S. Raghuwanshi
Department of Agricultural and Food Engineering, Indian Institute of Technology, Kharagpur (W.B.) – 721302, India

S. Khalkho
Indira Gandhi Krishi Vishwavidyalaya, Raipur (Chhattisgarh) – 492012, India

R. Singh
Department of Agricultural and Food Engineering, Indian Institute of Technology, Kharagpur (W.B.) – 721302, India

Yaser Mohammadi
Department of Agricultural Development and Management, College of Agricultural Development and Economic, University of Tehran, Tehran, Iran

Hussein Shabanali Fami
Department of Agricultural Development and Management, College of Agricultural Development and Economic, University of Tehran, Tehran, Iran

Ali Asadi
Department of Agricultural Development and Management, College of Agricultural Development and Economic, University of Tehran, Tehran, Iran

Ndiaye Malick
Laboratoire de Biotechnologies des Champignons, Département de Biologie Végétale, Faculté des Sciences et Techniques, Université Cheikh Anta Diop, BP. 5005 Dakar-Fann, Sénégal

Cavalli Eric
Laboratoire de Nanomédecine, Imagerie et Thérapeutique, EA 4662, UFR Sciences Médicales et Pharmaceutiques, Université de Franche-Comté, 19 rue Ambroise Paré, 25030 Besançon cedex, France

Leye El Hadji Malick
Laboratoire National de Recherche sur les Productions Végétales/ Institut Sénégalais de Recherche Agricole, 'LNRPV/ISRA', Dakar, Sénégal

Diop Tahir Abdoulaye
Laboratoire de Biotechnologies des Champignons, Département de Biologie Végétale, Faculté des Sciences et Techniques, Université Cheikh Anta Diop, BP. 5005 Dakar-Fann, Sénégal

Biswajit Mondal
Central Soil and Water Conservation Research and Training Institute, Research Centre, Bellary, Karnataka, India

N. Loganandhan
Central Soil and Water Conservation Research and Training Institute, Research Centre, Bellary, Karnataka, India

ADEDOKUN, Mathew Adewale
Fisheries Technology Department, Oyo State College of Agriculture, P. M. B. 10 Igboora, Oyo State, Nigeria

O. O. FAWOLE
Pure and Applied Biology Department, Ladoke Akintola University of Technology, P.M.B 4000 Ogbomoso, Oyo State, Nigeria

T. A. AYANDIRAN
Pure and Applied Biology Department, Ladoke Akintola University of Technology, P.M.B 4000 Ogbomoso, Oyo State, Nigeria

Hossein Samadi-Boroujeni
Department of Water Engineering, Faculty of Agriculture, Shahrekord University, Shahrekord, Iran

Mehri Saeedinia
Department of Irrigation and Drainage, Faculty of Water Sciences Engineering, Shahid Chamran University, Ahwaz, Iran

S. Muthu Kumar
Horticultural College and Research Institute, Tamil Nadu Agricultural University, Periyakulam, Theni District-625604, Tamil Nadu, India

V. Ponnuswami
Horticultural College and Research Institute, Tamil Nadu Agricultural University, Periyakulam, Theni District-625604, Tamil Nadu, India

Nigus Demelash
Gondar Agricultural Research Center, P. O. Box 1337, Ethiopia

Naser Mohammadian Roshan
Department of Agriculture, Lahijan Branch, Islamic Azad University, Lahijan, Iran

Maral Moradi
Department of Agriculture, Lahijan Branch, Islamic Azad University, Lahijan, Iran

Ebrahim Azarpour
Department of Agriculture, Lahijan Branch, Islamic Azad University, Lahijan, Iran

Hamid Reza Bozorgi
Department of Agriculture, Lahijan Branch, Islamic Azad University, Lahijan, Iran

Chongliang Sun
State Key Laboratory of Resources and Environmental Information System, Institute of Geographical Sciences and Natural Resources Research, Chinese Academy of Sciences, China

Dong Jiang
State Key Laboratory of Resources and Environmental Information System, Institute of Geographical Sciences and Natural Resources Research, Chinese Academy of Sciences, China

Juanle Wang
State Key Laboratory of Resources and Environmental Information System, Institute of Geographical Sciences and Natural Resources Research, Chinese Academy of Sciences, China

Yunqiang Zhu
State Key Laboratory of Resources and Environmental Information System, Institute of Geographical Sciences and Natural Resources Research, Chinese Academy of Sciences, China

S. S. WANDRE
Department of Soil and Water Engineering, College of Agricultural Engineering and Technology, Junagadh Agricultural University, Junagadh - 362001, India

H. D. RANK
Department of Soil and Water Engineering, College of Agricultural Engineering and Technology, Junagadh Agricultural University, Junagadh - 362001, India

R. E. Okonji
Department of Biochemistry, Obafemi Awolowo University, Ile-Ife, Nigeria

O. O. Komolafe
Department of Zoology, Obafemi Awolowo University, Ile-Ife, Nigeria

M. O. Popoola
Department of Zoology, Obafemi Awolowo University, Ile-Ife, Nigeria

A. Kuku
Department of Biochemistry, Obafemi Awolowo University, Ile-Ife, Nigeria

K. Pushpa
Department of Agronomy, College of Agriculture, GKVK, Bangalore- 560 065

N. Krishna Murthy
Department of Agronomy, College of Agriculture, GKVK, Bangalore- 560 065

R. Krishna Murthy
Soil and Water Management, Zonal Agricultural Research Station, V.C. Farm, Mandya -571 405, Karnataka, India

S. Anitta Fanish
Department of Agronomy, Tamil Nadu Agricultural University, Coimbatore, Tamil Nadu– 641 003, Indian

José Antonio Rodrigues De Souza
Department of Agricultural Engineering, Instituto Federal Goiano – Câmpus Urutaí, Urutaí – GO, Brazil

Débora Astoni Moreira
Department of Agricultural Engineering, Instituto Federal Goiano – Câmpus Urutaí, Urutaí – GO, Brazil

Antonio Texeira De Matos
Department of Agricultural Engineering, Universidade Federal de Viçosa, Viçosa - MG, Brazil

Aline Sueli De Lima Rodrigues
Department of Environmental Geology, Instituto Federal Goiano – Câmpus Urutaí, Urutaí – GO, Brazil

H. Ravi Sankar
Central Tobacco Research Institute, Rajahmundry, Andhra Pradesh – 533 105, India

C. Chandrasekhararao
Central Tobacco Research Institute, Rajahmundry, Andhra Pradesh – 533 105, India

K. Sivaraju
Central Tobacco Research Institute, Rajahmundry, Andhra Pradesh – 533 105, India

Alberto Jesús Perea Moreno
Department of Applied Physics, University of Cordoba, Campus Rabanales, 14071, Cordoba, Spain

Federico Manzano Agugliaro
Grupo Tragsa, Avda. Imperio Argentina no. 19, 29004, Málaga, Spain

K. Krishna Surendar
Vanavarayar Institute of Agriculture, Pollachi, India

V. Rajendran
Vanavarayar Institute of Agriculture, Pollachi, India

D. Durga Devi
Department of Crop PhysiologyTamil Nadu Agricultural University Coimbatore, India

P. Jeyakumar
Department of Crop PhysiologyTamil Nadu Agricultural University Coimbatore, India

I. Ravi
National Research Centre for Banana (ICAR), Thiruchirapalli, India

K. Velayudham
Director of CSCMS, TNAU, Coimbatore-641 003, India

Mellissa Ananias Soler da Silva
Embrapa Rice and Beans, Rodovia GO-462, km 12 Zona Rural P. O. Box 179, CEP 75375-000 Santo Antônio de Goiás, GO, Brazil
Federal University of Goias, Campus Samambaia - Rodovia, km 0 – PO-box 131, CEP 74690-900, Goiânia, GO, Brazil

Huberto José Kliemann
Federal University of Goias, Campus Samambaia - Rodovia, km 0 – PO-box 131, CEP 74690-900, Goiânia, GO, Brazil

Alfredo Borges De-Campos
Federal University of Goias, Campus Samambaia - Rodovia, km 0 – PO-box 131, CEP 74690-900, Goiânia, GO, Brazil

Beáta Emöke Madari
Embrapa Rice and Beans, Rodovia GO-462, km 12 Zona Rural P. O. Box 179, CEP 75375-000 Santo Antônio de Goiás, GO, Brazil

Jácomo Divino Borges
Federal University of Goias, Campus Samambaia - Rodovia, km 0 – PO-box 131, CEP 74690-900, Goiânia, GO, Brazil

Janine Mesquita Gonçalves
Federal University of Goias, Campus Samambaia - Rodovia, km 0 – PO-box 131, CEP 74690-900, Goiânia, GO, Brazil

Junshu Jia
Key Laboratory of Water Cycle and Related Land Surface Processes, Institute of Geographic Sciences and Natural Resources Research, CAS, Beijing, China

Yaohu Kang
Key Laboratory of Water Cycle and Related Land Surface Processes, Institute of Geographic Sciences and Natural Resources Research, CAS, Beijing, China

Shuqin Wan
Key Laboratory of Water Cycle and Related Land Surface Processes, Institute of Geographic Sciences and Natural Resources Research, CAS, Beijing, China

Mohammed Karrou
ICARDA, P. O. Box 6299, Rabat – Instituts, Morocco

Hechmi Chehab
Institute of Olivier of Sousse, Tunisia

Manel Issaoui
Laboratoire de Biochimie UR 'Nutrition Humaine et Altérations Métaboliques' Faculté de Médecine, Rue Avicenne, 5019 Monastir, Tunisie

Guido Flamini
Dipartimento di Chimica Bioorganica e Biofarmacia, via
Bonanno 33, 56126 Pisa, Italy

Beligh Mechri
Laboratoire de Biochimie UR 'Nutrition Humaine et
Altérations Métaboliques' Faculté de Médecine, Rue
Avicenne, 5019 Monastir, Tunisie

Faouzi Attia
Agronutrition Parc Activestre-3 avenue de l'Orchidée
31390 carbonne- France

Cioni Pier Luigi
Dipartimento di Chimica Bioorganica e Biofarmacia, via
Bonanno 33, 56126 Pisa, Italy

Dalenda Boujnah
Institute of Olivier of Sousse, Tunisia

Mohamed Hammami
Laboratoire de Biochimie UR 'Nutrition Humaine et
Altérations Métaboliques' Faculté de Médecine, Rue
Avicenne, 5019 Monastir, Tunisie